Lecture Notes in Artificial Intelligence 1794

Subseries of Lecture Notes in Computer Science
Edited by J. G. Carbonell and J. Siekmann

Lecture Notes in Computer Science

Edited by G.Goos, J. Hartmanis, and J. van Leeuwen

T0223278

Berlin
Heidelberg
New York
Barcelona
Hong Kong
London
Milan
Paris
Singapore
Tokyo

Hélène Kirchner
Christophe Ringeissen (Eds.)

Frontiers of Combining Systems

Third International Workshop, FroCoS 2000
Nancy, France, March 22-24, 2000
Proceedings

Series Editors

Jaime G. Carbonell, Carnegie Mellon University, Pittsburgh, PA, USA
Jörg Siekmann, University of Saarland, Saarbrücken, Germany

Volume Editors

Hélène Kirchner
LORIA - CNRS, Campus scientifique
BP 239, 54506 Vandoeuvre-les-Nancy Cedex, France
E-mail: Helene.Kirchner@loria.fr

Christophe Ringeissen
LORIA - INRIA, Campus scientifique
615, rue du Jardin Botanique
54602 Villers-les-Nancy, France
E-mail: Christophe.Ringeissen@loria.fr

Cataloging-in-Publication Data applied for

Die Deutsche Bibliothek - CIP-Einheitsaufnahme

Frontiers of combining systems : third international workshop / FroCoS
2000, Nancy, France, March 2000. Hélène Kirchner ; Christophe
Ringeissen. - Berlin ; Heidelberg ; New York ; Barcelona ; Hong Kong ;
London ; Milan ; Paris ; Singapore ; Tokyo : Springer, 2000
  (Lecture notes in computer science ; Vol. 1794 : Lecture notes in
  artificial intelligence) ISBN 3-540-67281-8

CR Subject Classification (1991): I.2.3, F.4.1, F.4

ISBN 3-540-67281-8 Springer-Verlag Berlin Heidelberg New York

Springer is a company in the BertelsmannSpringer publishing group.
© Springer-Verlag Berlin Heidelberg 2000
Printed in Germany

Typesetting: Camera-ready by author, data conversion by PTB-Berlin, Stefan Sossna
Printed on acid-free paper   SPIN: 10720084   06/3142   5 4 3 2 1 0

Preface

This volume contains the proceedings of FroCoS 2000, the 3rd International Workshop on Frontiers of Combining Systems, held March 22–24, 2000, in Nancy, France. Like its predecessors organized in Munich (1996) and in Amsterdam (1998), FroCoS 2000 is intended to offer a common forum for research activities related to the combination and the integration of systems in the areas of logic, automated deduction, constraint solving, declarative programming, and artificial intelligence.

There were 31 submissions of overall high quality, authored by researchers from countries including Australia, Brasil, Belgium, Chili, France, Germany, Japan, Ireland, Italy, Portugal, Spain, Switzerland, The Netherlands, the United Kingdom, and the United States of America. All submissions were thoroughly evaluated on the basis of at least three referee reports, and an electronic program committee meeting was held through the Internet. The program committee selected 14 research contributions. The topics covered by the selected papers include: combination of logics; combination of constraint solving techniques, combination of decision procedures; modular properties for theorem proving; combination of deduction systems and computer algebra; integration of decision procedures and other solving processes into constraint programming and deduction systems.

We welcomed five invited lectures by Alexander Bockmayr on "Combining Logic and Optimization in Cutting Plane Theory", Gilles Dowek on "Axioms vs. Rewrite Rules: From Completeness to Cut Elimination", Klaus Schulz on "Why Combined Decision Problems Are Often Intractable", Tomas Uribe on "Combinations of Theorem Proving and Model Checking", and Richard Zippel on "Program Composition Techniques for Numerical PDE Codes". Full papers of these lectures, except the last one, are also included in this volume.

Many people and institutions have contributed to the realization of this conference. We would like to thank the members of the program committee and all the referees for their care and time in reviewing and selecting the submitted papers; the members of the organizing committee for their help in the practical organization of the conference; and all institutions that supported FroCoS 2000: CNRS, Communauté Urbaine du Grand Nancy, Conseil Général de Meurthe et Moselle, Conseil Régional de Lorraine, France Telecom CNET, GDR ALP, INPL, INRIA, LORIA, Université Henri-Poincaré–Nancy 1, and Université Nancy 2.

February 2000

Hélène Kirchner
Christophe Ringeissen

Conference Chairs

Hélène Kirchner (Nancy)
Christophe Ringeissen (Nancy)

Program Committee

Franz Baader (Aachen)
David Basin (Freiburg)
Frédéric Benhamou (Nantes)
Thom Fruehwirth (Munich)
Fausto Giunchiglia (Trento)
Bernhard Gramlich (Vienna)
Hélène Kirchner (Nancy)

Christoph Kreitz (Ithaca, NY)
Till Mossakowski (Bremen)
Jochen Pfalzgraf (Salzburg)
Marteen de Rijke (Amsterdam)
Christophe Ringeissen (Nancy)
Tony Scott (Paderborn)
Mark Wallace (London)

Local Organization

Christelle Bergeret
Anne-Lise Charbonnier
Armelle Demange

Hélène Kirchner
Christophe Ringeissen
Laurent Vigneron

External Reviewers

L. Bachmair	J. Harm	G. Salzer
B. Beckert	F. Jacquemard	F. Saubion
O. Bournez	R. B. Kearfott	Ch. Scholl
B. Buchberger	Ch. Lueth	W. Schreiner
C. Castro	Ch. Lynch	L. Schroeder
M. Cerioli	L. Mandel	K. U. Schulz
H. Christiansen	C. Marché	W. M. Seiler
P. Crégut	D. Méry	K. Stokkermans
E. Dantsin	F. Messine	F. Stolzenburg
F. Drewes	M. Milano	G. Struth
H. Dubois	E. Monfroy	L. Vigano
C. Dupre	P.-E. Moreau	L. Vigneron
Ch. Fermueller	M. Richters	A. Voronkov
F. Goualard	R. Rioboo	B. Wolff
L. Granvilliers	M. Roggenbach	H. Zhang
G. Hains	M. Rusinowitch	

Table of Contents

Invited Paper

Session 4

Invited Paper

Session 5

Combining Logic and Optimization in Cutting Plane Theory

Alexander Bockmayr[1] and Friedrich Eisenbrand[2]

[1] Université Henri Poincaré, LORIA
B.P. 239, F-54506 Vandœuvre-lès-Nancy, France
`bockmayr@loria.fr`
[2] Max-Planck-Institut für Informatik
Im Stadtwald, D-66123 Saarbrücken, Germany
`eisen@mpi-sb.mpg.de`

Abstract. Cutting planes were introduced in 1958 by Gomory in order to solve integer linear optimization problems. Since then, they have received a lot of interest, not only in mathematical optimization, but also in logic and complexity theory. In this paper, we present some recent results on cutting planes at the interface of logic and optimization. Main emphasis is on the length and the rank of cutting plane proofs based on the Gomory-Chvátal rounding principle.

1 Introduction

Cutting planes were introduced in a seminal paper of Gomory (1958) in order to solve linear optimization problems over the integer numbers. In mathematical optimization, cutting planes are used to tighten the linear programming relaxation of an integer optimization problem. From the viewpoint of logic, a cutting plane is a linear inequality $c^T x \leq \delta$ derived from a system of linear inequalities $Ax \leq b$ such that the implication $Ax \leq b \rightarrow c^T x \leq \delta$ is valid in the ring of integer numbers $\mathbb{Z}$, but not in the field of real or rational numbers $\mathbb{R}$ resp. $\mathbb{Q}$. Geometrically speaking, a cutting plane cuts off a part of the real solution set P of $Ax \leq b$, while leaving invariant the set S of integer solutions (see Fig. 1).

Cutting plane principles developed in mathematical optimization have nice logical properties. They provide sound and complete inference systems for reasoning with linear inequalities in integer variables. Any clause in propositional logic

$$x_1 \vee \ldots \vee x_m \vee \overline{y}_1 \vee \ldots \vee \overline{y}_k \tag{1}$$

can be translated into a *clausal inequality*

$$
\begin{aligned}
x_1 + \cdots + x_m + (1 - y_1) + \cdots + (1 - y_k) &\geq 1 \qquad \text{or} \\
x_1 + \cdots + x_m \quad - y_1 - \quad \cdots \quad - y_k &\geq 1 - k.
\end{aligned}
\tag{2}
$$

A clause set is satisfiable if and only if the corresponding system of clausal inequalities together with the bounds $0 \leq x_i, y_j \leq 1$ has an integer solution.

H. Kirchner and C. Ringeissen (Eds.): FroCoS 2000, LNAI 1794, pp. 1–17, 2000.

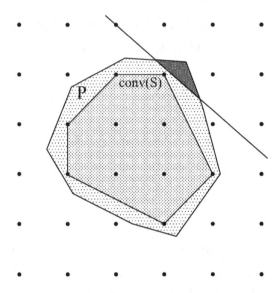

Fig. 1. Cutting plane

This shows that reasoning in propositional logic can be seen as a special case of reasoning with linear inequalities in integer variables.

Looking at logical inference from an optimization perspective allows one to apply tools from algebra and geometry to study the complexity of a logical inference problem, and to represent the set of all possible inferences in an algebraic way. This can be used to select one or more strongest inferences according to some quality measure, which leads to new algorithms for both satisfiability and optimization.

The structure of this paper is as follows. In Sect. 2, we use inference rules to present some basic results on solving equations and inequalities over real or rational numbers. We also recall the complexity of the underlying algorithmic problems. In Sect. 3, we introduce cutting planes from the viewpoint of integer linear optimization. Then we focus on their logical properties and present a soundness and completeness result for the corresponding inference system. Sect. 4 is about the complexity of the elementary closure, which is obtained by adding to a linear inequality system all possible Gomory-Chvátal cutting planes. Sect. 5 contains a number of important results on the complexity of cutting plane proofs in propositional logic. Finally, Sect. 6 presents logic-based cutting planes for extended clauses.

The results presented in this paper are by no way exhaustive. We selected some recent results at the interface of logic and optimization that are particularly relevant for cutting plane theory. For a comprehensive treatment of optimization methods for logical inference, we refer the reader to the beautiful monograph (Chandru & Hooker 1999).

2   Linear Equations and Inequalities

2.1   Linear Equations over $\mathbb{Q}$ or $\mathbb{R}$

Let $\mathbb{F}$ denote the field of real or rational numbers and consider a system of linear equations

$$
\begin{array}{c}
a_1^T x = \beta_1 \\
\vdots \quad \vdots \\
a_m^T x = \beta_m
\end{array}
\quad \Longleftrightarrow \quad Ax = b, \tag{3}
$$

with vectors $a_1, \ldots, a_m \in \mathbb{F}^n$, numbers $\beta_1, \ldots, \beta_m \in \mathbb{F}$, a matrix $A \in \mathbb{F}^{m \times n}$, and a vector $b \in \mathbb{F}^m$. Moreover, $\cdot^T$ denotes transposition of a vector or a matrix.

The basic inference principle to derive from this system a new equation $c^T x = \delta$, with $c \in \mathbb{F}^n, \delta \in \mathbb{F}$, is taking a *linear combination* of the given equations. More precisely, we can multiply each equation $a_i^T x = \beta_i$ by a number $u_i \in \mathbb{F}$ and add up the resulting equations $u_i a_i^T x = u_i \beta_i$, for $i = 1, \ldots, m$. In matrix notation, this can be expressed by the inference rule

$$
\texttt{lin\_com} : \quad \frac{Ax = b}{(u^T A)x = u^T b} \quad \text{if } u \in \mathbb{F}^m \tag{4}
$$

If $Ax = b$ is solvable in $\mathbb{F}^n$, the rule $\texttt{lin\_com}$ allows us to derive any implied equation $c^T x = \delta$. If $Ax = b$ is not solvable, we can get the contradiction $0 = 1$.

Proposition 1. *The system $Ax = b$ is not solvable in $\mathbb{F}^n$ if and only if there exists $u \in \mathbb{F}^m$ such that $u^T A = 0$ and $u^T b = 1$. If $Ax = b$ is solvable, an equation $c^T x = \delta$ is implied by $Ax = b$ in $\mathbb{F}^n$ if and only if there exists $u \in \mathbb{F}^m$ such that $u^T A = c^T$ and $u^T b = \delta$.*

The solution set of a single linear equation $a^T x = \beta$ defines a *hyperplane* in the n-dimensional affine space $\mathbb{F}^n$. The solution set S of a system of linear equations $Ax = b$ corresponds to an intersection of hyperplanes, i.e., an *affine subspace*. For $x^0 \in S$ the set $U = \{x - x^0 \mid x \in S\}$ defines a linear subspace of the n-dimensional vector space $\mathbb{F}^n$. If $\dim U = k$, then U is generated by k linearly independent vectors $x^1, \ldots, x^k \in \mathbb{F}^n$.

Using Gaussian elimination, systems of linear equalities over the rational numbers can be solved in polynomial time. The basic idea of this algorithm is successive variable elimination by repeated application of rule $\texttt{lin\_com}$.

2.2   Linear Inequalities over $\mathbb{Q}$ or $\mathbb{R}$

Given a system of linear inequalities $Ax \leq b, x \in \mathbb{F}^n$, the basic principle to derive a new inequality $c^T x \leq \delta$ is taking a *non-negative* linear combination of the given inequalities. Another possibility is *weakening* the right-hand side. We capture this by the following two inference rules:

$$\texttt{nonneg\_lin\_com}: \frac{Ax \leq b}{(u^T A)x \leq u^T b} \quad \text{if} \quad \begin{matrix} u \in \mathbb{F}^m \\ u \geq 0 \end{matrix} \tag{5}$$

$$\texttt{weak\_rhs}: \frac{a^T x \leq \beta}{a^T x \leq \beta'} \quad \text{if} \quad \beta \leq \beta' \tag{6}$$

Again there is a soundness and completeness result, which in the theory of linear optimization is known as *Farkas' Lemma*. Any implied inequality of a solvable system can be obtained by one application of the rule $\texttt{nonneg\_lin\_com}$ followed by one application of $\texttt{weak\_rhs}$.

Proposition 2 (Farkas' Lemma). *The system $Ax \leq b$ has no solution $x \in \mathbb{F}^n$ if and only if there exists $y \in \mathbb{F}^m, y \geq 0$ such that $y^T A = 0$ and $y^T b = -1$. If $Ax \leq b$ is solvable, then an inequality $c^T x \leq \delta$ is implied by $Ax \leq b$ in $\mathbb{F}^n$ if and only if there exists $y \in \mathbb{F}^m, y \geq 0$ such that $y^T A = c$ and $y^T b \leq \delta$.*

The solution set of a linear inequality $a^T x \leq \beta, a \neq 0$, defines a *halfspace* in n-dimensional affine space $\mathbb{F}^n$, which will be denoted by $(a^T x \leq \beta)$. The solution set P of a system of linear inequalities $Ax \leq b$, i.e., an intersection of finitely many halfspaces, is called a *polyhedron*. The polyhedron P is *rational* if $A \in \mathbb{Q}^{m \times n}$ and $b \in \mathbb{Q}^m$. A bounded polyhedron is called a *polytope*. An inequality $c^T x \leq \delta$ is called *valid* for a polyhedron P if $P \subseteq (c^T x \leq \delta)$. A set $F \subseteq P$ is called a *face* of P if there exists an inequality $c^T x \leq \delta$ valid for P such that $F = P \cap \{x \in P \mid c^T x = \delta\}$. A point v in P such that $\{v\}$ is a face of P is called a *vertex* of P. A polyhedron is called *pointed* if it has a vertex. A non-empty polyhedron $P = \{x \mid Ax \leq b\}$ is pointed if and only if A has full column rank. A *facet* of P is an inclusionwise maximal face F with $\emptyset \neq F \neq P$. Equivalently, a facet is a nonempty face of P of dimension $\dim(P) - 1$. If a polyhedron P is full-dimensional then P has a representation $P = \{x \in \mathbb{F}^n \mid Ax \leq b\}$ such that each inequality in the system $Ax \leq b$ defines a facet of P and such that each facet of P is defined by exactly one of these inequalities.

In addition to the outer description of a polyhedron as the solution set of a linear inequality system, there is also an inner description, which is based on vertices and extreme rays. A vector $r \in \mathbb{R}^n$ is called a *ray* of the polyhedron P if for each $x \in P$ the set $x + \mathbb{F}_{\geq 0} \cdot r$ is contained in P. A ray r of P is *extreme* if there do not exist two linearly independent rays r^1, r^2 of P such that $r = \frac{1}{2}(r^1 + r^2)$. For each polyhedron $P \subseteq \mathbb{F}^n$ there exist finitely many points $p^1, \ldots, p^k$ in P and finitely many rays $r^1, \ldots, r^l$ of P such that

$$P = \text{conv}(p^1, \ldots, p^k) + \text{cone}(r^1, \ldots, r^l). \tag{7}$$

Here $\text{conv}(p^1, \ldots, p^k) = \{\lambda_1 p^1 + \cdots + \lambda_k p^k \mid \lambda_1, \ldots, \lambda_k \geq 0, \lambda_1 + \cdots + \lambda_k = 1\}$ denotes the *convex hull* of $p^1, \ldots, p^k$, and $\text{cone}(r^1, \ldots, r^l) = \{\mu_1 r^1 + \cdots + \mu_l r^l \mid \mu_1, \ldots, \mu_l \geq 0\}$ the *conic hull* of $r^1, \ldots, r^l$. If the polyhedron P is pointed then the p^i are the uniquely determined vertices of P, and the r^j are representatives of the up to scalar multiplication uniquely determined extreme rays of P. This yields a parametric description of the solution set of a system of linear inequalities. In particular, every polytope can be described as the convex hull of its vertices.

For a long time, it was an open question whether there exists a polynomial algorithm for solving linear inequalities over the rational numbers. Only in 1979, Khachiyan showed that the *ellipsoid method* for nonlinear programming can be adapted to solve linear inequalities and linear programming problems in polynomial time.

Theorem 1 (Khachiyan 79). *The following problems are solvable in polynomial time:*

- *Given a matrix $A \in \mathbb{Q}^{m \times n}$ and a vector $b \in \mathbb{Q}^m$, decide whether $Ax \leq b$ has a solution $x \in \mathbb{Q}^n$, and if so, find one.*
- *Given a rational matrix $A \in \mathbb{Q}^{m \times n}$ and a rational vector $b \in \mathbb{Q}^m$, decide whether $Ax = b$ has a nonnegative solution $x \in \mathbb{Q}^n, x \geq 0$, and if so, find one.*
- *(Linear programming problem) Given a matrix $A \in \mathbb{Q}^{m \times n}$ and vectors $b \in \mathbb{Q}^m, c \in \mathbb{Q}^n$, decide whether $\max\{c^T x \mid Ax \leq b, x \in \mathbb{Q}^n\}$ is infeasible, finite, or unbounded. If it is finite, find an optimal solution. If it is unbounded, find a feasible solution x_0, and find a vector $z \in \mathbb{Q}^n$ with $Az \leq 0$ and $c^T z > 0$.*

2.3   Linear Inequalities over $\mathbb{Z}$

The problems that we have considered so far can all be solved in polynomial time. In this section, we consider systems of linear inequalities over the integer numbers

$$Ax \leq b, x \in \mathbb{Z}^n, \tag{8}$$

and, as a special case, systems of linear equations over the non-negative integer numbers $Ax = b, x \in \mathbb{N}^n$. The solution set of (8) may be infinite. However, there is again a finite parametric description.

Theorem 2. *Let $P = \{x \in \mathbb{R}^n \mid Ax \leq b\}$, with $A \in \mathbb{Z}^{m \times n}$ and $b \in \mathbb{Z}^m$, be a rational polyhedron and let $S = P \cap \mathbb{Z}^n$ be the set of integer points contained in P. Then there exist finitely many points $p^1, \ldots, p^k$ in S and a finitely many rays $r^1, \ldots, r^l \in \mathbb{Z}^n$ of P such that*

$$S = \{\sum_{i=1}^{k} \lambda_i p^i + \sum_{j=1}^{l} \mu_j r^j \mid \lambda_i, \mu_j \in \mathbb{N}, \sum_{i=1}^{k} \lambda_i = 1\} \tag{9}$$

The absolute value of all p^i and r^j can be bounded by $(n+1)\Delta$, where Δ is the maximum absolute value of the subdeterminants of the matrix $(A \mid b)$.

In particular, this theorem implies that if a system of linear inequalities has an integral solution, then it has one whose encoding length is polynomially bounded by the encoding length of the input $(A \mid b)$. Therefore, we can conclude:

Theorem 3. *The problem of deciding for a rational matrix $A \in \mathbb{Q}^{m \times n}$ and a rational vector $b \in \mathbb{Q}^m$ whether the system of linear inequalities $Ax \leq b$ has an integral solution $x \in \mathbb{Z}^n$ is NP-complete.*

In contrast to the polynomial solvability of 2SAT, the problem stays NP-complete if we admit only two variables in each constraint (Lagarias 1985). The following problems are also NP-complete:

- Given $A \in \mathbb{Q}^{m \times n}, b \in \mathbb{Q}^m$, does $Ax = b$ have a *nonnegative* integral solution $x \in \mathbb{Z}^n, x \geq 0$?
- Given $A \in \mathbb{Q}^{m \times n}, b \in \mathbb{Q}^m$, does $Ax = b$ have a 0-1 solution $x \in \{0,1\}^n$?
- Given $a \in \mathbb{Q}^n, \beta \in \mathbb{Q}$, does $a^T x = \beta$ have a *nonnegative* integral solution $x \in \mathbb{Z}^n, x \geq 0$?
- Given $a \in \mathbb{Q}^n, \beta \in \mathbb{Q}$, does $a^T x = \beta$ have a 0-1 solution $x \in \{0,1\}^n$?
- Knapsack problem: Given $a, c \in \mathbb{Q}^n$ and $\beta, \delta \in \mathbb{Q}$, is there a 0-1 vector $x \in \{0,1\}^n$ with $a^T x \leq \beta$ and $c^T x \geq \delta$?

While solving linear diophantine inequalities in general is NP-hard, a celebrated result of Lenstra (1983) states that this can be done in polynomial time, if the dimension n is fixed. Before Lenstra's theorem was obtained, this was known only for the cases $n = 1, 2$.

Theorem 4 (Lenstra 83). *For each fixed natural number n, there exists a polynomial algorithm which finds an integral solution $x \in \mathbb{Z}^n$ of a given system $Ax \leq b, A \in \mathbb{Q}^{m \times n}, b \in \mathbb{Q}^n$ in n variables, or decides that no such solution exists.*

3   Cutting Planes

Consider the *integer linear optimization problem*

$$\max\{c^T x \mid Ax \leq b, x \in \mathbb{Z}^n\}, \quad A \in \mathbb{Z}^{m \times n}, b \in \mathbb{Z}^m \tag{10}$$

An important consequence of Theorem 2 is that for a rational polyhedron $P = \{x \in \mathbb{R}^n \mid Ax \leq b\}$, the convex hull $\mathrm{conv}(S)$ of the set of integer points $S = P \cap \mathbb{Z}^n$ in P is again a rational polyhedron. Therefore, in principle, the integer optimization problem (10) can be reduced to an ordinary linear optimization problem

$$\max\{c^T x \mid x \in \mathrm{conv}(S) \subseteq \mathbb{R}^n\} \tag{11}$$

over the rational polyhedron $\mathrm{conv}(S)$. If (11) has a bounded optimal value, then it has an optimal solution, namely an extreme point of $\mathrm{conv}(S)$, which is an optimal solution of (10). The objective value of (10) is unbounded from above if and only if the objective value of (11) is unbounded from above.

If we had a linear inequality system $Cx \leq d$ defining the polyhedron $\mathrm{conv}(S)$, we could solve (11) with a polynomial linear programming algorithm. In general, the polyhedron conv(S) will have exponentially many facets. Therefore, such a description cannot be obtained in polynomial time. However, in order to find an optimal solution of (10), an approximation of conv(S) might be sufficient. This idea leads to *cutting plane algorithms* for solving the integer linear optimization problem. We start with the *linear programming relaxation* $P = \{x \in \mathbb{R}^n \mid Ax \leq b\}$ of the original problem. Then we add new inequalities which are satisfied by all points in S but which cut off at least one fractional vertex of P. These are called *cutting planes* (Fig. 1). This is repeated until an integer vertex is obtained, which belongs to the integer solution set S and maximizes $c^T x$.

3.1 Gomory-Chvátal Cutting Planes

One basic inference rule for deriving cutting planes is the Gomory-Chvátal rounding principle:

$$\text{Rounding} : \frac{a^T x \leq \beta}{a^T x \leq \lfloor \beta \rfloor} \quad \text{if } a \in \mathbb{Z}^n \tag{12}$$

If $a \in \mathbb{Z}^n$ is an integer vector, then for all integer vectors $x \in \mathbb{Z}^n$, the left-hand side of the inequality $a^T x \leq \beta$ is an integer. Therefore, we do not loose any integer solution if we round down the right-hand side β to the next integer number $\lfloor \beta \rfloor$. Geometrically, this means the following. Assume that the entries of a are relative prime integer numbers and that β is fractional. Then rounding β means that we translate the hyperplane $a^T x = \beta$ until it meets an integer point. Given a system of linear inequalities $Ax \leq b$, a *Gomory-Chvátal cutting plane* is obtained by combining the rules Nonneg_lin_com and Rounding.

$$\text{GC\_cp} : \frac{Ax \leq b}{u^T Ax \leq \lfloor u^T b \rfloor} \quad \text{if } \begin{array}{l} u \geq 0, \\ u^T A \in \mathbb{Z}^n \end{array} \tag{13}$$

A *cutting plane proof* of *length* l of an inequality $c^T x \leq \delta$ from $Ax \leq b$ is a sequence of inequalities $c_1^T x \leq \delta_1, \ldots, c_l^T x \leq \delta_l$ such that

$$\text{GC\_cp} : \frac{Ax \leq b, c_1^T x \leq \delta_1, \ldots, c_{i-1}^T x \leq \delta_{i-1}}{c_i^T x \leq \delta_i} \tag{14}$$

for each $i = 1, \ldots, l$, and

$$\text{weak\_rhs} : \frac{c_l^T x \leq \delta_l}{c^T x \leq \delta} \tag{15}$$

Gomory-Chvátal cutting planes provide a complete inference system for linear diophantine inequalities.

Theorem 5 (Chvátal 73, Schrijver 80). *Let* $P = \{x \in \mathbb{F}^n \mid Ax \leq b\}$ *be a non-empty polyhedron that is rational or bounded and let* $S = P \cap \mathbb{Z}^n$.

- *If* $S \neq \emptyset$ *and* $c^T x \leq \delta$ *is valid for* S, *then there is a cutting plane proof of* $c^T x \leq \delta$ *from* $Ax \leq b$.

- *If $S = \emptyset$, then there is a cutting plane proof starting from $Ax \leq b$ yielding the contradiction $0 \leq -1$.*

In the given form, the Gomory-Chvátal principle is not effective. It is not clear how to choose the weights in the non-negative linear combination of the given constraint matrix. However, several well-known inference rules are subsumed by the Gomory-Chvátal principle, see Section 6. The first cutting plane procedure for solving integer linear optimization problems was proposed by Gomory (1958). Here, the cutting planes are generated from a basic feasible solution in the Simplex algorithm. If v is a vertex of the polyhedron P such that the i-th component v_i is fractional, then by a simple rounding operation one can obtain a valid inequality for $P \cap \mathbb{Z}^n$ that cuts off v. Gomory showed that by systematically adding these cuts, and using the dual simplex method with appropriate anticycling rules, one can obtain a finitely terminating cutting plane algorithm for general integer linear optimization problems. The practical performance of Gomory's original algorithm is not good. However, recent work in computational integer programming has shown that Gomory's cutting planes may be very useful when used in a branch-and-cut algorithm, see e.g. (Balas, Ceria, Cornuéjols & Natraj 1996).

3.2 Disjunctive Cutting Planes

We briefly mention a second principle to derive cutting planes, which is based on disjunctive programming (Balas 1979, Sherali & Shetty 1980). A *disjunctive optimization problem* is of the form

$$\max\{c^T x \mid \bigvee_{h \in H} A^h x \leq b^h, x \geq 0\} \tag{16}$$

for H finite, $A^h \in \mathbb{F}^{m_h \times n}$, and $b^h \in \mathbb{F}^{m_h}$. The feasible set of the disjunctive optimization problem (16) is the union of the polyhedra $P^h = \{x \in \mathbb{R}^n \mid A^h x \leq b^h, x \geq 0\}, h \in H$. *Disjunctive cutting planes* are obtained by applying the rule

$$\texttt{disj\_cp}: \quad \frac{\bigvee_{h \in H} A^h x \leq b^h, x \geq 0}{a^T x \leq \beta} \quad \text{if} \quad \begin{cases} u_h \in \mathbb{F}^{m_h}, u_h \geq 0, \\ a^T \leq u_h^T A^h, \\ u_h^T b^h \leq \beta, \\ \text{for all } h \in H \end{cases} \tag{17}$$

We first take a non-negative linear combination of each disjunct and then weaken the left-hand and the right-hand side of all the inequalities obtained. The constraint $x \geq 0$ allows us to weaken also the left-hand side. The completeness of the disjunctive cutting plane principle was studied by several authors, see e.g. (Blair & Jeroslow 1978, Jeroslow 1977). To apply the inference rule $\texttt{disj\_cp}$ to integer programming problems, one may, e.g., consider disjunctions of the form

$$\begin{array}{cc} Ax \leq b, \ x \geq 0 \\ x_j \leq d \end{array} \quad \bigvee \quad \begin{array}{cc} Ax \leq b, \ x \geq 0 \\ x_j \geq d+1, \end{array} \tag{18}$$

for some $j \in \{1, \ldots, n\}$ and $d \in \mathbb{Z}$. A very successful branch-and-cut algorithm for general linear 0-1 optimization, which is based on disjunctive cutting planes, is the *lift-and-project method* developed by Balas, Ceria & Cornuéjols (1993, 1996).

4   The Elementary Closure

The set of vectors satisfying all Gomory-Chvátal cutting planes for P is called the *elementary closure* P' of P. Thus, if P is defined by the system $Ax \leq b$, where $A \in \mathbb{Z}^{m \times n}$ and $b \in \mathbb{Z}^m$, then P' is defined as

$$P' = \bigcap_{\substack{\lambda \in \mathbb{R}^m_{\geq 0} \\ \lambda^T A \in \mathbb{Z}^n}} (\lambda^T A x \leq \lfloor \lambda^T b \rfloor). \tag{19}$$

A crucial observation made by Schrijver (1980) is that the weight-vectors λ can be chosen such that $0 \leq \lambda \leq 1$. This holds because an inequality $\lambda^T A x \leq \lfloor \lambda^T b \rfloor$ with $\lambda \in \mathbb{R}^m_{\geq 0}$ and $\lambda^T A \in \mathbb{Z}^n$ is implied by $Ax \leq b$ and $(\lambda - \lfloor \lambda \rfloor)^T A x \leq \lfloor (\lambda - \lfloor \lambda \rfloor)^T b \rfloor$, since

$$\lambda^T A x = (\lambda - \lfloor \lambda \rfloor)^T A x + \lfloor \lambda \rfloor^T A x \leq \lfloor (\lambda - \lfloor \lambda \rfloor)^T b \rfloor + \lfloor \lambda \rfloor^T b = \lfloor \lambda^T b \rfloor. \tag{20}$$

Thus P' is of the form

$$P' = \bigcap_{\substack{0 \leq \lambda \leq 1 \\ \lambda^T A \in \mathbb{Z}^n}} (\lambda^T A x \leq \lfloor \lambda^T b \rfloor). \tag{21}$$

An integer vector $c \in \mathbb{Z}^n$ of the form $c^T = \lambda^T A$, where $0 \leq \lambda \leq 1$, has infinity norm $\|c\|_\infty \leq \|A^T\|_\infty \|\lambda\|_\infty \leq \|A^T\|_\infty$. Therefore the elementary closure P' of a rational polyhedron is a rational polyhedron again, as observed by Schrijver (1980).

Proposition 3 (Schrijver 80). *If P is a rational polyhedron, then P' is a rational polyhedron.*

4.1   Complexity of the Elementary Closure

An important question is, whether one can optimize in polynomial time over the elementary closure of a polyhedron. Grötschel, Lovász & Schrijver (1988) have shown that the optimization problem over a polyhedron Q is equivalent to the *separation problem* for Q, which is defined as follows: Given some $\hat{x}$, decide whether $\hat{x}$ is in Q and if $\hat{x} \notin Q$, provide an inequality which is valid for Q but not for $\hat{x}$. Equivalence means that there is a polynomial algorithm for optimization if and only if there is a polynomial algorithm for separation.

Thus equivalent to the optimization over the elementary closure of P' is the separation problem over P'. A subproblem of the separation problem for P' is the *membership problem* for P'. It can be stated as:

Given $A \in \mathbb{Z}^{m \times n}, b \in \mathbb{Z}^m$ and a rational vector $\hat{x} \in \mathbb{Q}^n$, decide whether $\hat{x}$ is outside the elementary closure P' of the polyhedron $P = \{x \in \mathbb{R}^n \mid Ax \leq b\}$.

Here you are only faced with the decision problem of whether $\hat{x}$ is outside P' and not with the additional task of computing a separating hyperplane. Eisenbrand (1999) proved that the membership problem for the elementary closure of a polyhedron is NP-complete. In other words, it is NP-complete to decide whether there is an inference using rule GC_cp that separates the point $\hat{x}$ from P'. Therefore the optimization problem over P' is NP-hard.

4.2   The Elementary Closure in Fixed Dimension

The upper bound of $\|A^T\|_\infty$ on the number of inequalities defining P' is exponential in the size of the input P, even in fixed dimension. Integer programming in fixed dimension, however, is polynomial. Also, there is a structural result concerning the number of vertices of the integer hull of a polyhedron in fixed dimension. If P is defined by an integral system $Ax \leq b$, then the number of vertices of P_I is polynomial in the encoding length size(P) of A and b in fixed dimension. This follows from a generalization of a result by Hayes & Larman (1983), see (Schrijver 1986). A tight upper bound was provided by Cook, Hartmann, Kannan & McDiarmid (1992). For the elementary closure a similar result holds. Bockmayr & Eisenbrand (1999) show that in fixed dimension only a polynomial number of inequalities is needed to describe P'.

Theorem 6. *The number of inequalities needed to describe the elementary closure of a rational polyhedron* $P = \{x \in \mathbb{R}^n \mid Ax \leq b\}$ *with* $A \in \mathbb{Z}^{m \times n}$ *and* $b \in \mathbb{Z}^m$, *is* O$(m^n \mathrm{size}(P)^n)$ *in fixed dimension* n.

A *simplicial cone* is a polyhedron $P = \{x \in \mathbb{R}^n \mid Ax \leq b\}$, where $A \in \mathbb{R}^{n \times n}$ is a nonsingular matrix. If A is in addition an integral matrix with $d = |\det(A)|$, then the cutting planes of the simplicial cone P are of the form

$$\frac{\mu^T A}{d} x \leq \left\lfloor \frac{\mu^T b}{d} \right\rfloor, \text{ where } \mu \in \{0, \ldots, d\}^n, \text{ and } \mu^T A \in (d \cdot \mathbb{Z})^n. \quad (22)$$

Thus cutting planes of P are represented by integral points in the *cutting plane polytope* Q which is the set of all (μ, y, z) in $\mathbb{R}^{2n+1}$ satisfying the inequalities

$$\begin{aligned}
\mu &\geq 0 \\
\mu &\leq d \\
\mu^T A &= d\,y \\
(\mu^T b) - d + 1 &\leq d\,z \\
(\mu^T b) &\geq d\,z.
\end{aligned} \quad (23)$$

If (μ, y, z) is integral, then $\mu \in \{0, \ldots, d\}^n$, $y \in \mathbb{Z}^n$ enforces $\mu^T A \in (d \cdot \mathbb{Z})^n$ and z is the only integer in the interval $[(\mu^T b + 1 - d)/d, \mu^T b/d]$. Bockmayr & Eisenbrand (1999) show that the facets of P' are represented by vertices of Q_I. Theorem 6 then follows from triangulation.

4.3   Choosing Cutting Planes for Simplicial Cones

In practice, Gomory-Chvátal cutting planes are normally used in a branch-and-cut framework. Therefore, one is faced with the problem of finding efficiently cutting planes that tighten the relaxation as much as possible. It would be most useful to cut with facets of the elementary closure of P'. However this is so far not possible. If P is a simplicial cone defined by $Ax \le b$, then the vertices of the integer hull Q_I of Q defined by (23) include the facets of P'. Bockmayr & Eisenbrand (1999) present a polynomial algorithm which computes cutting planes for a simplicial cone that correspond to vertices of Q_I. These cutting planes can be used to separate $A^{-1}b$.

Theorem 7. *Let $P = \{x \in \mathbb{R}^n \mid Ax \le b\}$ be a rational simplicial cone, where $A \in \mathbb{Z}^{n \times n}$ is of full rank, $b \in \mathbb{Z}^n$ and $d = |\det(A)|$. Then one can compute in $O(n^\omega)$ basic operations of $\mathbb{Z}_d$ a vertex of Q_I corresponding to a cutting plane $(\mu/d)^T Ax \le \lfloor (\mu/d)^T b \rfloor$ separating $A^{-1}b$ with maximal possible amount of violation $\nu_{\max}/d$.*

Here ω is a constant such that $n \times n$ matrix multiplication can be carried out in $O(n^\omega)$.

4.4   The Chvátal-Gomory Procedure

The elementary-closure operation can be iterated. Let $P^{(0)} = P$ and $P^{(i+1)} = (P^{(i)})'$, for $i \ge 0$. Then the *Chvátal rank* of P is the smallest number t such that $P^{(t)} = P_I$, where $P_I = \operatorname{conv}(P \cap \mathbb{Z}^n)$ is the *integer hull* of P.

The *rank* of an inequality $c^T x \le \delta$ is the smallest t such that $c^T x \le \delta$ is valid for $P^{(t)}$. The following theorem clarifies the relation between the rank and the length of a cutting plane proof.

Theorem 8 (Chvátal, Cook and Hartmann 89). *Let $Ax \le b$, with $A \in \mathbb{Z}^{m \times n}$ and $b \in \mathbb{Z}^m$, have an integer solution, and let $c^T x \le \delta$ have rank at most d relative to $Ax \le b$. Then there is a cutting plane proof of $c^T x \le \delta$ from $Ax \le b$ of length at most $(n^{d+1} - 1)/(n - 1)$.*

Chvátal (1973) showed that every bounded polyhedron $P \subseteq \mathbb{R}^n$ has finite rank. Schrijver (1980) extended this result to possibly unbounded, but rational polyhedra $P = \{x \in \mathbb{R}^n \mid Ax \le b\}$, where $A \in \mathbb{Q}^{m \times n}$, $b \in \mathbb{Q}^m$. Both results are implicit in (Gomory 1958). Cook, Gerards, Schrijver & Tardos (1986) and Gerards (1990) proved that for every matrix $A \in \mathbb{Z}^{m \times n}$ there exists $t \in \mathbb{N}$ such that for all right-hand sides $b \in \mathbb{Z}^m$, the Chvátal rank of $P_b = \{x \in \mathbb{R}^n \mid Ax \le b\}$ is bounded by t.

Already in dimension 2, there exist rational polyhedra of arbitrarily large Chvátal rank (Chvátal 1973). To see this, consider the polytopes

$$P_k = \operatorname{conv}\{(0,0), (0,1)(k, \tfrac{1}{2})\}.$$

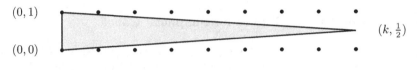

$(0,1)$

$(0,0)$

$(k, \frac{1}{2})$

Fig. 2.

One can show that $P_{k-1} \subseteq P'_k$. For this, let $c^T x \leq \delta$ be valid for P_k with $\delta = \max\{c^T x \mid x \in P_k\}$. If $c_1 \leq 0$, then the point $(0,0)$ or $(0,1)$ maximizes $c^T x$, thus $(c^T x = \delta)$ contains integral points. If $c_1 > 0$, then $c^T(k, \frac{1}{2}) \geq c^T(k-1, \frac{1}{2}) + 1$. Therefore the point $(k-1, \frac{1}{2})$ is in the half space $(c^T x \leq \delta - 1) \subseteq (c^T x \leq \lfloor \delta \rfloor)$. Unfortunately, the lower bound k is exponential in the encoding length of P_k which is $O(\log(k))$.

4.5   Bounds on the Chvátal-Rank in the 0/1 Cube

Of particular interest in propositional logic and combinatorial optimization, however, are polytopes in the 0/1 cube. Note that combinatorial optimization problems can often be modeled as an integer program in 0/1-variables.

Bockmayr, Eisenbrand, Hartmann & Schulz (1999) studied the Chvátal rank of polytopes contained in the n-dimensional cube $[0,1]^n$. They showed that the Chvátal rank of a polytope $P \subseteq [0,1]^n$ is $O(n^3 \log n)$ and prove the linear upper and lower bound n for the case $P \cap \mathbb{Z}^n = \emptyset$. The basic proof technique here is scaling. Each integral 0/1 polytope can be described by a system of integral inequalities $Ax \leq b$ such that each absolute value of an entry in A is bounded by $n^{n/2}$, see e.g. (Padberg & Grötschel 1985). The sequence of integral vectors obtained from a^T by dividing it by decreasing powers of 2 followed by rounding gives a better and better approximation of a^T itself. One estimates the number of iterations of the Chvátal-Gomory rounding procedure needed until the face given by some vector in the sequence contains integer points, using the fact that the face given by the previous vector in the sequence also contains integer points. Although the size of the vector is doubled every time, the number of iterations of the Chvátal-Gomory rounding procedure in each step is at most quadratic.

This $O(n^3 \log n)$ upper bound was later improved by Eisenbrand & Schulz (1999) down to $O(n^2 \log n)$. Lower bounds that are asymptotically larger than linear are not known. Eisenbrand & Schulz (1999) show that the case $P_I = \emptyset$ and $\text{rank}(P) = n$ can only be modeled with an exponential number of inequalities.

Theorem 9. *Let $P \subseteq [0,1]^n$ be a polytope in the 0/1-cube with $P_I = \emptyset$ and $\text{rank}(P) = n$. Any inequality description of P has at least 2^n inequalities.*

This implies that an unsatisfiable set of clauses in n variables can only model a polytope of rank n, if the number of clauses is exponential in n.

5   Cutting Plane Proofs

By using the translation of propositional clauses into clausal inequalities (cf. Sect. 1) cutting planes yield a proof system for propositional logic. It is easy to see that any resolvent of a set of propositional clauses can also be obtained as a Gomory-Chvátal cutting plane from the corresponding clausal inequalities and the bound inequalities $0 \leq x_i, y_j \leq 1$.

Proposition 4. *If the unsatisfiability of a set of clauses has a resolution proof of length l, then it has a cutting plane proof of the same length.*

Let Pig_n be a set of clauses representing the *pigeon hole problem* "Place $n+1$ pigeons into n holes". Using a propositional variable x_{ij} to indicate whether or not pigeon i is placed into hole j, Pig_n is given by the clauses

$$x_{i1} \vee \cdots \vee x_{in}, \text{ for } i = 1, \ldots, n+1$$
$$\bar{x}_{i_1 j} \vee \bar{x}_{i_2 j}, \qquad \text{for } 1 \leq i_1 < i_2 \leq n+1, j = 1, \ldots, n. \tag{24}$$

Theorem 10 (Haken 85). *There is no polynomial upper bound on the length of a shortest resolution proof of the unsatisfiability of Pig_n.*

Proposition 5 (Cook, Coullard and Turán 87). *The unsatisfiability of Pig_n has a cutting plane proof of length $O(n^3)$.*

This shows that cutting planes are strictly more powerful than resolution. There exists a polynomial proof system for propositional logic if and only if NP = coNP. A longstanding open problem was to find a set of formulae for which there exist no polynomial cutting plane proofs. This was finally solved Pudlák in 1995.

Theorem 11 (Pudlák 97). *There exist unsatisfiable propositional formulae φ_n, for which there is no polynomial upper bound on the length of a shortest cutting plane proof.*

Pudlák's proof is based on an effective interpolation theorem that allows one to reduce the lower bound problem on the length of proofs to a lower bound problem on the size of some circuits that compute with real numbers and use non-decreasing functions as gates. Pudlák obtained an exponential lower bound for these circuits by generalizing Razborov's lower bound for monotone Boolean circuits (Razborov 1985). Another exponential lower bound for monotone real circuits has been given independently by Haken & Cook (1999).

The formulae φ_n used in Pudlák's proof express that a graph on n vertices contains a clique of size m and that it is $(m-1)$-colorable. We introduce variables $p_{ij}, 1 \leq i < j \leq n$, to encode the graph, variables $q_{ki}, 1 \leq k \leq m, 1 \leq i \leq n$, to encode a one-to-one mapping from an m-element set into the vertex set of the graph, and variables $r_{il}, 1 \leq i \leq n, 1 \leq l \leq m-1$, to encode an $(m-1)$-coloring of the graph. The inequalities are as follows:

$$
\begin{aligned}
\sum_{i=1}^{n} q_{ki} \geq 1, && \text{for } 1 \leq k \leq m \\
\sum_{k=1}^{m} q_{ki} \leq 1, && \text{for } 1 \leq i \leq n \\
\sum_{l=1}^{m-1} r_{il} \geq 1, && \text{for } 1 \leq i \leq n \\
q_{ki} + q_{k'j} \leq 1 + p_{ij}, && \text{for } 1 \leq i < j \leq n, 1 \leq k \neq k' \leq m \\
p_{ij} + r_{il} + r_{jl} \leq 2, && \text{for } 1 \leq i < j \leq n, 1 \leq l \leq m-1
\end{aligned}
\tag{25}
$$

Pudlák shows that any cutting plane proof of $0 \geq 1$ from these inequalities has at least $2^{\Omega((n/\log n)^{1/3})}$ steps.

There exist several other proof systems for linear 0-1 inequalities whose complexity has been investigated, see (Pudlák 1999) for an overview.

6   Generalized Resolution

Hooker (1988) and (1992) generalized resolution from propositional clauses to arbitrary linear 0-1 inequalities. Here, we present his method for the case of extended clauses. An *extended clause* is a linear 0-1 inequality of the form

$$
L_1 + \cdots + L_m \geq d,
\tag{26}
$$

where $d \geq 1$ is a positive integer number and $L_i, i = 1, \ldots, m$, is either a *positive literal* x_i or a *negative literal* $\overline{x}_i = 1 - x_i$. The intuitive meaning of (26) is that at least d out of the m literals L_i have to be true. Classical clauses or clausal inequalities correspond to the case $d = 1$. If $x_i, i = 1, \ldots, l$, are the positive literals and $\overline{y}_j, j = 1, \ldots, k$, are the negative literals, then the extended clause (26) can also be written in the form

$$
x_1 + \cdots + x_l - y_1 - \cdots - y_k \geq d - k.
\tag{27}
$$

In many cases, extended clauses give a more compact representation of a set $S \subseteq \{0,1\}^n$ than classical clauses. For example, the extended clause $L_1 + \cdots + L_m \geq d$ is equivalent to the conjunction $\bigwedge_{I \subseteq \{1,\ldots,m\}:|I|=m-d+1} \sum_{i \in I} L_i \geq 1$ of $\binom{m}{m-d+1}$ classical clauses.

A classical clause $L_1 + \cdots + L_m \geq 1$ implies another clause $L'_1 + \cdots + L'_k \geq 1$ if and only if $L = \{L_1, \ldots, L_m\} \subseteq L' = \{L'_1, \ldots, L'_k\}$. A similar result holds for extended clauses. Abbreviate an extended clause $L_1 + \cdots + L_m \geq d$ by $L \geq d$ and view L as a set of literals $\{L_1, \ldots, L_m\}$ of cardinality $|L| = m$. Then $L \geq d$ implies $L' \geq d'$ if and only if $|L \setminus L'| \leq d - d'$. This means that the implication problem for extended clauses is easy, while for arbitrary linear 0-1 inequalities it is coNP-complete. *Generalized resolution* is described by the inference rules:

$$\text{Impl\_Ext\_Cl}: \qquad \frac{L \geq d}{L' \geq d'} \quad \text{if } |L \setminus L'| \leq d - d'$$

$$\text{Resolution}: \qquad \frac{L_i + M \geq 1, \overline{L}_i + N \geq 1}{M + N \geq 1}$$

$$\text{Diagonal\_Sum}: \qquad
\begin{array}{l}
\square \quad L_2 \; + \cdots \; + L_{m-1} + L_m \geq \quad d \\
\overline{L}_1 \; \square \; + \cdots \; + L_{m-1} + L_m \geq \quad d \\
\quad \vdots \\
\underline{L_1 \; + L_2 + \cdots \; + L_{m-1} \; \square \; \geq \quad d} \\
L_1 \; + L_2 + \cdots \; + L_{m-1} + L_m \geq \; d + 1
\end{array}
\tag{28}$$

Generalizing a classical result from Quine, Hooker (1992) showed that by applying these rules one can compute all prime extended clauses for a set of extended clauses C. An extended clause $L \geq d$ is *prime* for C if $L \geq d$ is implied by C and if there is no other extended clause implied by C that implies $L \geq d$. A set of extended clauses is unsatisfiable if and only one can derive by generalized resolution the empty clause $0 \geq 1$. Barth (1996) introduced *simplifying* resolvents and diagonal sums and showed how these can be computed efficiently. He also devised an algorithm to compute for an arbitrary linear 0-1 inequality an equivalent set of extended clauses. (Bockmayr 1993, Bockmayr 1995, Barth & Bockmayr 1995, Barth & Bockmayr 1996) study logic-based and polyhedral 0-1 constraint solving in the context of constraint logic programming.

References

Balas, E. (1979), 'Disjunctive programming', *Annals of Discrete Mathematics* **5**, 3 – 51.

Balas, E., Ceria, S. & Cornuéjols, G. (1993), 'A lift-and-project cutting plane algorithm for mixed 0-1 programs', *Mathematical Programming* **58**, 295 – 324.

Balas, E., Ceria, S. & Cornuéjols, G. (1996), 'Mixed 0-1 programming by lift-and-project in a branch-and-cut framework', *Management Science* **42**(9), 1229 – 1246.

Balas, E., Ceria, S., Cornuéjols, G. & Natraj, N. R. (1996), 'Gomory cuts revisited', *Operations Research Letters* **19**.

Barth, P. (1996), *Logic-based 0-1 constraint programming*, Operations Research/Computer Science Interfaces Series, Kluwer.

Barth, P. & Bockmayr, A. (1995), Finite domain and cutting plane techniques in CLP($\mathcal{PB}$), *in* L. Sterling, ed., 'Logic Programming. 12th International Conference, ICLP'95, Kanagawa, Japan', MIT Press, pp. 133 – 147.

Barth, P. & Bockmayr, A. (1996), Modelling 0-1 problems in CLP($\mathcal{PB}$), *in* 'Practical Application of Constraint Technology, PACT'96, London', The Practical Application Company Ltd, pp. 1 – 9.

Blair, C. & Jeroslow, R. G. (1978), 'A converse for disjunctive constraints', *Journal of Optimization Theory and Applications* **25**, 195 – 206.

Bockmayr, A. (1993), Logic programming with pseudo-Boolean constraints, *in* F. Benhamou & A. Colmerauer, eds, 'Constraint Logic Programming. Selected Research', MIT Press, chapter 18, pp. 327 – 350.

Bockmayr, A. (1995), Solving pseudo-Boolean constraints, *in* 'Constraint Programming: Basics and Trends', Springer, LNCS 910, pp. 22 – 38.

Bockmayr, A. & Eisenbrand, F. (1999), Cutting planes and the elementary closure in fixed dimension, Research Report MPI-I-99-2-008, Max-Planck-Institut für Informatik, Im Stadtwald, D-66123 Saarbrücken, Germany.

Bockmayr, A., Eisenbrand, F., Hartmann, M. & Schulz, A. S. (1999), 'On the Chvátal rank of polytopes in the 0/1 cube', *Discrete Applied Mathematics* **98**, 21 – 27.

Chandru, V. & Hooker, J. N. (1999), *Optimization Methods for Logical Inference*, Wiley.

Chvátal, V. (1973), 'Edmonds polytopes and a hierarchy of combinatorial problems', *Discrete Mathematics* **4**, 305 – 337.

Chvátal, V., Cook, W. & Hartmann, M. (1989), 'On cutting-plane proofs in combinatorial optimization', *Linear Algebra and its Applications* **114/115**, 455–499.

Cook, W., Coullard, C. R. & Turán, G. (1987), 'On the complexity of cutting plane proofs', *Discrete Applied Mathematics* **18**, 25 – 38.

Cook, W., Gerards, A. M. H., Schrijver, A. & Tardos, E. (1986), 'Sensitivity theorems in integer linear programming', *Mathematical Programming* **34**, 251 – 264.

Cook, W., Hartmann, M., Kannan, R. & McDiarmid, C. (1992), 'On integer points in polyhedra', *Combinatorica* **12**(1), 27 –37.

Eisenbrand, F. (1999), 'On the membership problem for the elementary closure of a polyhedron', *Combinatorica* **19**(2), 297–300.

Eisenbrand, F. & Schulz, A. S. (1999), Bounds on the Chvátal rank of polytopes in the 0/1 cube, *in* 'Integer Programming and Combinatorial Optimization, IPCO 99', Springer, LNCS 1610, pp. 137–150.

Gerards, A. M. H. (1990), On cutting planes and matrices, *in* W. Cook & P. D. Seymour, eds, 'Polyhedral Combinatorics', DIMACS, pp. 29–32.

Gomory, R. E. (1958), 'Outline of an algorithm for integer solutions to linear programs', *Bull. AMS* **64**, 275 – 278.

Grötschel, M., Lovász, L. & Schrijver, A. (1988), *Geometric algorithms and combinatorial optimization*, Vol. 2 of *Algorithms and Combinatorics*, Springer.

Haken, A. (1985), 'The intractability of resolution', *Theoretical Computer Science* **39**, 297 – 308.

Haken, A. & Cook, S. A. (1999), 'An exponential lower bound for the size of monotone real circuits', *Journal of Computer and System Sciences* **58**, 326 – 335.

Hayes, A. C. & Larman, D. G. (1983), 'The vertices of the knapsack polytope', *Discrete Applied Mathematics* **6**, 135 – 138.

Hooker, J. N. (1988), 'Generalized resolution and cutting planes', *Annals of Operations Research* **12**, 217 – 239.

Hooker, J. N. (1992), 'Generalized resolution for 0-1 linear inequalities', *Annals of Mathematics and Artificial Intelligence* **6**, 271–286.

Jeroslow, R. G. (1977), 'Cutting-plane theory: disjunctive methods', *Annals of Discrete Mathematics* **1**, 293 – 330.

Khachiyan, L. G. (1979), 'A polynomial algorithm in linear programming', *Soviet Math. Doklady* **20**, 191 – 194.

Lagarias, J. C. (1985), 'The computational complexity of simultaneous diophantine approximation problems', *SIAM J. Computing* **14**(1), 196 – 209.

Lenstra, H. W. (1983), 'Integer programming with a fixed number of variables', *Mathematics of Operations Research* **8**(4), 538 – 548.

Padberg, M. W. & Grötschel, M. (1985), Polyhedral computations, *in* E. L. Lawler, J. K. Lenstra, A. Rinnoy Kan & D. B. Shmoys, eds, 'The Traveling Salesman Problem', John Wiley & Sons, pp. 307–360.

Pudlák, P. (1997), 'Lower bounds for resolution and cutting plane proofs and monotone computations', *Journal of Symbolic Logic* **62**(3), 981–988.

Pudlák, P. (1999), On the complexity of the propositional calculus, *in* 'Sets and Proofs, Invited papers from Logic Colloquium'97', Cambridge Univ. Press, pp. 197 – 218.

Razborov, A. A. (1985), 'Lower bounds on the monotone complexity of some Boolean functions', *Soviet Math. Dokl.* **31**(2), 354 – 357.

Schrijver, A. (1980), 'On cutting planes', *Annals of Discrete Mathematics* **9**, 291 – 296.

Schrijver, A. (1986), *Theory of Linear and Integer Programming*, John Wiley.

Sherali, H. D. & Shetty, C. M. (1980), *Optimization with disjunctive constraints*, Vol. 181 of *Lecture Notes in Economics and Mathematical Systems*, Springer.

Towards Cooperative Interval Narrowing

Laurent Granvilliers

IRIN, Université de Nantes
B.P. 92208, F-44322 Nantes Cedex 3, France
Laurent.Granvilliers@irin.univ-nantes.fr

Abstract. P. Van Hentenryck *et al.* have designed an efficient interval constraint solver combining box consistency and Gauss-Seidel iterations, that is the core of Numerica. F. Benhamou *et al.* have shown that hull consistency may be faster and more accurate than box consistency. Their algorithm merges both consistency techniques taking care of the constraints' expressions. This paper presents a new algorithm BC5 enforcing hull consistency, box consistency and the interval Gauss-Seidel method. The main idea is to weaken the local contractions and to let the propagation operate between all elementary solvers in order to accelerate the computation while preserving the same precision. Algorithm BC5 is finally compared with the constraint solving algorithm of Numerica.

1 Introduction

Numerical constraints over continuous domains are involved in many applications from robotics, chemistry, image synthesis, etc. Sets of constraints over continuous variables are called *numeric constraint satisfaction problems* (CSP). Constraint solving algorithms are of particular importance since they are often used by optimization or differentiation engines. Among them, branch and prune algorithms alternate domain pruning, enforcing local consistency techniques to remove from variables' domains some inconsistent values w.r.t. constraints, and branching to search for the solutions.

This paper focuses on pruning —filtering— algorithms enforcing local consistency techniques over numeric CSPs. Two consistency notions worth mentioning are *hull* [6] and *box* [5] consistency. The collective thought advised until recently that box consistency should be enforced over *complex* constraints and hull consistency over *simple* constraints (w.r.t. the length of constraints' expressions and the number of occurrences of variables). In [4], it has been shown that both must be combined to gain in precision and time. In [29], box consistency has been used with the interval Gauss-Seidel method where the numeric CSP is linearized through a first order Taylor expansion. Their interaction is a form of coarse-grained cooperation since both solvers only alternate after the computation of a fixed-point. Furthermore, constraint consistency techniques were shown to be more efficient when variables' domains are large, though Gauss-Seidel methods are at their best when the domains are tight. Thus, it is a trivial idea to combine both in the previously mentioned ordering.

H. Kirchner and C. Ringeissen (Eds.): FroCoS 2000, LNAI 1794, pp. 18–31, 2000.

This paper reviews the combination of constraint consistency techniques and Gauss-Seidel-based methods. Both algorithms presented in [29,4] are integrated in a new solver that takes advantage of their own characteristics. The main conclusions are the following:

- an effective cooperation is needed between all solvers; in particular, the computation of a fixed-point of each solver leads to slow convergence phenomena, though the application of another solver before the fixed-point is reached may contract more the domains at a lower cost;
- some knowledge of the solvers' behavior is required for accelerating the whole solving process; in particular some heuristics concerning their efficiencies w.r.t. the constraints' expressions, the size of domains, etc. lead to fix a partial ordering on their applications, that avoids some unnecessary computations. In other words, a pure concurrency or a coarse-grained cooperation are in general not efficient.

The conclusions are confirmed by some experimental results on the set of all benchmarks we have collected (but only the more significant ones are reported). Different algorithms we have implemented are compared with the constraint solving algorithm of Numerica. The best one is from five to ten times faster on average.

The rest of the paper is organized as follows: Section 2 introduces some materials from interval arithmetic and local consistency techniques. Section 3 presents the novel algorithm. Experimental results are discussed in Section 4. Directions for future research are sketched in Section 5.

2 Preliminary Notions

This section reviews some efficient techniques from interval analysis and artificial intelligence for solving numerical constraints over continuous domains.

2.1 Interval Arithmetic

Interval arithmetic was designed by R.E. Moore [24] for computing roundoff error bounds of numerical computations. Real quantities are represented by *intervals* of real numbers enclosing them.

Definition 1 (Interval). *Given two real numbers a and b such that $a \leqslant b$, the set $\{x \in \mathbb{R} \mid a \leqslant x \leqslant b\}$ is an* interval *denoted by* $[a, b]$.

Practical experiments are based on the set $\mathbb{I}$ of machine-representable intervals whose bounds are floating-point numbers. Given a relation $\rho \subseteq \mathbb{R}$, $\mathsf{Hull}(\rho)$ denotes the intersection of all elements of $\mathbb{I}$ containing ρ. An interval is said *canonical* if it contains at most two floating-point numbers. The *width* of an interval $[a, b]$ is the quantity $b - a$.

Interval arithmetic operations are set theoretic extensions of the real ones. They are usually implemented by floating point operations[1] over the bounds of intervals; for example: $[a, b] + [c, d] = [a + c, b + d]$ and $[a, b] - [c, d] = [a - d, b - c]$. Composition of operations defines a fast and reliable method for computing an *outer approximation* of the range of a continuous real function, also called *interval evaluation* of the function: for example, the range of $f(x) = 1 + 2 \times x - x$ over $[0, 1]$ is necessarily included in $[1, 1] + [2, 2] \times [0, 1] - [0, 1] = [0, 3]$. This process corresponds to the evaluation of the interval function $F(X) = \mathsf{Hull}(\{1\}) + \mathsf{Hull}(\{2\}) \times X - X$ called the *natural interval extension* of f. More generally, we have the following definition:

Definition 2 (Interval extension). *A function $F : \mathbb{I}^n \to \mathbb{I}$ is an* interval extension *of $f : \mathbb{R}^n \to \mathbb{R}$ if the range of f over I is included in $F(I)$ for every I in the domain of f.*

2.2   Constraint Satisfaction

In the following, a *constraint* is an atomic formula built from a real-based structure $\langle \mathbb{R}, \mathcal{O}, \{=, \leqslant, \geqslant\} \rangle$ —where $\mathcal{O}$ is a set of operations— and a set of real-valued variables. We denote by ρ_c the relation associated with any constraint c (considering the usual interpretation). A *numeric CSP* is defined by a set of constraints and the interval vector $I = (I_1, \dots, I_n)$ of variables' domains. A solution of the CSP is an element of the Cartesian product $I_1 \times \cdots \times I_n$ —called a *box*— verifying all constraints.

The limitations of machine arithmetic only permit the computation of approximations —here, interval vectors— of the solutions. An *approximate solution* of the CSP is a vector $J \subseteq I$ verifying all constraints. A vector J verifies an elementary constraint c of the form $\mathsf{r}(f, g)$ if $\mathsf{R}(F(J), G(J))$ holds[2], where F (resp. G) is an interval extension of f (resp. g), and R is the interval symbol corresponding to r. Interval symbols are interpreted as follows: given $I = [a, b]$ and $J = [a', b']$, we have:

$$I = J \text{ if } I \cap J \neq \varnothing \qquad I \leqslant J \text{ if } a \leqslant b' \qquad I \geqslant J \text{ if } b \geqslant a'$$

In other words, the constraint is verified if the terms' interval evaluations are in the interval interpretation of the corresponding relation symbol. If J does not satisfy c, then the fundamental property of interval arithmetic [24] guarantees that it does not contain any solution of c.

2.3   Constraint Narrowing and Propagation

The process of solving a CSP is defined as the computation of precise (in the sense of domains' widths) approximate solutions of all solutions. A generic branch and

---

[1] Results of operations are rounded towards the infinities on machines supporting IEEE binary floating-point arithmetic [16].

[2] Constraints' relation symbols considered in this paper are binary.

prune algorithm solves a CSP by iterating two steps: the pruning of domains while preserving the solution set; and the splitting of domains —branching step— to generate more precise sub-domains when some solutions cannot be separated. The pruning step is generally implemented by a filtering algorithm —constraint propagation process— that iteratively applies some elementary solvers associated with constraints, called *constraint narrowing operators*, until no domain can be further tightened (see [26,3,2] for more details).

Definition 3 (Constraint narrowing operator). *A* constraint narrowing operator (CNO) *for the constraint c is a contracting* $(N(\boldsymbol{B}) \subseteq \boldsymbol{B})$ *and monotonic* $(\boldsymbol{B} \subseteq \boldsymbol{B}' \Rightarrow N(\boldsymbol{B}) \subseteq N(\boldsymbol{B}'))$ *function N verifying* $\rho_c \cap \boldsymbol{B} \subseteq N(\boldsymbol{B})$ *for every boxes* $\boldsymbol{B}, \boldsymbol{B}'$.

In the framework of numeric CSPs, a CNO usually checks a consistency property of a set of variables' domains with respect to a (local) set of constraints. This paper focuses on approximations of *arc consistency* over intervals, namely *hull* and *box consistency* (see [10] for a comparison of partial consistencies over intervals), verifying at a time the consistency of one interval domain with respect to one constraint of the CSP. These notions are defined as instances of *2B consistency* proposed by O. Lhomme [18]. The main idea is to check only the consistency of the bounds of domains, that avoids the combinatorial explosion arising when all values are considered as in the discrete case.

The standard filtering algorithm for 2B consistency —a variation of AC3 [19]— maintains a queue that verifies the following invariant: the domains are necessarily a fixed-point of every CNO not in that queue. If the queue becomes empty then a global fixed-point is reached, and if the product of domains becomes empty then the CSP's inconsistency is established.

Box Consistency and Related Algorithms

Box consistency was defined by F. Benhamou *et al.* in [5].

Definition 4 (Box consistency). *A constraint c is said* box consistent w.r.t. *a box* $I_1 \times \cdots \times I_n$ *and a given interval extension if and only if* $I_i = \mathsf{Hull}(\{r_i \in I_i \mid (I_1, \ldots, I_{i-1}, \mathsf{Hull}(\{r_i\}), I_{i+1}, \ldots, I_n)$ *satisfies* $c\})$ *for all* $i \in \{1, \ldots, n\}$.

Roughly speaking, c is box consistent if the canonical intervals at bounds of I_i satisfy it. Consequently, a CNO implementing box consistency typically removes from the bounds of one domain some inconsistent canonical intervals. An efficient implementation operates a binary or ternary splitting of I_i for searching for the leftmost and rightmost consistent canonical intervals. Furthermore, all constant terms (that do not contain the i-th variable) can be evaluated once before the search. Algorithm BC3revise [5] combines this process with an *interval Newton method* (an interval version of the real one). Algorithm BC3 [5] is the filtering algorithm derived from AC3 that applies CNOs implemented by some variations of BC3revise.

In practice, it is often more efficient to stop propagation and narrowing processes before the computation of a fixed-point in BC3 and BC3revise, and to apply another CNO or filtering algorithm. The precision of BC3revise is called here *narrowing precision factor*, that estimates the maximal distance between the output domain and the fixed-point domain (only some of the inconsistent canonical intervals may be removed). This idea has been proposed in [13] in order to speed-up BC3 by enforcing an algorithm weaker than BC3revise while preserving the same final precision (computation of box consistency). The precision of BC3 is also influenced by the *domain improvement factor* that estimates the smallest width of domains' contractions that may stop the propagation (in this case box consistency may not be computed). In practice, a good tuning of both precision parameters is a condition for efficiency.

Hull Consistency and Related Algorithms

Hull consistency originates from a paper of J.G. Cleary [9].

Definition 5 (Hull consistency). *A constraint c is said* hull consistent w.r.t. a box $\boldsymbol{B}$ *if and only if $\boldsymbol{B} = \mathsf{Hull}(\rho_c \cap \boldsymbol{B})$.*

In other terms, a constraint is hull consistent if the domains cannot be strictly contracted without removing some elements of the associated relation. The different narrowing algorithms that enforce hull consistency are based on the inversion of constraints' expressions. Algorithm HC3revise [9] takes as input a *primitive* constraint (in practice, a constraint having at most one operation symbol) though Algorithm HC4revise [4] operates on more general constraints. HC4revise is able to contract the domains of all variables occurring in any constraint c by performing two traversals in the tree-structured representation of c. The main difference is that HC4revise is not idempotent (it has to be reinvoked for the computation of a fixed-point) unlike HC3revise.

The computation of HC3revise for the constraint $c : x + y = z$ and domains I_x, I_y, and I_z, is as follows:

$$
\begin{aligned}
I_x &\leftarrow I_x \cap (I_z - I_y) \\
I_y &\leftarrow I_y \cap (I_z - I_x) \\
I_z &\leftarrow I_z \cap (I_x + I_y)
\end{aligned}
$$

Algorithm HC4 is the filtering algorithm derived from AC3 that applies CNOs implemented by HC4revise. The precision of domains returned by HC4 depends on the domain improvement factor as in BC3.

2.4   Gauss-Seidel Method

The *interval Gauss-Seidel method* is an algorithm for bounding the solution set of interval linear systems [14,1,25,15]. In the case of square systems (n variables, n constraints) of nonlinear equations $\boldsymbol{F}(\boldsymbol{X}) = \boldsymbol{0}$, the linear system is obtained through a first-order Taylor expansion around the center of variables' domains.

Given an interval vector I, $\mathsf{m}(I)$ the midpoint of I, $F'(I)$ an interval extension[3] of the Jacobian matrix of F over I, Equation (1) holds:

$$F(\mathsf{m}(I)) + F'(I)(I - \mathsf{m}(I)) = 0 \qquad (1)$$

Let $V = I - \mathsf{m}(I)$ then Equation (1) becomes:

$$F'(I)V = -F(\mathsf{m}(I)) \qquad (2)$$

E.R. Hansen [15] has shown that if $F'(I)$ is not *diagonally dominant*, it is more efficient to apply a *preconditioning* —multiplication by a matrix preserving the solution set— transforming Equation (2) into:

$$PF'(I)V = -PF(\mathsf{m}(I)) \qquad (3)$$

where P —a preconditioner— is the inverse of the real matrix of midpoints of $F'(I)$. The resulting linear system $AV = B$ (Equation (3)) is then solved through Gauss-Seidel iterations: the initial value of V is $W = I - \mathsf{m}(I)$. For i taking the values from 1 to n, V_i is computed by the following formula, assuming that the value of A_{ii} is different from 0 (otherwise V_i is not modified):

$$V_i = (B_i - \sum_{j=1}^{i-1} A_{ij}V_j - \sum_{j=i+1}^{n} A_{ij}W_j)/A_{ii}$$

The resulting variables' domains are then $I \cap V + \mathsf{m}(I)$. This process, commonly called *multidimensional interval Newton method* in the interval analysis community, can be seen as a non-idempotent filtering algorithm that may be applied on a square subpart of the initial CSP.

E.R. Hansen has proposed to iterate some Gauss-Seidel-based algorithms until no sufficient domain modifications happen. In the framework of numeric CSPs, we argue that iterating Gauss-Seidel is less efficient than applying other (constraint) solving techniques that may contract the variables' domains at a lower cost.

3   Cooperative Algorithm

We propose to design a new interval constraint solver combining two existing methods: box consistency combined with the Gauss-Seidel method [29], implemented in Numerica, and box consistency combined with hull consistency [4]. As the latter was shown to be more efficient than box consistency, the idea of combining both methods is immediate. Nevertheless, finding an efficient strategy is not so easy and needs an exhaustive experimental study. Heuristics must decide on: fine-grained integration of all methods or coarse-grained cooperation; ordering of application of methods; computation or not of a local fixed-point for each method; computation or not of a global fixed-point; tunings for the precision of CNOs and filtering algorithms in order to prevent slow convergences phenomena. The quality criterion will be the balance between pruning and computation time.

---

[3] Each coefficient (ij) of the matrix $F'(I)$ is an outer approximation of the range of the partial derivative $\partial f_i / \partial x_j$.

3.1   Knowledge of Existing Methods

The design of the new algorithm is based on the knowledge of existing methods we are aware of, summarized below:

- The computation of hull consistency using HC4revise is fast since the contraction of the domains of all variables in a constraint only needs two traversals in the constraint's expression [4].
- Hull consistency is more precise than box consistency when variables occur once in constraints' terms. Thus, BC3revise is only used in the case of variables occurring more than once [4].
- Gauss-Seidel is efficient when variables' domains are tight [15], since the error term of the Taylor formula is bounded by the evaluation of the derivative over the whole domains. As a consequence, constraint processing techniques should beforehand contract the domains as much as possible [29].
- The computation of a fixed-point for box consistency may lead to slow convergence phenomena, though stopping before the fixed-point may lead to the generation of a huge number of sub-domains in a bisection algorithm. An algorithm presented in [13] enforces a weak box consistency and gradually increases the precision during the search of the fixed-point. Other narrowing/propagation strategies have been presented in [22].

3.2   Algorithm BC5

Table 1 presents the filtering algorithm BC5 for contracting the domains of a numeric CSP. BC5 combines three filtering algorithms: HC4 processing all constraints having some simple occurrences of variables; BC3 taking as input all constraints' projections over variables with multiple occurrences in the associated constraints; and Gauss-Seidel iterations —procedure GS— applied on a square system resulting from the application of any formal method —algorithm SquareSystem (see [8] for some implementations of that procedure). Let remark that SquareSystem must generate a CSP that defines a superset of the set of solutions of the initial CSP. The ordering of application of these methods is motivated by both remarks that HC4 is faster (but eventually less precise) than BC3, and that tighter are the input domains (that may be obtained by enforcing consistency techniques), more precise are Gauss-Seidel iterations.

Algorithm HC4 stops when there is no sufficient contraction of the domain of a variable occurring once in a constraint from $\mathcal{C}$ —the width of domain is reduced by less than 10% (parameter γ being the domain improvement factor)— since box consistency is supposed to be more precise for multiple occurrences of variables. BC3 stops when there is no sufficient contraction of any domain (parameter γ). Moreover, BC3revise is parameterized by a narrowing precision factor (parameter φ) and does not use any (costly) Newton's method. Consequently, no fixed-point of any filtering or narrowing algorithm is in general computed.

The combination of HC4 with BC3 has led to the design of Algorithm BC4 in [4]. The combination of BC3 on the whole set of projections with Gauss-Seidel has been presented in [29]. Algorithm BC5 does more than combining

Table 1. Filtering algorithm BC5.

BC5 (in: $\{C_1, \ldots, C_m\}$; inout: $\boldsymbol{I} = (I_1, \ldots, I_n)$)
let $\gamma = 10\%$ be the domain improvement factor
let $\varphi = 10^{-8}$ be the narrowing precision factor
begin

$\quad \mathcal{C} \leftarrow \{C_i \mid 1 \leqslant i \leqslant m$, there exists a variable occurring once in $C_i\}$

$\quad \mathcal{P} \leftarrow \{(C_i, x_j) \mid 1 \leqslant i \leqslant m,\ 1 \leqslant j \leqslant n,\ x_j$ occurs more than once in $C_i\}$

$\quad \mathcal{S} \leftarrow \mathsf{SquareSystem}(\{C_1, \ldots, C_m\}, n)$

$\quad$**if** $\mathcal{C} \neq \varnothing$

$\quad$**then** HC4($\mathcal{C},\boldsymbol{I},\gamma$)   % *constraint processing with hull consistency*

$\quad$**endif**

$\quad$**if** $\mathcal{P} \neq \varnothing$ **and** $\boldsymbol{I} \neq \varnothing$

$\quad$**then** BC3($\mathcal{P},\boldsymbol{I},\gamma,\varphi$)   % *constraint processing with box consistency*

$\quad$**endif**

$\quad$**if** $\mathcal{S} \neq \varnothing$ **and** $\boldsymbol{I} \neq \varnothing$

$\quad$**then** GS($\mathcal{S},\boldsymbol{I}$)   % *Gauss-Seidel iterations*

$\quad$**endif**

end

both methods since it is equivalent to HC4 combined with Gauss-Seidel if the set of projections $\mathcal{P}$ is empty, and to BC3 combined with Gauss-Seidel if the set of constraints $\mathcal{C}$ is empty. All these algorithms will be experimentally compared (see section 4).

Finally, we argue that it is more efficient to bisect the domains before reinvoking BC5 for solving a numeric CSP with a finite set of solutions (typically defined by a square set of equations). The implemented bisection strategy splits one domain in three equal parts. Domains to be bisected are chosen using a round-robin strategy [29] that is the more efficient on average.

The problem of characterizing the output domains from BC5 is important. In [4], it has been shown that any fixed-point strategy that combines HC4 with BC3 computes box consistency. In [29], the domains obtained from a fixed-point of Gauss-Seidel iterations have been shown to be box consistent with respect to the constraints' Taylor extensions (one line of Equation (1)). As a consequence, if BC5 is iteratively applied given $\gamma = \varphi = 0$, until no domain can be further tightened, then the output domains are box consistent with respect to a set of interval extensions of the input constraints. Nevertheless, the domains resulting from one application of BC5 are not in general box consistent. We have the following, expected, result:

Proposition 1. *Let $\langle C, \boldsymbol{I} \rangle$ be a CSP, and $\boldsymbol{J} = (J_1, \ldots, J_n)$ the interval vector resulting from the call of* BC5($C, \boldsymbol{I}$). *Then, we have:*

- *$\boldsymbol{J} \subseteq \boldsymbol{I}$,*
- *$J_1 \times \cdots \times J_n$ is a superset of the solution set of $\langle C, \boldsymbol{I} \rangle$.*

Proof. Follows directly from the property that a composition of narrowing operators defined from elementary constraints is still a narrowing operator for the conjunction of these constraints [2].

4   Experimental Results

Table 2 compares the computation times of several filtering algorithms embedded in a standard branch and prune —bisection— algorithm that splits the domains (in three parts) if the desired precision of 10^{-8} is not reached. HC4+BC3 corresponds to BC5 when the Gauss-Seidel method is disconnected; BC3+GS to BC5 when HC4 is disconnected and $\mathcal{P}$ is composed of all constraints' projections; HC4+GS to BC5 when BC3 is disconnected and $\mathcal{C}$ is composed of all constraints. The last column contains the results from Numerica on the same machine (SUN Sparc Ultra 2/166MHz). The figures in the column labelled by v are the number of variables of the CSPs. Computation times exceeding one hour are replaced by a question mark. A blank space means that the data is not available (the results of Numerica are extracted from [30]).

Benchmarks in Table 2, extracted from [30,31] and corresponding to challenging problems for the state-of-the-art computer algebra, continuation or interval methods, are classified in four categories with respect to the more efficient method solving them: respectively box consistency combined with Gauss Seidel, hull consistency, and hull consistency combined with Gauss-Seidel; the last ones are easily solved by all methods. Moreover, problem Transistor is solved by embedding the filtering algorithms into a *3B consistency*-based search process.

The results are analyzed now:

1. BC5 is the best algorithm on average. HC4+BC3 suffers from the impossibility to remove some solutions that can be handled by Gauss-Seidel. BC3+GS is slow for simple occurrences of variables. HC4+GS may be imprecise for multiple occurrences of variables. Furthermore BC5 outperforms all other algorithms for problems Nbody and Moré-Cosnard composed of variables occurring once and twice in constraints (since HC4+BC3 is faster than BC3 and more precise than HC4).

2. Hull consistency is always faster and more precise than box consistency in the case of simple occurrences of variables. The local search performed by BC3revise may be necessary in the case of multiple occurrences of variables, for example for problems Broyden-Banded and Chemistry. Surprisingly, hull consistency is more efficient than box consistency on some problems composed of multiple occurrences of variables as Kinematics 2, Rouiller Robot, etc. But unfortunately, this behavior cannot be anticipated.

3. Gauss-Seidel is efficient when the domains are tight. It permits us to separate the solutions and then to avoid the generation of a huge number of parasite domains: this is the case for Nbody, Rose and all benchmarks from the third category (see the rather bad results in column HC4+BC3).

4. HC4+GS is trivially faster than BC3+GS when CSPs only contain variables occurring once in constraints. This is the case for DC Circuit, Pentagon, Dessin d'Enfant 2, Kinematics 1, Neurophysiology, Transistor, I1 and I4. In this case, BC5 is equivalent to HC4+GS.

5. When hull consistency is combined with box consistency, HC4revise is stopped when no domain of a variable occurring once is contracted (since box

Table 2. Comparison of Filtering Algorithms (times in seconds).

Benchmark	v	BC5	HC4+BC3	BC3+GS	HC4+GS	Numerica
Nbody	6	**341.40**	?	764.70	1185.00	
Rose	3	**2.65**	?	3.10	15.70	
Broyden-Banded	10	**0.10**	0.25	0.10	3.40	0.50
	20	**0.25**	0.50	0.25	?	1.70
	40	**0.50**	1.05	0.50	?	4.00
	80	**1.10**	2.10	1.10	?	9.20
	160	**2.20**	4.10	2.20	?	30.00
	320	**4.75**	8.35	4.75	?	151.80
	640	**9.85**	17.35	10.25	?	
	1280	**20.75**	34.70	21.10	?	
Chemistry	5	**0.35**	38.05	0.50	35.35	1.10
Caprasse	4	5.80	25.10	**4.10**	5.00	
DC circuit	10	0.20	**0.15**	1.15	0.20	
Pentagon One	11	**0.05**	0.15	0.15	0.05	
Pentagon All	11	**1.75**	6.00	5.60	1.75	
Moré-Cosnard	10	**0.05**	0.15	0.10	0.10	1.80
	20	**0.15**	0.40	0.70	0.40	17.30
	40	**0.55**	1.70	5.20	1.45	204.90
	80	**2.45**	6.95	41.40	6.40	
Moré-Cosnard Aux	10	**0.20**	0.20	0.25	0.20	0.40
	20	0.60	**0.55**	0.70	0.60	1.30
	40	2.45	**1.95**	2.90	2.45	7.40
	80	13.75	**7.05**	14.20	13.75	51.80
	160	96.10	**12.75**	96.60	96.10	397.00
Dessin d'Enfant 1	8	1062.25	?	1486.55	**1047.50**	
Transistor	9	**444.80**	?	1677.65	444.80	2359.80
Seyfert Filter	9	378.70	?	1093.25	**328.85**	
Noonburg Network	5	**80.55**	?	170.25	80.55	
Bellido Kinematics	9	84.35	?	245.55	**81.55**	
Rouiller Robot	9	66.50	?	121.90	**53.50**	
Dessin d'Enfant 2	10	**57.50**	?	154.05	57.50	
Neurophysiology	8	**46.15**	178.95	141.95	46.15	108.00
Kinematics 2	8	25.05	1175.20	53.35	**19.20**	243.30
Katsura Magnetism	6	7.40	321.30	14.75	**7.20**	
Trinks	6	3.10	37.00	9.10	**3.00**	
Wood function	4	2.50	79.05	6.60	**2.10**	
Solotarev	4	1.40	?	3.40	**0.40**	
Cyclohexane	3	1.05	?	1.05	**0.75**	
Brown	5	**0.85**	55.80	3.35	0.85	
Kinematics 1	12	**0.75**	110.90	2.00	0.75	7.20
Wright	5	**0.40**	1.65	1.90	0.40	
Bifurcation	3	**0.15**	0.55	0.25	2.55	
Combustion	10	**0.00**	0.05	0.10	0.00	0.00
I1	10	**0.00**	0.00	0.05	0.00	0.10
I4	10	**13.65**	16.90	23.50	13.65	

consistency is supposed to be more precise), and after the computation of a fixed-point when used alone. This is efficient for Moré-Cosnard: BC5 is two times faster than HC4+GS. Nevertheless, this is not always efficient, in particular for Caprasse and Cyclohexane where the improvement ratio is inversed.

6. Moré-Cosnard Aux problem is a counterexample to the concept of variable introduction proposed in [30] in order to share the common expressions between constraints. This process considerably increases the system's dimension (from n^2 to about $9n^2$) that dramatically slows down Gauss-Seidel, that is of no need for solving the problem. This can be seen in Table 2 for Moré-Cosnard Aux with 160 variables: the cost of Gauss-Seidel is about seven times the cost of HC4+BC3. Furthermore, hull consistency is very efficient on the original Moré-Cosnard problem: the filtering process takes about 6s for 80 variables, and 13s for Moré-Cosnard Aux.

   Nevertheless, variable introduction permits us to speed-up the solving of Kinematics 1 and Transistor (as is proposed in [30]) and the overcost of Gauss-Seidel is insignificant.

7. The application of BC5 where $\mathcal{P}$, $\mathcal{C}$ and $\mathcal{S}$ correspond to all datas from the CSP is not efficient since some unnecessary computations are performed.

8. BC3+GS outperforms Numerica, that implements the same algorithms. We may identify two reasons: Numerica computes a fixed-point of BC3 and of Gauss-Seidel before bisection; and the implementation of BC3revise in our system does not use any Newton's method and does not compute exact box consistency due to the nonzero narrowing precision factor. The improvement is remarkable for problem Moré-Cosnard since the cost of the interval Newton method grows according to the length of constraints' expressions.

5   Conclusion and Perspectives

The solving time for numeric CSPs is improved through cooperation of narrowing algorithms. Algorithm BC5 herein described makes cooperate algorithms enforcing hull consistency, box consistency, along with a Gauss-Seidel method to compute box consistency for a set of different interval extensions of some constraints. To date, four algorithms are devoted to the computation of box consistency, namely: BC3, the algorithm used by Numerica, BC4, and BC5.

The research reported in this paper may be extended in many ways, some of which are described hereafter. First, the accuracy of narrowing algorithms may be improved through the use of several interval extensions for the considered constraints. Consequently, we plan to implement *Bernstein* and *nested* forms [28] for polynomial constraints, as well as centered forms like Taylor extensions of order $k \geqslant 2$ [20]. Therefrom, a challenging task is the isolation of heuristics to efficiently combine these extensions.

High-order consistencies [18] are powerful but their implementation needs refinement. There is a room for improvement concerning the efficient combination of bisections and interval tests for searching for zeros of conjunctions of constraints. Furthermore, choosing the right consistency level to be used for a given CSP requires heuristics that still need to be devised. More generally, given (parts of) a CSP, a difficult task is to know the lowest consistency level that permits eventually reaching some given precision for the solutions.

Gaps in variables' domains could be handled as proposed in [15]. For example, consider Algorithm HC4 where union of intervals are computed, and then immediately connected into intervals for complexity reasons: some improvement could be obtained by keeping track of the information provided by the gaps, to be used during the bisection process; another idea [17] is to extract solutions from domains as soon as they are isolated during the narrowing process.

Algorithm BC5 uses a basic strategy. Though the tuning of precision parameters thus allowed appears efficient, a clever strategy would probably be to merge computations from HC4 and BC3, then gradually decreasing the parameters' values.

Two wider research themes to improve interval narrowing algorithms may also be identified: cooperation of linear/non-linear solvers [11,27], and cooperation of symbolic/numeric techniques [21,23,12]. Achievements from these cooperations are, for instance, the symbolic generation of redundant constraints (e.g. by means of Gaussian elimination, the Simplex algorithm, or Gröbner bases), a first step towards our final goal: the processing of a CSP as a whole rather than by combining elementary computations from several constraints considered independently.

Finally, we report two problems that we consider difficult since the solving time by BC5 is much long. The first one is a square, scalable to any arbitrary order n, system from Reimer appearing in [7]. The variables lie in the domain $[-10^8, 10^8]$. The equations of the system have the generic form:

$$0.5 = \sum_{j=1}^{i} (-1)^{j+1} x_j^{n-i+2} \quad 1 \leqslant i \leqslant n$$

The second system, from C. Nielsen and B. Hodgkin—called here Dipole—and also described in [7], corresponds to the "*determination of magnitudes, directions, and locations of two independent dipoles in a circular conducting region from boundary potential measurements*". The variables are initially known to lie in $[-10, 10]$. The system is as follows:

$$\begin{cases} 0.6325400 = a + b \\ -1.3453400 = c + d \\ -0.8365348 = ta + ub - vc - wd \\ 1.7345334 = va + wb + tc + ud \\ 1.3523520 = at^2 - av^2 - 2ctv + bu^2 - bw^2 - 2duw \\ -0.8434530 = ct^2 - cv^2 + 2atv + du^2 - dw^2 + 2buw \\ -0.9563453 = at^3 - 3atv^2 + cv^3 - 3cvt^2 + bu^3 - 3buw^2 + dw^3 - 3dwu^2 \\ 1.2342523 = ct^3 - 3ctv^2 - av^3 + 3avt^2 + du^3 - 3duw^2 - bw^3 + 3bwu^2 \end{cases}$$

The solving of these problems will probably sound the knell of Algorithm BC5.

Acknowledgements. This work was partly supported by the LOCO project of INRIA Rocquencourt and the franco-russian Liapunov Institute.

The author thanks Frédéric Benhamou and Frédéric Goualard for interesting discussions on these topics. The author is grateful to the referees, Martine Ceberio and Frédéric Goualard whose remarks have led to significant improvements of this paper.

References

1. G. Alefeld and J. Herzberger. *Introduction to Interval Computations.* Academic Press, 1983.
2. K.R. Apt. The Essence of Constraint Propagation. *Theoretical Computer Science*, 221(1-2):179–210, 1999.
3. F. Benhamou. Heterogeneous Constraint Solving. In *Proc. of International Conference on Algebraic and Logic Programming*, 1996.
4. F. Benhamou, F. Goualard, L. Granvilliers, and J-F. Puget. Revising Hull and Box Consistency. In *Proc. of International Conference on Logic Programming*, 1999.
5. F. Benhamou, D. McAllester, and P. Van Hentenryck. CLP(Intervals) Revisited. In *Proc. of International Logic Programming Symposium*, 1994.
6. F. Benhamou and W. J. Older. Applying Interval Arithmetic to Real, Integer and Boolean Constraints. *Journal of Logic Programming*, 32(1):1–24, 1997.
7. D. Bini and B. Mourrain. Handbook of Polynomial Systems. 1996.
8. M. Ceberio. Preconditioning and Solving Techniques for Constrained Global Optimization. Master's thesis, University of Nantes, France, 1999.
9. J.G. Cleary. Logical Arithmetic. *Future Computing Systems*, 2(2):125–149, 1987.
10. H. Collavizza, F. Delobel, and M. Rueher. Comparing Partial Consistencies. *Reliable Computing*, 5(3):213–228, 1999.
11. A. Colmerauer. Naive solving of non-linear constraints. In Frédérric Benhamou and Alain Colmerauer, editors, *Constraint Logic Programming: Selected Research*, pages 89–112. MIT Press, 1993.
12. L. Granvilliers. A Symbolic-Numerical Branch and Prune Algorithm for Solving Non-linear Polynomial Systems. *Journal of Universal Computer Science*, 4(2):125–146, 1998.
13. L. Granvilliers, F. Goualard, and F. Benhamou. Box Consistency through Weak Box Consistency. In *Proc. of International Conference on Tools with Artificial Intelligence*, 1999.
14. E.R. Hansen. On the Solution of Linear Algebraic Equations with Interval Coefficients. *Linear Algebra and its Applications*, 2:153–165, 1969.
15. E.R. Hansen. *Global Optimization using Interval Analysis.* Marcel Dekker, 1992.
16. IEEE. IEEE standard for binary floating-point arithmetic. Technical Report IEEE Std 754-1985, Institute of Electrical and Electronics Engineers, 1985. Reaffirmed 1990.
17. Y. Lebbah and O. Lhomme. Consistance énumérante. In *Proc. of French Conference on Practical Solving of NP-Complete Problems*, 1999.
18. O. Lhomme. Consistency Techniques for Numeric CSPs. In *Proc. of International Joint Conference on Artificial Intelligence*, 1993.
19. A.K. Mackworth. Consistency in Networks of Relations. *Artificial Intelligence*, 8(1):99–118, 1977.

20. K. Makino and M. Berz. Efficient Control of the Dependency Problem based on Taylor Model Method. *Reliable Computing*, 5:3–12, 1999.
21. P. Marti and M. Rueher. A Distributed Cooperating Constraint Solving System. *International Journal on Artificial Intelligence Tools*, 4(1):93–113, 1995.
22. E. Monfroy. Using Weaker Functions for Constraint Propagation over Real Numbers. In *Proc. of ACM Symposium on Applied Computing*, 1999.
23. E. Monfroy, M. Rusinowitch, and R. Schott. Implementing Non-Linear Constraints with Cooperative Solvers. In *Proc. of ACM Symposium on Applied Computing*, 1996.
24. R.E. Moore. *Interval Analysis*. Prentice-Hall, Englewood Cliffs, NJ, 1966.
25. A. Neumaier. *Interval Methods for Systems of Equations*. Cambridge University Press, 1990.
26. W. Older and A. Vellino. Constraint Arithmetic on Real Intervals. In Frédéric Benhamou and Alain Colmerauer, editors, *Constraint Logic Programming: Selected Research*. MIT Press, 1993.
27. M. Rueher and C. Solnon. Concurrent Cooperating Solvers over Reals. *Reliable Computing*, 3(3):325–333, 1997.
28. V. Stahl. *Interval Methods for Bounding the Range of Polynomials and Solving Systems of Nonlinear Equations*. PhD thesis, University of Linz, Austria, 1995.
29. P. Van Hentenryck, D. McAllester, and D. Kapur. Solving Polynomial Systems Using a Branch and Prune Approach. *SIAM Journal on Numerical Analysis*, 34(2):797–827, 1997.
30. P. Van Hentenryck, L. Michel, and Y. Deville. *Numerica: a Modeling Language for Global Optimization*. MIT Press, 1997.
31. J. Verschelde. Database of Polynomial Systems. Michigan State University, USA, 1999. http://www.math.msu.edu/j̃an/demo.html.

Integrating Constraint Solving into Proof Planning

Erica Melis[1], Jürgen Zimmer[2], and Tobias Müller[3]

[1] Fachbereich Informatik, Universität des Saarlandes,
D-66041 Saarbrücken
melis@ags.uni-sb.de
[2] Fachbereich Informatik, Universität des Saarlandes,
D-66041 Saarbrücken
jzimmer@ags.uni-sb.de
[3] Programming Systems Lab, Postfach 15 11 50, Universität des Saarlandes,
D-66041 Saarbrücken
tmueller@ps.uni-sb.de

Abstract. In proof planning mathematical objects with theory-specific properties have to be constructed. More often than not, mere unification offers little support for this task. However, the integration of constraint solvers into proof planning can sometimes help solving this problem. We present such an integration and discover certain requirements to be met in order to integrate the constraint solver's efficient activities in a way that is correct and sufficient for proof planning. We explain how the requirements can be met by n extension of the constraint solving technology and describe their implementation in the constraint solver $CoSIE$.

In automated theorem proving, mathematical objects satisfying theory-specific properties have to be constructed. More often than not, unification offers little support for this task and logic proofs, say of linear inequalities, can be very long and infeasible for purely logical theorem proving. This situation was a reason to develop theory reasoning approaches, e.g., in theory resolution [19], constrained resolution [6], and constraint logic programming [8] and to integrate linear arithmetic decision procedures into provers such as Nqthm [4]. Boyer and Moore, e.g., report how difficult such as integration may be.

In knowledge-based proof planning [12] external reasoners can be integrated. In particular, a domain-specific constraint solver can help to construct mathematical objects that are elements of a specific domain. As long as these mathematical objects are still unknown during the proof planning process they are represented by place holders, also called *problem variables*. In [11] we described a first hand-tailored constraint solver Lineq that incrementally restricts the possible object values. It checks for the inconsistency of constraints and thereby influences the search for a proof plan.

This paper presents the integration of an extended standard constraint solver into proof planning and describes several generally necessary extensions of off-the-shelf constraint solvers for their correct use in proof planning. As a result

H. Kirchner and C. Ringeissen (Eds.): FroCoS 2000, LNAI 1794, pp. 32–46, 2000.

more theorems from three investigated mathematical areas (convergence of real-valued functions, convergent sequences, and continuous functions) can be proved by our proof planner.

The paper is organized as follows: First we introduce knowledge-based proof planning as it is realized in the mathematical assistant system ΩMEGA [3] and its concrete integration of constraint solving into proof planning. In section 2 we summarize the requirements that the integration into proof planning causes for constraint solving. In section 3, we discuss the essential extensions of constraint solving for proof planning. Finally, we illustrate the proof planning and particularly $CoSIE$'s work with a concrete proof planning example. In the following, Δ, Φ, and Ψ denote sets of formulas.

1   Integration of Constraint Solving into Proof Planning

Proof planning, introduced by A.Bundy [5], differs from traditional search-based techniques by searching for appropriate proof steps at abstract levels and by a global guidance of the proof search. Knowledge-based proof planning [12] extends this idea by allowing for domain-specific operators and heuristics, by extending the means of heuristic guidance, and by integrating domain-specific external reasoning systems.

Proof planning can be described as an application of classical AI-planning where the initial state consists of the two proof assumptions represented by sequents[1] and of the goal which is a sequent representing the theorem to be proved. For instance, for proving the theorem LIM+ which states that the limit of the sum of two real-valued functions f and g at a point $a \in \mathbb{R}$ (a real number a) is the sum of their limits the initial planning state consists of the goal

$$\emptyset \vdash \lim_{x \to a} f(x) + g(x) = L_1 + L_2$$

and of the proof assumptions

$$\emptyset \vdash \lim_{x \to a} f(x) = L_1 \quad \text{and}$$

$$\emptyset \vdash \lim_{x \to a} g(x) = L_2.$$

After the expansion of the definiton of $\lim_{x \to a}$ the resulting planning goal is

$$\emptyset \vdash \forall \epsilon(\epsilon > 0 \to \exists \delta(\delta > 0 \land \forall x((|x - a| < \delta \land x \neq a) \to |(f(x) + g(x)) - (L_1 + L_2)| < \epsilon).$$

Proof planning searches for a sequence of operators that transfers the initial state into a state with no open planning goals. The proof plan operators represent complex inferences that correspond to mathematical proof techniques. These operators are usually more abstract than the rules of the basic logic calculus. Thus, a proof of a theorem is planned at an abstract level and a plan is an outline

---

[1] A sequent $(\Delta \vdash F)$ consists of a set of formulas Δ (the hypotheses) and a formula F and means that F is derivable from Δ.

of the proof. This plan can be recursively expanded to the calculus-level where it can be checked for correctness by a proof checker.[2]

In the following, we briefly introduce knowledge-based proof planning as it is realized in the ΩMEGA system.

1.1   Proof Planning in ΩMEGA

The operators in ΩMEGA have a frame-like representation. As a first example for planning operators, we explain `TellCS` which plays an important role in the integration of constraint solving into proof planning:

operator: `TellCS(CS)`	
premises	L1
conclusions	$\ominus$L2
appl-cond	`is-constraint`$(c,$CS$)$ AND `var-in`(c) AND `tell(L2, CS)`
proof schema	L1. Δ_1 $\vdash$ $\mathcal{C}$ () L2. $\Delta,\ \mathcal{C}$ $\vdash$ c (solveCS;L1)

`TellCS` has the constraint solver `CS` as a parameter. The application of `TellCS` works on goals c that are constraints. When `TellCS` is matched with the current planning state, c is bound to this goal. This is indicated by the conclusion L2. The $\ominus$ in $\ominus$L2 indicates that the planning goal is removed from the planning state when `TellCS` is applied. The operator introduces no new subgoals because there are no $\oplus$-*premises*. An operator is applied only if the application condition, *appl-cond*, evaluates to *true*. The application condition of `TellCS` says that the operator is applicable, if the following conditions are fulfilled. Firstly, the open goal that is matched with the c in line L2 of `TellCS` has to be a constraint, i.e., a formula of the constraint language of the constraint solver that instantiates `CS`. Secondly, the goal should contain at least one problem variable whose value is restricted by c. Last but not least, the constraint goal must be consistent with the constraints accumulated by `CS` so far. The latter is checked by `tell(L2,CS)` which evaluates to true, if `CS` does not find an inconsistency of the instantiated c with the constraints accumulated so far. The constraint solver is accessed via the `tell` function.

The *proof schema* of `TellCS` contains a meta-variable $\mathcal{C}$ that is a place holder for the conjunction of all constraints accumulated (also called *answer constraint*). The instantiation of $\mathcal{C}$ is relevant for line L2 in the *proof schema* that suggests that the constraint can be logically derived from the yet unknown answer constraint.

The control mechanism of our proof planner prefers the operator `TellCS`, if the current planning goal is an inequality or an equation.

---

[2] The basic calculus of the ΩMEGA system is natural deduction (ND) [17].

Another planning operator is `ExistsIntro` [3] which eliminates an existential quantification in a planning goal:

operator: `ExistsIntro`	
premises	$\oplus$L1
conclusions	$\ominus$L2
appl-cond	$M_x :=$`new-meta-var`(x)
proof schema	L1. Δ $\vdash$ $\varphi[M_x/x]$ (OPEN) L2. Δ $\vdash$ $\exists x.\varphi$ (ExistsI;L1)

`ExistsIntro` closes an existentially quantified planning goal that matches L2 by removing the quantifier and replacing the variable x by a new problem variable M_x. The formula $\varphi[M_x/x]$ is introduced as a new subgoal which is indicated by the $\oplus$-*premise* $\oplus$L1. The function `new-meta-var` in the application condition computes a new problem variable with the type of x. The *proof schema* is introduced into the partial proof plan when the operator is expanded. `ExistsIntro` is often applied iteratively for a number of quantifiers when normalizing a goal.

Even if only one operator is applicable, there may be infinitely many branches at a choice point in proof planning. This problem occurs, for example, when existentially quantified variables have to be instantiated. In a complete proof x in $\exists x.\varphi$ has to be replaced by a term t, a *witness* for x. Since usually t is still unknown when `ExistsIntro` is applied, one solution would be to guess a witness for x and to backtrack in search, if no proof can be found with the chosen witness. This approach yields unmanageable search spaces. We have chosen the approach to introduce M_x as a place-holder for the term t and to search for the instantiation of M_x when all constraints on t are known only.

Melis [10] motivates the use of domain-specific constraint solvers to find witnesses for existentially quantified variables. The key idea is to delay the instantiations as long as possible and let the constraint solver incrementally restrict the admissible object values.

1.2 The Integration

Constraint solvers employ domain-specific data structures and algorithms. The constraint solver $\mathcal{CoSIE}$, described later, is a propagation-based real-interval solver. It is integrated as a mathematical service into the distributed architecture of the ΩMEGA system.

Fig. 1 schematically depicts the interface between the proof planner of ΩMEGA and our constraint solver. The constraint solver can be accessed directly by the proof planner and by interface functions that are called in the application conditions of certain planning operators. The proof planner's application of the

---

[3] `ExistsIntro` encapsulates the ND-calculus rule ExistsI which is the rule $\frac{\Delta \vdash F[t/x]}{\Delta \vdash \exists x.F}$, where t is an arbitrary term.

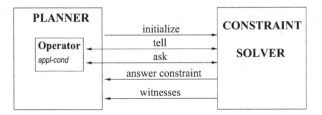

Fig. 1. Interface between constraint solving and proof planning.

operator `InitializeCS` initializes the constraint solver at the beginning of each planning process. During proof planning the main interface is provided via the planning operators `TellCS` and `AskCS`. `TellCS` sends new constraints to the solver by calling the `tell` function and `AskCS` tests entailment of constraints from the constraints collected so far by calling the `ask` function. At the end of the planning process, the proof planner directly calls the constraint solver to compute an answer constraint formula and to search for witnesses for problem variables.

A constraint solver can help to reduce the search during planning because it checks the validity of the application conditions of certain operators by checking for the inconsistency of constraints. When such an inconsistency is detected, the proof planner backtracks rather than continuing the search at that point in the search space.

2   Requirements of Constraint Solving in Proof Planning

For an appropriate integration of constraint solving into proof planning, several requirements have to be satisfied. The most relevant ones are discussed in the following.

1. Proof planning needs to process constraints containing terms, e.g., $E_1 \leq \epsilon/(2.0 * M)$. These terms may contain names of elements of a certain domain (e.g., 2.0) as well as variables (e.g., M, E_1) and symbolic constants (e.g., ϵ). So, as opposed to systems for variables constrained by purely numeric terms, the constraint representation and inference needs to include non-numeric (we say "symbolic") terms in order to be appropriate for proof planning.

In the following, we always use the notion "numeric" to indicate that a certain value or inference is related to a certain domain, although this domain does not necessarily have to contain natural, rational, or real numbers.

2. Since in the planning process not every variable occurs in the sequents of the initial state, the set of problem variables may be growing. In particular, proof planning operators may produce new auxiliary variables that are not contained in the original problem. Moreover, the set of constraints is incrementally growing and typically reaches a stable state at the end of the planning process only. Therefore, dynamic constraint solving [14] is needed.

3. Since backtracking is possible in proof planning constraints that have already been added to the constraint store may be withdrawn again.

4. In proof planning a constraint occurs in a sequent $(\Delta \vdash c)$ that consists of a set Δ of hypotheses and the actual constraint formula c. The hypotheses provide the *context* of a constraint and must be taken into account while accumulating constraints, in computing the answer constraint, and in the search for instantiations of problem variables. Therefore, we refer to a sequent $\Delta \vdash c$ as a *constraint* in the rest of this paper. Importantly, certain problem variables, called *shared variables*, occur in different - possibly contradicting - contexts. For instance, the new contexts $\Delta \cup \{X = a\}$ and $\Delta \cup \{X \neq a\}$ result from introducing a case split $(X = a \lor X \neq a)$ into a proof plan, where Δ is the set of hypotheses in the preceding plan step. When a new constraint $\Delta \cup \{X = a\} \vdash c$ is processed in the $X = a$ branch of the proof plan, its consistency has to be checked with respect to all constraints with a context Φ which is a subset of $\Delta \cup \{X = a\}$.

5. In order to yield a logically correct ND-proof when the operators are expanded, those constants that are introduced by the ND-rules $\forall I$ and $\exists E$ [4] have to satisfy the E*igenvariable condition*, i.e., they must not occur in other formulas beforehand. That is, they must not occur in witnesses that will be instantiated for place holders in the formulas. This condition must be satisfied by the search for witnesses of problem variables.

3   Constraint Solving for Proof Planning

Many off-the-shelf constraint solvers are designed to tackle combinatorial (optimization) problems. For them all problem variables are introduced at the beginning and the solver submits the problem to a monolithic search engine that tries to find a solution without any interference from outside.

An established model for (propagation-based) constraint solving [18] involves numeric constraint inference over a *constraint store* holding so-called *basic* constraints over a domain as, for example, the domain of integers, sets of integers, or real numbers. A basic constraint is of the form $X = v$ (X is bound to a value v of the domain), $X = Y$ (X is equated to another variable Y), or $X \in B$ (X takes its value in B, where B is an approximation of a value of the respective domain). Attached to the constraint store are *non-basic* constraints. Non-basic constraints, as for example "$X + Y = Z$" over integers or real numbers, are more expressive than basic constraints and, hence, require more computational effort. A non-basic constraint is realized by a computational agent, a *propagator*, observing the basic constraints of its parameters which are variables in the constraint store (in the example X, Y, and Z). The purpose of a propagator is to infer new basic constraints for its parameters and add them to the store. That happens until no further basic constraints can be inferred and written to the store, i.e., until a fix-point is reached. Inference can be resumed by adding new constraints

---

[4] these are the rules $\forall I \frac{\Delta \vdash F[a/x]}{\Delta \vdash \forall x.F}$ and $\exists E \frac{\Delta \vdash \exists x.F \quad \Delta, F[a/x] \vdash G}{\Delta \vdash G}$, where a must not occur in any formula in $\Delta \cup \{F, G\}$.

either basic or non-basic. A propagator terminates if it is either inconsistent with the constraint store or explicitly represented by the basic constraints in the store, i.e., entailed by the store.

The common functionalities of these constraint solvers are consistency check, entailment check, reflection, and search for instantiations. (In)consistency check includes the propagation of constraints combined with the actual consistency algorithm, e.g., with arc-consistency AC3 [9].

No previous solver satisfies all the above mentioned requirements and therefore we developed an extended constraint solver that can be safely integrated into proof planning. In the following, we describe the extensions of this solver and the implementation of these extensions.

3.1   Extensions of Constraint Solving

In order to meet requirement 1, a symbiosis of numeric inference techniques as well as domain specific term rewriting rules are needed. To meet the requirements 2,3, and 4, we introduce so called *context trees* which store constraints wrt. their context and enable an efficient test for subset relations between contexts. The context tree is also used to compute a logically correct answer constraint formula and to build the initial constraint store for the search for witnesses.

Constraint Inference. We employ two different kinds of constraint inference in order to detect inconsistencies as fast as possible and to symbolically solve and simplify symbolic constraints. One algorithm efficiently tests a set of constraints for inconsistencies by inspecting and handling the numeric bounds of variables. We refer to this algorithm as *numeric inference*. Another algorithm for *symbolic inference* uses term rewrite rules to simplify the symbolic representation of constraints and constraint simplification rules to transform a set of constraints into a satisfiability equivalent one which is in a unique *solved form*.

A typical constraint solver for (in)equalities in in real numbers $\mathbb{R}$ that represents constraints by numerical lower and upper bounds has to be extended because otherwise in some cases unique bounds cannot be determined. For example, if a problem variable D has two upper bounds, δ_1 and δ_2 which are symbolic constants. These bounds cannot be replaced by a unique upper bound unless a functions *min* is employed. Constraint simplification rules help to determine and to reduce the sets of upper (lower) bounds of a problem variable and to detect inconsistencies which cannot be found efficiently by purely numeric inference. For instance, the constraint $X < Y + Z \land Y + Z < W \land W < X$ is obviously inconsistent, but numeric inference cannot detect this inconsistency efficiently. This requires a constraint representation that can be handled by numeric and symbolic inference. The extension of a constraint solver needs to integrate both inference mechanisms into a single solver and benefit from the results of the respective other inference.

Context Trees. Context trees consist of nodes, the *context nodes*. Each such node N_Φ consists of a set Φ of hypotheses (the context) and a set $S_\Phi = \{c \mid \Delta \vdash c$ is constraint and $\Delta \subseteq \Phi\}$.

A context tree is a conjunctive tree representing the conjunction of all constraints stored in the nodes. Fig. 2 shows the structure of such a context tree.

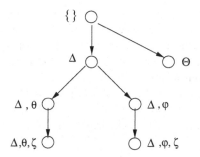

Fig. 2. A Context Tree with node annotations. Δ and Θ are sets of formulas. φ, θ, and ζ are formulas. Δ, φ stands for $\Delta \cup \{\varphi\}$.

The root node is annotated with the empty context $\{\ \}$. A directed edge from a node N_Δ to a child N'_Φ implies $\Delta \subset \Phi$. A subtree T_Φ of a context tree consists of all nodes with a context Ψ for which $\Phi \subseteq \Psi$ holds.

A new constraint $(\Delta \vdash c)$ must be consistent with the constraint sets S_Φ with $\Delta \subset \Phi$. The constraint solver has to check for consistency with the sets S_Φ in the leaf nodes only because the sets of constraints grow from the root node to the leaves. In other words $\Delta \subset \Phi$ implies $S_\Delta \subset S_\Phi$. If an inconsistency occurs in at least one leaf, the constraint $(\Delta \vdash c)$ is not accepted by the constraint solver. Otherwise, c is added to all sets S_Φ in the subtree T_Δ. If the subtree T_Δ is the empty tree, i.e., the context Δ is new to the constraint solver, new nodes N_Δ are created and inserted into the context tree as shown in Fig. 2. This operation preserves the subset relations in the context tree.

When a constraint $(\Delta \vdash c)$ has to be withdrawn because of backtracking in the proof planning, c is simply removed from all nodes in the subtree T_Δ. Empty context nodes are removed from the tree.

The Answer Constraint. At the end of the planning process, the constraint solver uses the structure of the context tree to compute the answer constraint formula. Let $\Delta_1, \ldots, \Delta_n$ be the contexts of all nodes in the context tree and $C_1, \ldots, C_n$ be the conjunctions of all formulas which are new in $S_{\Delta_1}, \ldots, S_{\Delta_n}$ respectively, i.e., $C_i := S_{\Delta_i} - \{c \mid c \in S_{\Delta_j} \text{ with } \Delta_j \subset \Delta_i\}$. Then the answer constraint formula is $\bigwedge_i (\Delta_i \to C_i)$.

Search for Witnesses. Since the context tree is a conjunctive tree witnesses of the problem variables have to satisfy all constraints in the context tree if the respective context is satisfied. The constraint solver searches for a solution for each problem variable which satisfies all constraints. In particular, the search for witnesses of *shared variables* which occur in different contexts has to take into account all constraints of these variables. Therefore, the constraint solver creates a single set with all constraints from the leaf nodes at the beginning of the search process

The search algorithm uses numeric inference and term rewriting to compute an interval constraint $max(L) \leq X \leq min(U)$ for every problem variable X, where $L(U)$ is a list whose first element is the numeric lower(upper) bound $l(u)$ and the rest of $L(U)$ consists of the symbolic lower(upper) bounds. An element is dropped from a bound list as soon as it is found to be not maximal (minimal). Eventually, the maximal lower bound $max(L)$ and the minimal upper bound $min(U)$ are used to compute a witness for X. The search algorithm must not compute witnesses which contain Eigenvariables of the respective problem variable.

3.2   Implementation

This section describes the constraint solver $\mathcal{CoSIE}$ (Constraint Solver for Inequalities and Equations over the field of real numbers). The constraint language of $\mathcal{CoSIE}$ consists of arithmetic (in)equality constraints over the real numbers, i.e., constraints with one of the relations $<,\leq,=,\geq$, and $>$. Terms in formulas of this language are built from real numbers, symbolic constants and variables, and the function symbols $+,-,*,/,min,$ and max. Terms may also contain *ground alien terms*, i.e. ground terms which contain function symbols unknown to $\mathcal{CoSIE}$, i.e., alien. For instance, $|f'(a)|$ is a ground alien term containing the two function symbols $|.|$ and f'. $\mathcal{CoSIE}$ handles these alien terms by variable abstraction, i.e., for constraint inference these terms are replaced by variables and later on instantiated again.

$\mathcal{CoSIE}$ is implemented in the concurrent constraint logic programming language Mozart Oz [16]. $\mathcal{CoSIE}$ builds a context tree whose nodes are *computation spaces* annotated with contexts. A computation space is a Mozart Oz data structure that encapsulates data, e.g., constraints, and any kind of computation including constraint inference. After constraint inference has reached a fix-point, a computation space may have various states: the constraints are inconsistent, all propagators vanished since they are represented by the basic constraints in the constraint store, or the space contains disjunctions, i.e., constraint inference will proceed in different directions.

When a new constraint $(\Delta \vdash c)$ is sent to the solver by TellCS, it has to be added to certain computation spaces in the context tree. Therefore, a new computation space s_c containing c only is created and merged with all computation spaces in the leaf nodes of the subtree T_Δ. In each of these computation spaces, the symbolic inference procedure tries to simplify constraints and detect non-trivial inconsistencies. Propagation, i.e. numeric inference, is triggered by the symbolic inference procedure as described in the next paragraph. When a fix-point is reached in numeric and symbolic inference, the resulting computation spaces are asked for their state to detect inconsistencies. If no inconsistency is detected c is inserted into every computation space of the subtree T_Δ by merging with the space s_c.

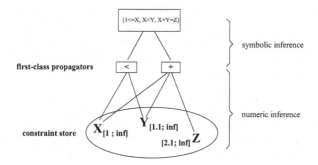

Fig. 3. Combining symbolic and numeric inference.

Symbolic and Numeric Constraint Inference. In $\mathcal{C}o\mathcal{SIE}$, numeric inference is based on the off-the-shelf Real-Interval (RI-) module coming with the Mozart Oz system. The RI-module provides *RI-variables* (constraint variables attributed with intervals of real numbers). As an extension, now the RI-module provides first-class propagators for all relations and functions from $\mathcal{C}o\mathcal{SIE}$'s constraint language. Because of being a first-class data structure these propagators can be inspected, started, and terminated, e.g., by the symbolic inference procedure and at the same time work on the constraint store in the usual way.

The symbiosis of symbolic and numeric inference is based on a shared representation of constraints and by the first-class propagators. Every variable and every symbolic constant occurring in a constraint processed by $\mathcal{C}o\mathcal{SIE}$ is connected to a corresponding RI-variable. The relations and non-alien functions of a constraint are connected to the first-class propagator of those relations and functions of the RI-module.

Fig. 3 illustrates the combination of symbolic and numeric inference. It shows $\mathcal{C}o\mathcal{SIE}$'s connections of the constraint $1 \leq X \wedge X < Y \wedge X + Y = Z$ to the first-class propagators for $<$ and $+$ and to the RI-variables for X, Y, and Z in the constraint store.

The symbolic inference procedure applies (conditional) rewrite rules and constraint simplification rules from the theory of real numbers to (symbolic) constraints in order to transform these constraints into an equivalent normal form. Since the symbolic inference changes the term structure of constraints, it directly influences the corresponding first-class propagators. It starts or terminates first-class propagators connected to the relations and non-alien functions of the terms changed by the application of rewrite and constraint simplification rules. One of the rewrite rules used by $\mathcal{C}o\mathcal{SIE}$ is the following.

$$(t_1 \cdot t_2)/(t_1 \cdot t_3) \quad [t_1 > 0] \Rightarrow t_2/t_3 \qquad (1)$$

If the condition $t_1 > 0$ holds, then the rule cancels out a common factor t_1 in a fraction. When the symbolic inference procedure receives, for instance, the constraint $a > 0 \wedge E \le (a \cdot \epsilon)/(a \cdot M)$, it creates new RI-variables for E, M, ϵ, and a (in case they do not exist yet) and computes new first-class propagators for the relations $>$ and $\le$ and for all occurrences of the functions $/$ and $\cdot$. The rule (1) is applied, to the term $(a \cdot \epsilon)/(a \cdot M)$, which is transformed to the normal form ϵ/M. Thus, the first-class propagators for $\cdot$ in $(a \cdot \epsilon)$ and $(a \cdot M)$ are terminated.

The symbolic inference applies constraint simplification rules to detect inconsistencies as early as possible, e.g.,

$$(t_1 < t_2) \wedge (t_2 < t_3) \wedge (t_3 < t_1) \;\Rightarrow\; \bot \qquad (2)$$

For instance, the constraint $X < Y + Z \;\wedge\; Y + Z < W \;\wedge\; W < X$, already mentioned above, is instantly simplified to $\bot$ by the application of rule (2). With pure numeric inference it would take several minutes to detect this inconsistency.

Search. The search procedure of $\mathcal{CoSIE}$ collects all constraints of the leaf nodes of the context tree in a single computation space, the *root space* of the search tree. As described below, the search may create new computation spaces. The search procedure checks recursively for each space whether it is inconsistent or contains a solution. For each computation space, propagation reduces the domains of the variables. Additionally, the symbolic inference applies term rewrite rules and constraint simplification rules to transform the constraint store into a solved form, to compute a unique symbolic smallest(greatest) upper(lower) bound for each variable, and to detect inconsistencies as early as possible. A set of constraints in solved form does not contain any redundant or trivially valid constraints, e.g., $0 < 1$. One of the simplification rules is

$$(X \le t_1) \wedge (X \le t_2) \;\Rightarrow\; X \le \min\{t_1, t_2\},$$

where the t_i are arithmetic terms and X is a problem variable. When propagation has reached a fix-point and no rewrite and constraint simplification rules are applicable, the space whose state is not failed is said to be stable. For a stable space with undetermined variables a distribution algorithm computes alternatives for the values of a carefully chosen variable X. The search algorithm uses these alternatives to create new disjunctive branches in the search tree, i.e., new computation spaces for every alternative for the domain of X. The new computation spaces contain exactly one of the alternatives and are submitted to recursive exploration again. The entire process is aborted as soon as a solution is found. For instance, if a variable X is constrained by $0 < X \wedge X < \epsilon$, three alternatives for X are computed, expressed by the new constraints $X = \frac{\epsilon}{2}$, $X < \frac{\epsilon}{2}$, and $X > \frac{\epsilon}{2}$.

4   Worked Example

ΩMEGA's proof planner and the integrated constraint solver $\mathcal{CoSIE}$ could find proof plans for many theorems, examples, and exercises from two chapters of the introductory analysis textbook [2]. The now extended constraint solver allows for correctly handling proofs plans that involve a case split. A case split produces alternative contexts of constraints.

A proof that requires a case split is, e.g., the proof of the theorem ContIfDeriv. This theorem states that if a function $f \colon \mathbb{R} \to \mathbb{R}$ has a derivative $f'(a)$ at a point $a \in \mathbb{R}$, then it is continuous in a. In the following, we briefly describe those parts of the planning process for ContIfDeriv that are relevant for the integrated constraint solving. Let's assume a formalization of the problem that implies an initial planning state with the assumption (1):

$$\emptyset \vdash \forall \epsilon_1 (\epsilon_1 > 0 \to \exists \delta_1 (\delta_1 > 0 \to (\forall x_1 (|x_1 - a| < \delta_1 \to ((x_1 \neq a) \to (|\tfrac{f(x_1) - f(a)}{x_1 - a} - f'(a)| < \epsilon_1))))))$$

and the planning goal[5]

$$\emptyset \vdash \forall \epsilon (\epsilon > 0 \to \exists \delta (\delta > 0 \to (\forall x (|x - a| < \delta \to |f(x) - f(a)| < \epsilon))))$$

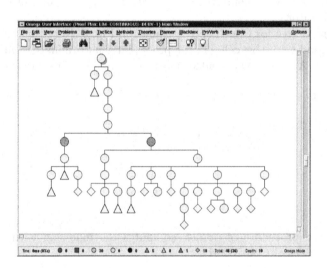

Fig. 4. The proof plan of ContIfDeriv.

The proof planner finds a proof plan for ContIfDeriv as depicted in the screen shot in Fig. 4. During proof planning, the following constraints are passed to the constraint solver $\mathcal{CoSIE}$:

$$
\begin{array}{ll}
\Delta \vdash E_1 > 0 & \Delta \vdash \delta_1 > 0 \\
\Delta \vdash D \leq \delta_1 & \Delta, (X_1 \neq a) \vdash 0 < M \\
\Delta, (X_1 \neq a) \vdash 0 < M' & \Delta, (X_1 \neq a) \vdash D \leq M \\
\Delta, (X_1 \neq a) \vdash |f'(a)| \leq M' & \Delta, (X_1 \neq a) \vdash D \leq \epsilon/(4 * M') \\
\Delta, (X_1 \neq a) \vdash E_1 \leq \epsilon/(2 * M) & \Delta, (X_1 = a) \vdash X_1 = x \ ,
\end{array}
$$

---

[5] In this formalization the definitions of *limit* and *derivative* have already been expanded but this is not crucial for the purpose of this paper.

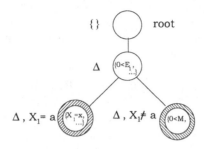

Fig. 5. The context tree for ContIfDeriv.

where Δ consists of the proof assumption (1) and the constraints $\epsilon > 0$ and $D > 0$. The problem variables D, X_1, and E_1 correspond to δ, x_1, and ϵ_1 in the formalization of the problem. M and M' are auxiliary variables introduced by a planning operator.

The context tree for ContIfDeriv is shown in Fig. 5. Note that the two branches correspond to the branches of the proof plan that originate from a case split on $(X_1 = a \vee X_1 \neq a)$. The shaded nodes correspond to the shaded plan nodes in Fig. 4.

At the end of the planning process, $\mathcal{CoSIE}$ computes the following answer constraint:

$$
\begin{aligned}
E_1 > 0 \;\wedge\; \delta_1 > 0 \;\wedge\; D \le \delta_1 \;\wedge& \\
(X_1 \neq a \;\rightarrow\; 0 < M \;\wedge\; D \le M \;\wedge& \\
0 < M' \;\wedge\; |f'(a)| \le M' \;\wedge& \\
E_1 \le \epsilon/(2 \cdot M) \;\wedge& \\
D \le \epsilon/(4 \cdot M')) \;) \;\wedge& \\
(X_1 = a \;\rightarrow\; X_1 = x \;)).&
\end{aligned}
$$

The search procedure of $\mathcal{CoSIE}$ computes the following witnesses for the problem variables of ContIfDeriv:

$$
D = min\{\delta_1, \tfrac{\epsilon}{4 \cdot (|f'(a)|+1)}\} \;,\; X_1 = x \;,\; E_1 = 2 \cdot (|f'(a)| + 1) \;,\; M = D \;,\\
M' = (|f'(a)| + 1).
$$

These witnesses satisfy the Eigenvariable conditions $forbidden(E_1) = \{\delta_1\}$ and $forbidden(D) = \{x\}$.

5   Conclusion

The main theme of this paper is the integration of constraint solvers into proof planning and the nonstandard requirements caused by proof planning. Since off-the-shelf constraint solvers are typically geared towards other applications, we address generic extensions of a standard constraint solver that may also extend the potential application areas of constraint solving.

The reasons for the extensions are manifold: the constraint solver's service has to be integrated into the proof planner in a logically correct way, the constraints are usually not purely numeric, and the control of proof planning, e.g., backtracking, has to be matched on the constraint solver's side.

The programming language Oz is well-suited for the extensions reported in this paper because it provides concurrent propagation-based constraint inference encapsulated in computation spaces. The development of first-class propagators in Oz has been initiated, among others, by our need to combine numeric and symbolic constraint inference. Additionally, Oz provides the means for building new constraint systems from scratch that are as efficient as the built-in ones.

Related Work. A few theorem proving systems directly include specially designed decision procedures for constraint domains, e.g., [4], or a constraint solver [20]. All these systems tightly integrate the constraint solving into theorem proving rather than integrating an external, stand-alone constraint solver. And, of course, none of them does proof planning.

Our previous work [11] mainly dealt with interfacing and integrating the specially designed external constraint solver Lineq into proof planning by designing (Tell and Ask) operators, interface functions, and instantiation procedures. We also investigated with the merits/benefits such an integration can have for proof planning if applied appropriately and correctly [13]. We knew that additional features of the constraint solver are needed but did not elaborate on this. Now *CoSIE* has been developed based on our previous experiences with symbolic constraint solving and based on the RI-module constraint solver of Mozart.

SoleX [15] is a general scheme for the extension of the constraint language of an existing constraint solvers preserving soundness and completeness properties. It combines symbolic and numeric inference in a sequential way. We used the SoleX approach to handle so-called alien terms in the constraint language of *CoSIE*. Constraint handling rules [7] define constraint theories and implement constraint solvers at the same time.

A context is used in the constraint logic programming language CAL [1] to handle guarded clauses. Running a CAL program results in a context tree. Therefore, context trees in CAL are conceptually different to the context trees presented in this paper.

References

1. A. Aiba and R. Hasegawa. Constraint Logic Programming System - CAL, GDCC and Their Constraint Solvers. In *Proc. of the Conference on Fifth Generation Computer Systems.*, pages 113–131. ICOT, 1992.
2. R.G. Bartle and D.R. Sherbert. *Introduction to Real Analysis.* John Wiley& Sons, New York, 1982.
3. C. Benzmueller, L. Cheikhrouhou, D. Fehrer, A. Fiedler, X. Huang, M. Kerber, M. Kohlhase, K. Konrad, A. Meier, E. Melis, W. Schaarschmidt, J. Siekmann, and V. Sorge. OMEGA: Towards a Mathematical Assistant. In W. McCune, editor, *Proc. of CADE-14.* Springer, 1997.

4. R. S. Boyer and J S. Moore. Integrating Decision Procedures into Heuristic Theorem Provers: A Case Study of Linear Arithmetic. *Machine Intelligence (Logic and the Acquisition of Knowledge)*, 11, 1988.
5. A. Bundy. The Use of Explicit Plans to Guide Inductive Proofs. In E. Lusk and R. Overbeek, editors, *Proc. CADE-9*, LNCS 310, Argonne, 1988. Springer.
6. H.-J. Bürckert. A Resolution Principle for Constrained Logics. *Artificial Intelligence*, 66(2), 1994.
7. T. Frühwirth. Constraint Handling Rules. In A. Podelski, editor, *Constraint Programming: Basics and Trends*, LNCS 910. Springer, 1995.
8. J. Jaffar and J-L. Lassez. Constraint Logic Programming. In *Proc. 14th ACM Symposium on Principles of Programming Languages*, 1987.
9. A. K. Mackworth. Consistency in Networks of Relations. *Artificial Intelligence*, 8:99–118, 1977.
10. E. Melis. AI-Techniques in Proof Planning. In *European Conference on Artificial Intelligence*. Kluwer Academic, 1998.
11. E. Melis. Combining Proof Planning with Constraint Solving. In *Proceedings of Calculemus and Types'98*, 1998. Electronic Proceedings http://www.win.tue.nl/math/dw/pp/calc/proceedings.html.
12. E. Melis and J.H. Siekmann. Knowledge-based Proof Planning. *Artificial Intelligence*, 115(1):65–105, 1999.
13. E. Melis and V. Sorge. Employing External Reasoners in Proof Planning. In A. Armando and T. Jebelean, editors, *Calculemus'99*, 1999.
14. S. Mittal and B. Falkenhainer. Dynamic Constraint Satisfaction Problems. In *Proceedings of the 10th National Conference on Artificial Intelligence, AAAI-90*, pages 25–32, Boston, MA, 1990.
15. E. Monfroy and Ch. Ringeissen. SoleX: a Domain-Independent Scheme for Constraint Solver Extension. In J. Calmet and J. Plaza, editors, *Artificial Intelligence and Symbolic Computation AISC'98*, LNAI 1476. Springer, 1998.
16. The Mozart Consortium. *The Mozart Programming System*. http://www.mozart-oz.org/.
17. D. Prawitz. *Natural Deduction - A Proof Theoretical Study*. Almquist and Wiksell, Stockholm, 1965.
18. G. Smolka. The Oz programming model. In Jan van Leeuwen, editor, *Current Trends in Computer Science*. Springer, 1995.
19. M.K. Stickel. Automated Deduction by Theory Resolution. In *Proc. of the 9th International Joint Conference on Artificial Intelligence*, 1985.
20. F. Stolzenburg. Membership Constraints and Complexity in Logic Programming with Sets. In F. Baader and U. Schulz, editors, *Frontiers in Combining Systems*. Kluwer Academic, 1996.

Termination of Constraint Contextual Rewriting

Alessandro Armando and Silvio Ranise

DIST – Università di Genova, Viale Causa 13 – 16145 Genova, Italia
{armando, silvio}@dist.unige.it

Abstract. The effective integration of decision procedures in formula simplification is a fundamental problem in mechanical verification. The main source of difficulty occurs when the decision procedure is asked to solve goals containing symbols which are interpreted for the prover but uninterpreted for the decision procedure. To cope with the problem, Boyer & Moore proposed a technique, called *augmentation*, which extends the information available to the decision procedure with suitably selected facts. Constraint Contextual Rewriting (CCR, for short) is an extended form of contextual rewriting which generalizes the Boyer & Moore integration schema. In this paper we give a detailed account of the control issues related to the termination of CCR. These are particularly subtle and complicated since augmentation is mutually dependent from rewriting and it must be prevented from indefinitely extending the set of facts available to the decision procedure. A proof of termination of CCR is given.

1   Introduction

The effective integration of decision procedures in formula simplification is one of the key problems in mechanical verification. Unfortunately the problem is not easy: while it is relatively straightforward to plug a decision procedure inside a prover, to obtain an effective integration can be a challenge. The main source of difficulty occurs when the decision procedure is asked to solve goals containing symbols which are interpreted for the prover but uninterpreted for the decision procedure. In such situations it is often the case that the decision procedure can not solve the problem and therefore it is of no help to the prover.

To cope with the problem, Boyer & Moore [4] devised and implemented a heuristics, called *augmentation*, which extends the information available to the decision procedure with facts encoding properties of the symbols the decision procedure does not know anything about. In the Boyer & Moore experience the heuristics is crucial to obtain an effective integration: they found out that without the heuristics the decision procedure is of limited use, whereas its introduction improves dramatically the performance of the prover (both in speed and in decreased user interaction).

* We are indebted to Michael Rusinowitch and Sorin Stratulat for many helpful discussions. We are grateful to Enrico Giunchiglia for very useful feedback on an early draft of this paper. The authors are supported in part by *Conferenza dei Rettori delle Università Italiane (CRUI)* in collaboration with *Deutscher Akademischer Austaunschdienst (DAAD)* under the *Vigoni Programme*.

The problem with the augmentation heuristics is that it greatly complicates the integration schema and sophisticated control strategies are needed to control the augmentation activity. The complexity of the resulting integration schema (together with the fact that in the 40 page long description available in [4] high level design decisions are intermixed with optimizations) has probably discouraged many, thereby preventing a wide use of the approach.[1]

Constraint Contextual Rewriting [1] (CCR(X) for short) is a generalized form of contextual rewriting [21, 20] which incorporates the functionalities provided by a decision procedure. The services of the decision procedure are characterized abstractly and the notation CCR(X) (by analogy with the CLP(X) notation used to denote the Constraint Logic Programming paradigm [11]) is used to stress the independence of CCR(X) from the theory decided by the decision procedure. In the same paper we showed that contextual rewriting as well as the integration schemas employed in the simplifiers of NQTHM [4] and Tecton [14] are instances of CCR(X).

This paper extends [1] by providing a detailed account of the control issues of CCR(X). These are particularly subtle and complicated since *(i)* augmentation is mutually dependent from rewriting and *(ii)* augmentation must be prevented from indefinitely extending the set of facts available to the decision procedure. In this paper we refine CCR(X) so to ensure its termination. Notice that the problem of termination is not even mentioned in [4].

Similarly to [4] our work is addressed to those interested in the effective integration of decision procedures in larger provers or in the design of decision procedures intended for such a purpose. By using CCR(X) as a reference model, the problem of the integration of decision procedures in formula simplification is reduced to the implementation of a decision procedure for the fragment of choice whose interface functionalities are as specified in the paper.

It is worth emphasizing that we do not address the problem of *combining* decision procedures as done, e.g., in [19, 17, 6]. Instead, we focus on the complementary problem of *integrating* (or *incorporating*) decision procedures (either compound or not) within formula simplification.

Plan of the paper. We begin in Section 2 by means of an introductory discussion on contextual rewriting. Section 3 presents in details (the refined version of) CCR(X) and includes a thorough discussion on augmentation and the control problems it introduces. The soundness and termination of CCR(X) are then formally stated in Section 4 and the proof of termination is given in the same section. The related work is discussed in Section 5 and some conclusions are drawn in Section 6.

2   Contextual Rewriting

Contextual rewriting is an extended form of conditional rewriting whereby information contained in the context of the expression being rewritten is used by

---

[1] To our knowledge—if we exclude NQTHM and its descendant ACL2 [15]—Tecton [13] is the only prover based on the Boyer & Moore's original ideas.

the rewriting activity. To illustrate, let us consider the problem of rewriting the literal p occurring in the clause $\{p\} \cup E$ via a set of conditional rewrite rules, say R. The key idea of contextual rewriting is that while rewriting p (the *focus* literal) it is legal to assume the truth of the set of negations of the literals in E (the *(rewriting) context*).

Example 1. Let R contain the conditional rule

$$f(U) = f(V) \rightarrow r(g(U, V), U) = U \tag{1}$$

and let $\{r(g(y, z), x) = x, g(x, y) \neq g(y, z), y \neq x\}$ be the clause to be simplified. If we take $r(g(y, z), x) = x$ as focus literal, then the context is $\{g(x, y) = g(y, z), y = x\}$. Notice that the left hand side of (1) does not match with the left hand side of the focus literal. However, the first element of the context, namely $g(x, y) = g(y, z)$, allows us to rewrite the focus literal to $r(g(x, y), x) = x$. The left hand side of (1) now matches with (the rewritten version of) the focus. (1) can then be applied (leaving us with the identity $x = x$) provided that the condition $f(x) = f(y)$ can be established. $f(x) = f(y)$ is not a consequence of R, however it readily follows from the fact $y = x$ occurring in the context by the properties of equalities. $\star$

As the previous example illustrates, the context is used both *(i)* to rewrite the focus literal (possibly enabling the application of rewrite rules) and *(ii)* to establish the conditions of conditional rewrite rules. In both cases the kind of reasoning applied amounts to reasoning about the properties of ground equalities. Indeed a decision procedure for ground equalities, as described for instance in [18], suffices to support contextual reasoning and hence to mechanize the reasoning of Example 1. The key observation here is that contextual rewriting can be obtained by combining traditional conditional rewriting with a "decision procedure" capable of deciding whether any given literal is entailed by the context and normalizing expressions w.r.t. the information available in the context.[2] Furthermore the pattern of interaction between rewriting and the decision procedure does not depend on the theory decided by the decision procedure. CCR(X) is then the result of abstracting contextual rewriting from the theory decided by the decision procedure employed. The traditional notion of contextual rewriting therefore becomes an instance of CCR(X) whereby X is instantiated by a decision procedure for ground equalities and new forms of contextual rewriting can be obtained by instantiating X to decision procedures for different decidable theories.

To illustrate, let us assume that X is instantiated to a decision procedure for total orders and let us consider the following variant of Example 1.

Example 2. Let R contain the conditional rule

$$g(U, V) \leq U \rightarrow r(g(U, V), U) = U \tag{2}$$

and let $r(g(y, z), x) = x$ and $\{x > g(y, z), g(x, y) \leq g(y, z), g(x, y) \geq g(y, z)\}$ be the focus and the context respectively. Similarly to Example 1, the left hand

---

[2] Consistently with the literature we use the term decision procedure in a very liberal sense. A precise definition is given in Section 3.3.

side of (2) does not match with the left hand side of the focus literal. However, since $g(x, y) = g(y, z)$ is a logical consequence of the context we assume that the decision procedure rewrites the focus literal to $r(g(x, y), x) = x$. This step enables the matching of the left hand side of (2) and therefore we are left with the problem of establishing the condition $g(x, y) \leq x$ which is easily found to be a consequence of the context by the decision procedure. $\star$

It is worth pointing out that contextual rewriting differs from conventional rewriting in two essential ways. Firstly proofs do not have a linear structure. This feature (inherited from conditional rewriting) is due to the presence of subsidiary proofs needed to establish the conditions of conditional rewrite rules. This results in a hybrid notion of proof which combines the linear structure of reduction proofs with the tree-structured proofs of sequent calculi. Secondly, due to the dependency from the context, rewriting is a ternary relation and not a binary relation as in conventional rewriting. These considerations motivate the definition of contextual reduction systems introduced in Section 3.2.

3   Constraint Contextual Rewriting

3.1   Preliminaries

By Σ, Π (possibly subscripted) we denote finite sets of function and predicate symbols (with their arity), respectively. A *signature* is a pair of the form (Σ, Π). V (possibly subscripted) denotes a finite set of variables.[3] $\tau(\Sigma, V)$ is the set of *terms* built on Σ and V and defined in the usual way. $\tau(\Sigma)$ abbreviates $\tau(\Sigma, \emptyset)$, i.e. the set of *ground terms*. For the sake of simplicity we consider quantifier-free first-order languages and we assume the usual conceptual machinery (e.g. the notion of substitution and the definition of *position* in an expression) as given, e.g., in [7]. A (Σ, Π, V)-*atom* is either an expression $q(t_1, \ldots, t_n)$ where $q \in \Pi$ and $t_i \in \tau(\Sigma, V)$ $(i = 1, \ldots, n)$ or one of the propositional constants true and false denoting truth and falsity respectively. (Σ, Π, V)-*formulae* are built in the obvious way using the standard logical connectives (i.e. $\neg, \wedge, \vee, \rightarrow, \leftrightarrow$). A (Σ, Π, V)-*literal* is either a (Σ, Π, V)-atom, $p(t_1, \ldots, t_n)$, or a negated (Σ, Π, V)-atom, $\neg p(t_1, \ldots, t_n)$, and $t_1, \ldots, t_n$ are the *arguments* of the literal. The set of (Σ, Π, V)-*expressions* is the union of $\tau(\Sigma, V)$ and the set of (Σ, Π, V)-formulae. We write (Σ, Π)-atom (-literal, -expression) instead of $(\Sigma, \Pi, \emptyset)$-atom (-literal, -expression). A (Σ, Π, V)-*clause* is a disjunction of literals which we indicate as finite set of (Σ, Π, V)-literals. We will write $s \neq t$, $s \not< t$, ... in place of $\neg(s = t)$, $\neg(s < t)$, ... (resp.). $e \sim e'$ stands for $e = e'$ $(e \leftrightarrow e')$ if e and e' are terms (formulae, resp.). If a is an atom, then $\bar{a}$ abbreviates $\neg a$ and $\overline{\neg a}$ stands for a. If Q is a set of literals, then $\overline{Q}$ abbreviates $\{\bar{q} : q \in Q\}$, $Q \rightarrow p$ abbreviates the clause $\overline{Q} \cup \{p\}$, and $\bigwedge Q$ abbreviates any conjunction of all the literals in Q.

---

[3] We use different fonts to emphasize the distinction between the variables and the symbols occurring in signatures and the symbolism used in the informal presentation. For instance we use the notation "x" for a symbol which is part of a given signature, "X" (capital letter) for a variable, and use the notations "x" and "X" in the informal metamathematics. Similarly we use the notation $=, <, \leq, >, \geq$ when these symbols belong to a given signature, and use the notation $=, <, \leq, >, \geq$ when they belong to the informal metamathematics.

If ϕ is a (Σ, Π, V)-formula and Γ is a set of (Σ, Π, V)-formulae, then ϕ is a *logical consequence* of Γ iff $\Gamma \models \phi$, where $\models$ denotes entailment in classical logic. A (Σ, Π, V)-*theory* is a set of (Σ, Π, V)-formulae closed under logical consequence. If T is a theory, then $\Gamma \models_T \phi$ abbreviates $T \cup \Gamma \models \phi$. ϕ is T-*consistent* iff there exists a model of $T \cup \{\phi\}$, and T-*inconsistent* otherwise. ϕ is T-*valid* iff ϕ is a logical consequence of T or, equivalently, iff $\phi \in T$; ϕ and ψ are T-*equivalent* iff $(\phi \leftrightarrow \psi)$ is T-valid.

In the following we consider two theories T_c and T_j of signature (Σ_c, Π_c) and (Σ_j, Π_j) respectively s.t. $\Sigma_c \subseteq \Sigma_j$, $\Pi_c \subseteq \Pi_j$, and $T_c \subseteq T_j$. The objective will be to simplify (Σ_j, Π_j)-expressions using a decision procedure for T_c (hence T_c is assumed to be decidable).

3.2 Contextual Reduction Systems

We now introduce the notion of contextual reduction system by generalizing that of abstract reduction system given in [16]. This will be useful to specify CCR(X) in a way that is *precise*, *concise*, and *incremental* at the same time.

Let $\mathcal{L}$ be a set of labels. For all $\ell \in \mathcal{L}$, let C_ℓ, E_ℓ be sets of expressions and $\mathcal{S}(C_\ell, E_\ell)$ be the set of sequents of the form $c :: e \xrightarrow{\ell} e'$ for all $c \in C_\ell$ and $e, e' \in E_\ell$. A *contextual reduction system* (CRS for short) is a structure $\langle \{\mathcal{S}(C_\ell, E_\ell)\}_{\ell \in \mathcal{L}}, \mathcal{R} \rangle$, where $\mathcal{R}$ is a set of *inference rules* of the form:[4]

$$\frac{c_1 :: e_1 \xrightarrow{\ell_1} e_1' \quad \cdots \quad c_n :: e_n \xrightarrow{\ell_n} e_n'}{c :: e \xrightarrow{r} e'} \quad \text{if } Cond \tag{3}$$

where the r above the arrow in the conclusion of the rule is the name of the inference rule. An ℓ-*reduction of e_0 to e_m in context c* is an expression of the form $c :: e_0 \xrightarrow{\Pi_1}{\ell} e_1 \xrightarrow{\Pi_2}{\ell} \ldots e_{m-1} \xrightarrow{\Pi_m}{\ell} e_m$ with $m \geq 1$ s.t. either $m = 1$, $e_1 = e_0$, and $\Pi_1 = []$ (called *trivial reduction*) or $c :: e_{i-1} \xrightarrow{\ell_i} e_i$ (for $i = 1, \ldots, m$) is the conclusion of an inference rule with premises $c_{i,j} :: e_{i,j} \xrightarrow{\ell_{i,j}} e_{i,j}'$ for $j = 1, \ldots, n_i$ and the j-th element of Π_i is an $\ell_{i,j}$-reduction of $e_{i,j}$ to $e_{i,j}'$ in context $c_{i,j}$. $c :: e_0 \xrightarrow{*}{\ell} e_m$ ($c :: e_0 \xrightarrow{+}{\ell} e_m$) denotes the existence of a (non trivial, resp.) ℓ-reduction of e_0 to e_m in context c.

Let δ and δ' be ℓ-reductions s.t. δ' properly occurs in δ. We say that δ' is a *maximal ℓ-reduction properly occurring in δ* iff there is no other ℓ-reduction δ'' properly occurring in δ and δ' properly occurs in δ''. A CRS is *terminating* iff there exists a family of well-founded relations $\{\prec_\ell\}_{\ell \in \mathcal{L}}$ such that for all $\ell \in \mathcal{L}$ and for all ℓ-reductions $c :: e \xrightarrow{\Pi}{\ell} e'$ we have that $\langle c, e' \rangle \prec_\ell \langle c, e \rangle$ and if $c_1 :: e_1 \xrightarrow{\Pi_1}{\ell} e_1'$ is a maximal ℓ-reduction properly occurring in Π then $\langle c_1, e_1 \rangle \prec_\ell \langle c, e \rangle$.

---

[4] For termination it is convenient to implicitly assume (as we do in the sequel) that the condition $e' \neq e$ is in *Cond*.

3.3   Reasoning Specialists

According to the usual definition, a decision procedure for T_c is a procedure which takes a (Σ_c, Π_c)-formula as input and returns a 'yes-or-no' answer indicating whether the input formula is T_c-consistent or not. Unfortunately, although simple and conceptually elegant, this definition is seldom adequate in practical applications. Efficiency considerations require the procedure to be *incremental*, i.e. capable of processing parts of the input problem as soon as they become available. Moreover, the procedure is often required to return more than a yes-or-no answer. For instance, it can be required the ability of "normalizing" any given expression w.r.t. the information available in its internal state (cf. Example 2). This generalized notion of decision procedure is captured by the notion of reasoning specialist. A *reasoning specialist* is a state-based procedure whose states (called *constraint stores*) are finite sets of (Σ_c, Π_c)-literals represented in some internal form and whose functionalities are abstractly characterized in the following way.[5]

Initialization of the Constraint Store.   The first functionality we consider is the relation cs-init(C) which characterizes the "empty" constraint stores. cs-init(C) is required to be a decidable relation s.t. cs-init(C) holds only if C is T_c-valid.

Detection of Inconsistency.   cs-unsat(C) characterizes a set of T_c-inconsistent constraint stores C whose T_c-inconsistency can be checked by means of a computationally inexpensive syntactic check. We require that cs-unsat(C) is decidable and that cs-unsat(C) implies the T_c-inconsistency of C.

Constraint Store Simplification.   The main functionality of the reasoning specialist is a transition relation over constraint stores, $P :: C \xrightarrow[\text{cs−simp}]{} C'$, which models the activity of adding a finite set of (Σ_j, Π_j)-literals P to C yielding a new constraint store C'. The relation is required to enjoy the following properties:

($sound.cs\text{-}simp$) if $P :: C \xrightarrow[\text{cs−simp}]{} C'$, then $P, \|C\| \models_{T_c} \bigwedge \|C'\|$;

($decreasing.cs\text{-}simp$) if $P :: C \xrightarrow[\text{cs−simp}]{} C'$, then $\langle P, C' \rangle \prec^{cs} \langle P, C \rangle$;

where $\prec^{cs}$ is a well-founded relation over pairs of the form $\langle P, C \rangle$.

Normalization. $C :: s \xrightarrow[\text{cs−normal}]{} s'$ computes a normal representation of s w.r.t. the information stored in C and must enjoy the following properties:

($sound.cs\text{-}normal$) if $C :: s \xrightarrow[\text{cs−normal}]{} s'$, then $\|C\| \models_{T_c} s \sim s'$;

($decreasing.cs\text{-}normal$) if $C :: s \xrightarrow[\text{cs−normal}]{} s'$, then $s' \prec^e s$;

---

[5] If C is a constraint store, then $\|C\|$ denotes the set of literals represented by C. To simplify the presentation we will often blur the distinction between constraint stores and the formulae they encode. For instance, we will talk about the consistency of C meaning the consistency of $\|C\|$.

c	$edges(c)$	c	$edges(c)$
$t_1 \leq t_2$	$\{\langle t_1, \leq, t_2\rangle\}$	$t_1 = t_2$	$\{\langle t_1, \leq, t_2\rangle, \langle t_2, \leq, t_1\rangle\}$
$t_1 \neq t_2$	$\{\langle t_1, \neq, t_2\rangle\}$	$t_1 < t_2$	$\{\langle t_1, \leq, t_2\rangle, \langle t_1, \neq, t_2\rangle\}$
$t_1 \nleq t_2$	$\{\langle t_2, \leq, t_1\rangle\}$	$t_1 \nleq t_2$	$\{\langle t_2, \leq, t_1\rangle, \langle t_1, \neq, t_2\rangle\}$

Table 1. Definition of $edges(c)$

where $\prec^e$ is a reduction ordering over (Σ_j, Π_j, V)-expressions (i.e. a well-founded relation closed under substitution and replacement) containing the sub-expression relation.

Example 3 (A reasoning specialist for total orders). Let $\Pi_c = \{=, <, \leq, >, \geq\}$. T_c is a (Σ_c, Π_c)-theory for total orders. We consider *labeled directed graphs* of the form $G = \langle N, E\rangle$ where $N \subseteq \tau(\Sigma_c)$ and $E \subseteq (N \times L \times N)$ where L is a finite set of edge labels. The set of nodes and the set of edges of G are denoted by $\mathcal{N}(G)$ and $\mathcal{E}(G)$ respectively. Constraint stores are labeled directed graphs (called *transitivity graphs*) $C = \langle N, E\rangle$ s.t. $N = \tau(\Sigma_c)$, $L = \{\leq, \neq\}$, and if $\langle x, \neq, y\rangle \in E$ then $\langle y, \neq, x\rangle \in E$. cs-init$(C)$ holds iff $\mathcal{E}(C) = \emptyset$. cs-unsat$(C)$ holds iff C contains an edge of the form $\langle t_1, \neq, t_2\rangle$ and t_1 and t_2 occur in a strongly $\leq$-connected subgraph of C. If P is a set of literals, then let $Edges(P) = \bigcup_{c \in P} edges(c)$ with $edges(c)$ defined as in Table 1 if c is a (Σ_c, Π_c)-literal and $edges(c) \hat{=} \emptyset$ otherwise. Furthermore, we define $edges(t_1 > t_2) \hat{=} edges(t_2 < t_1)$, $edges(t_1 \ngtr t_2) \hat{=} edges(t_1 < t_2)$, $edges(t_1 \geq t_2) \hat{=} edges(t_2 \leq t_1)$, and $edges(t_1 \ngeq t_2) \hat{=} edges(t_1 < t_2)$. $P :: C \xrightarrow[cs-simp]{} C'$ holds iff $Edges(P) \nsubseteq \mathcal{E}(C)$ and in such a case $C' = \langle \mathcal{N}(C), \mathcal{E}(C) \cup Edges(P)\rangle$. From this it readily follows that $Edges(P) \subseteq \mathcal{E}(C')$ and this suggests that a suitable definition for $\prec^{cs}$ is $\langle P', C'\rangle \prec^{cs} \langle P, C\rangle \hat{=} |Edges(P') \setminus \mathcal{E}(C')| < |Edges(P) \setminus \mathcal{E}(C)|$. The well-foundedness of $\prec^{cs}$ is a straightforward consequence of the well-foundedness of the relation of (proper) set inclusion over finite sets. (*decreasing.cs-simp*) readily follows from the above observations. $C :: e \xrightarrow[cs-normal]{} e'$ is defined to hold whenever $e = e[s]_u$, $e' = e[t]_u$, $e' \prec^e e$, and s and t occur in a strongly $\leq$-connected subgraph of C. $\|C\|$ is the set of atoms $r(s, t)$ s.t. there exists an edge $\langle s, r, t\rangle$ in C. It is easy to verify that $\xrightarrow[cs-simp]{}$ and $\xrightarrow[cs-normal]{}$ enjoy (*sound.cs-simp*) and (*sound.cs-normal*) respectively. $\star$

3.4   Clause Simplification and Rewriting

A simple form of clause simplification can be modeled as the binary relation over clauses, $\xrightarrow[simp]{}$, defined by the following rules:

$$E \cup \{\text{true}\} \xrightarrow[\text{simp}]{cl-true} \{\text{true}\} \qquad E \cup \{\text{false}\} \xrightarrow[\text{simp}]{cl-false} E$$

$$\frac{\overline{E} :: C_\circ \xrightarrow[\text{cs-extend}]{} C \quad C :: p \xrightarrow[\text{ccr}]{} p'}{E \cup \{p\} \xrightarrow[\text{simp}]{cl-simp} E \cup \{p'\}} \text{ if cs-init}(C_\circ)$$

The inference rule *cl-true* (*cl-false*) specifies how to simplify a clause when the constant true (false, resp.) is in it. *cl-simp* says that a literal p in a clause $E \cup \{p\}$ can be replaced by a new literal p' obtained by constraint contextually rewriting p in context C (premise $C :: p \xrightarrow[\text{ccr}]{} p'$), where C is obtained by extending the empty rewriting context $C_\circ$ (condition $\text{cs-init}(C_\circ)$) with the negated literals in E (premise $\overline{E} :: C_\circ \xrightarrow[\text{cs-extend}]{} C$). $\xrightarrow[\text{cs-extend}]{}$ is a ternary relation modeling the interface with the reasoning specialist that for the time being we assume it coincides with the $\xrightarrow[\text{cs-simp}]{}$ relation, i.e. it is defined by:

$$\frac{P :: C \xrightarrow[\text{cs-simp}]{} C'}{P :: C \xrightarrow[\text{cs-extend}]{cs-simp} C'}$$

where P is a finite set of (Σ_j, Π_j)-literals. The $\xrightarrow[\text{cs-extend}]{}$ relation is introduced for modularity reasons and will be extended in Section 3.5.

Let R be a finite set of T_j-valid clauses. The (ternary) *constraint contextual rewriting* relation $\xrightarrow[\text{ccr}]{}$ is defined by the following three rules. A literal p can be rewritten to true in rewrite context C if the result of extending the context with the negation of the literal being rewritten yields an inconsistent rewrite context. This is formalized by:

$$\frac{\{\overline{p}\} :: C \xrightarrow[\text{cs-extend}]{} C'}{C :: p \xrightarrow[\text{ccr}]{cxt-entails} \text{true}} \text{ if } \begin{array}{l} p \text{ is a } (\Sigma_j, \Pi_c)\text{-literal,} \\ \text{cs-unsat}(C') \end{array}$$

The activity of normalizing an expression w.r.t. the information stored in the rewriting context is modeled by:

$$\frac{C :: e \xrightarrow[\text{cs-normal}]{} e'}{C :: e \xrightarrow[\text{ccr}]{normal} e'}$$

Finally conditional rewriting is formalized by:[6]

$$\frac{C :: Q\sigma \xrightarrow[\text{ccr}]{} \emptyset \qquad (Q \to l = r) \in R,}{C :: s[l\sigma]_u \xrightarrow[\text{ccr}]{crew} s[r\sigma]_u} \text{ if } \begin{array}{l} \sigma \text{ is a ground substitution s.t.} \\ Q\sigma \cup \{s[r\sigma]_u\} \prec\!\!\prec^e \{s[l\sigma]_u\} \end{array}$$

---

[6] By abuse of notation we write $C :: Q \xrightarrow[\text{ccr}]{} \emptyset$ in place of $C :: q \xrightarrow[\text{ccr}]{} \text{true}$ for all $q \in Q$.

i.e. rewrite the sub-expression $l\sigma$ at position u in the expression s to the sub-expression $r\sigma$ in the rewriting context C if a clause of the form $(Q \rightarrow l = r)$ is in R, σ is a ground substitution s.t. all the expressions in $Q\sigma \cup \{s[r\sigma]_u\}$ are $\prec^e$-smaller than $s[l\sigma]_u$, and the instantiated conditions $Q\sigma$ can be recursively established in the same rewriting context C. $\prec\!\!\!\prec^e$ is the multiset extension of $\prec^e$.

3.5   Augmenting the Constraint Store

Although the notion of CCR(X) we have defined so far is already a significant improvement over contextual rewriting, there is obviously still room for improvement. A serious limitation is revealed by the situation in which the rewriting context is T_j-inconsistent but not T_c-inconsistent. When this is the case, the T_j-inconsistency of the rewriting context can not possibly be detected by the reasoning specialist. The occurrence in the rewriting context of (function) symbols interpreted in T_j but not in T_c is the main cause of the problem. The key idea of *augmentation* is to extend the rewriting context with T_j-valid facts, thereby informing the reasoning specialist about properties of function symbols it is not aware of. By adding T_j-valid facts to the rewriting context, the heuristics aims at generating a T_j-equivalent but T_c-inconsistent context whose T_j-inconsistency can therefore be detected by the reasoning specialist. The selection of suitable T_j-valid facts is done by looking through the available lemmas.

Example 4. Let T_j be a theory of natural numbers with the usual interpretation of the symbols. Let $\Sigma_c = \Sigma_j$ and let Π_c, T_c, and $\xrightarrow[\text{cs-simp}]{}$ be as in Example 3. Under such hypotheses, the following formula

$$X \geq 4 \rightarrow X^2 \leq 2^X \tag{4}$$

is in T_j, but not in T_c. Let us now consider the problem of establishing the T_j-inconsistency of the set of literals

$$P = \{a \geq 4, b \leq a^2, 2^a < b\} \tag{5}$$

If we add P to an initial constraint store, $C_\circ$, by means of the $\xrightarrow[\text{cs-simp}]{}$ relation, we get a new transitivity graph C comprising the edges $\langle 4, \leq, a \rangle$, $\langle b, \leq, a^2 \rangle$, $\langle 2^a, \leq, b \rangle$, and $\langle 2^a, \neq, b \rangle$. C (and hence P) is trivially T_c-consistent since no strongly $\leq$-connected subgraphs are in it. On the other hand it is also easy to verify that (5) is T_j-inconsistent. If (4) is an available lemma, then from it we can obtain the instance

$$a^2 \leq 2^a \tag{6}$$

by instantiating X to a and relieving the condition $a \geq 4$ using C as rewriting context. Now, it is easy to verify that if we add (6) to C we get a new transitivity graph C' in which 2^a and b occur in a $\leq$-connected subgraph and a $\neq$-labeled edge linking 2^a with b does exist. Therefore C' is T_c-inconsistent.

  Notice that the task of establishing the T_j-inconsistency of (5) falls largely beyond the scope of a decision procedure for total orders. The problem is nevertheless solved thanks to the use of the augmentation heuristics. $\star$

In the above example a single lemma (namely (4)) is available and a single instance of the lemma suffices to detect the inconsistency of the constraint store. In the general case, several lemmas may be available and multiple instances of each lemma can potentially be helpful in detecting the inconsistency of the constraint store. Also, in our example the condition $a \geq 4$ is immediately found to be a consequence of the constraint store whereas in the general case augmentation may also be necessary to establish the conditions of the lemmas. A form of augmentation capable to handle all these situations is obtained by extending the definition of $\xrightarrow[\text{cs-extend}]{}$ with the following rule:

$$\frac{C :: Q\sigma \xrightarrow[\text{ccr}]{} \emptyset \quad C :: c\sigma \xrightarrow[\text{ccr}]{} c' \quad \{c'\} :: C \xrightarrow[\text{cs-simp}]{} C'}{P :: C \xrightarrow[\text{cs-extend}]{naive-augment} C'} \quad \text{if } \begin{array}{l} (Q \to c) \in R, \\ c\sigma \text{ is a } (\Sigma_j, \Pi_c)\text{-literal}, \\ Q\sigma \not\ll^e \{c\sigma\} \end{array}$$

This inference rule states that the rewriting context can be extended with a new literal c' provided that a clause of the form $(Q \to c)$ is in R, σ is a substitution s.t. $Q\sigma \not\ll^e \{c\sigma\}$, the instantiated hypotheses $Q\sigma$ can be established, and c' is the result of (constraint contextual) rewriting $c\sigma$. The rationale for applying $\xrightarrow[\text{ccr}]{}$ to $c\sigma$ is that rewriting can replace some of the symbols in $c\sigma$ with symbols the reasoning specialist knows about (e.g. by unfolding definitions), therefore the addition of c' in place of c enhances the probability of detecting the inconsistency.

The problem with the above rule is that it compromises the termination of CCR(X). We recall from Section 3.2 that for CCR(X) to be terminating, there must exist a well-founded relation $\prec_{\text{cs-extend}}$ s.t. for all cs-extend-reductions $P ::$
$C \xrightarrow[\text{cs-extend}]{\Pi} C'$ we have that $\langle P, C' \rangle \prec_{\text{cs-extend}} \langle P, C \rangle$ and if $P_1 :: C_1 \xrightarrow[\text{cs-extend}]{\Pi'_1} \cdots$ is a maximal cs-extend-reduction properly occurring in Π then $\langle P_1, C_1 \rangle \prec_{\text{cs-extend}} \langle P, C \rangle$. But this is not the case in general. For instance, it is possible to have maximal cs-extend-reductions properly occurring in Π_1 of the form $\{p\} :: C \xrightarrow[\text{cs-extend}]{\Pi'} \cdots$ where p is s.t. $C :: \overline{p_i}\sigma \xrightarrow[\text{ccr}]{}{}^* p$ and $p \in Q$. For termination $\langle \{p\}, C \rangle \prec_{\text{cs-extend}} \langle P, C \rangle$ must hold, but this is not true in general (e.g. when $P = \{p\}$).

To retain termination the following, more elaborate, version of augmentation is needed:

$$\frac{C' :: Q\sigma \xrightarrow[\text{ccr}]{} \emptyset \quad C' :: c\sigma \xrightarrow[\text{ccr}]{} c' \quad \{c'\} :: C' \xrightarrow[\text{cs-simp}]{} C''}{P :: C \xrightarrow[\text{cs-extend}]{augment} C''} \quad \text{if } \begin{array}{l} (Q \to c) \in R, \\ c\sigma \text{ is a } (\Sigma_j, \Pi_c)\text{-literal}, \\ \langle P, C, c \rangle \mapsto \langle C', \sigma \rangle \\ Q\sigma \not\ll^e \{c\sigma\} \end{array}$$

where $\langle P, C, c \rangle \mapsto \langle C', \sigma \rangle$ models the functionality of computing a new constraint store C' and a ground substitution σ given a constraint store C, a finite set of (Σ_j, Π_j)-literals P, and a (Σ_j, Π_c, V)-literal c as input.

Selection of Constraints. The $\mapsto$ relation is required to enjoy the following properties:[7]

---

[7] The functionality modeled by $\mapsto$ is provided by the reasoning specialist and therefore it conceptually belongs to the list of functionalities given in Section 3.3.

(*sound.select*) if $\langle P, C, c \rangle \mapsto \langle C', \sigma \rangle$ then $\|C'\| = \|C\|$;
(*decreasing.select*) if $\langle P, C, c \rangle \mapsto \langle C', \sigma \rangle$ then $\{c\sigma\} \npreceq^e (P \cup \|C\|)$ and $\langle P', C' \rangle \prec^{cs} \langle P, C \rangle$
   for all $P' \npreceq^e (P \cup \|C\|)$.

(*sound.select*) says that C' has the same logical content of C.[8] (*decreasing.select*)
says that the selected constraint, $c\sigma$, is not $\prec^e$-bigger than any element in $P \cup \|C\|$
and that $\langle P', C' \rangle$ is $\prec^{cs}$-smaller than $\langle P, C \rangle$ whenever $P' \npreceq^e (P \cup \|C\|)$. This
property is fundamental for the termination of CCR(X).

The introduction of augmentation introduces additional constraints on $\prec^{cs}$
(we recall from Section 3.3 that $\prec^{cs}$ is a well-founded relation):

(*cs-prec1*) $\langle P, C \rangle \preceq^{cs} \langle P \cup P', C \rangle$;
(*cs-prec2*) if $\langle P, C \rangle \prec^{cs} \langle P, C' \rangle$ then $\langle P \cup P', C \rangle \preceq^{cs} \langle P \cup P', C' \rangle$;
(*cs-prec3*) if $P' \npreceq^e (P \cup \|C\|)$ then $\langle P \cup P', C \rangle \asymp^{cs} \langle P, C \rangle$.

where $\preceq^{cs}$ is a symmetric and transitive relation whose strict version coincides
with $\prec^{cs}$ and $\asymp^{cs}$ is the associated equivalence relation. (*cs-prec1*) and (*cs-prec2*)
state obvious properties related to the operation of extending P with new elements. (*cs-prec3*) states that by extending the first component of a pair $\langle P, C \rangle$
with literals which are no $\prec^e$-bigger than those already in $\langle P, C \rangle$ we get a $\asymp^{cs}$-
equivalent pair.

The following example shows that the above properties are easily realizable.
Notice that this obviously requires a good understanding of the inner workings of
the reasoning specialist, but no knowledge of the rewrite or simplification engines
is needed.

Example 5 (A reasoning specialist for total orders—continued). Let $L^\dagger = L \cup$
$\{\leq^\dagger, \neq^\dagger\}$ and define $Edges^\dagger(P) = \{\langle s, r^\dagger, t \rangle : \langle s, r, t \rangle \in Edges(P)\}$. $\langle P, C, c \rangle$
$\mapsto \langle C', \sigma \rangle$ holds iff $c\sigma$ is a (Σ_c, Π_c)-literal whose arguments are non-isolated
nodes of C or arguments of the literals in P s.t. $Edges^\dagger(\{c\sigma\}) \not\subseteq \mathcal{E}(C)$, and in
such a case $C' = \langle \mathcal{N}(C), \mathcal{E}(C) \cup Edges^\dagger(\{c\sigma\}) \rangle$. The rôle of the edges of the form
$\langle s, r^\dagger, t \rangle$ is to keep track of the edges currently being added by the augmentation
heuristics and to prevent the selection of candidate constraints which would lead
to the addition of the same edges. This prevents augmentation from looping.
cs-init(C), cs-unsat(C), and $\|C\|$ are defined as in Example 3 (i.e. they ignore
the edges of the form $\langle s, r^\dagger, t \rangle$). In the following, the *complementary graph* of a
graph $\langle N, E \rangle$ (where $E \subseteq (N \times L \times N)$) is the graph $\langle N, (N \times L \times N) \setminus E \rangle$.

(*sound.select*) trivially follows from the definition. To ensure termination
we must first show the existence of a well-founded relation $\prec^{cs}$ satisfying (*cs-
prec1*), (*cs-prec2*), and (*cs-prec3*), and then show that $\xrightarrow[\text{cs--simp}]{}$ and $\mapsto$ enjoy
(*decreasing.cs-simp*) and (*decreasing.select*) respectively. To this end we asso-
ciate each pair $\langle P, C \rangle$ with a finite transitivity graph $G(P, C)$ whose edges are
those of C and whose nodes are the terms s s.t. there exists a term t which occurs
as a non-isolated node of C or as an argument of the literals in P and $s \preceq^e t$.
Notice that the set of nodes of $G(P, C)$ is finite. We define $\langle P', C' \rangle \prec^{cs} \langle P, C \rangle$

---

[8] For the soundness of CCR(X) it would be sufficient to require the T_c-equivalence of C' with
C. However, as we will see in Section 4, (*sound.select*) plays also a rôle in the termination
argument.

iff G_1 is smaller (according to the lexicographic combination of $N' \prec\!\!\prec^e N$ and $E' \subset E$) than G_2, where G_1 and G_2 are the complementary graphs of $G(P', C')$ and $G(P, C)$, respectively. The well-foundedness of $\prec^{cs}$ readily follows from the well-foundedness of $\prec\!\!\prec^e$ and of the relation of (proper) set inclusion over finite sets. (*cs-prec1*), (*cs-prec2*), and (*cs-prec3*) directly follow from the definition. In order to show that $\xrightarrow[\text{cs-simp}]{}$ enjoys (*decreasing.cs-simp*) it suffices to observe that if $P :: C \xrightarrow[\text{cs-simp}]{} C'$ then $\mathcal{N}(G(P, C')) = \mathcal{N}(G(P, C))$ and that $\mathcal{E}(G(P, C)) \subset \mathcal{E}(G(P, C')) \subseteq (\mathcal{N}(G(P, C)) \times L^\dagger \times \mathcal{N}(G(P, C)))$. The proof of (*decreasing.select*) is a bit more involved and it is not given here for the lack of space.   ⋆

4   Properties of Constraint Contextual Rewriting

We are now able to state and prove two key properties of CCR(X): soundness and termination. Let $CCR(X)$ indicate the CRS containing axioms for $\xrightarrow[\text{cs-simp}]{}$ and $\xrightarrow[\text{cs-normal}]{}$ enjoying the properties stated in Section 3.3, the rules *cs-simp*, *cxt-entails*, *normal*, and *crew* presented in Section 3.4, and the rule *augment* presented in Section 3.5.

Theorem 1 (Soundness). $CCR(X)$ *is sound, i.e.*

1. if $C :: e \xrightarrow[\text{ccr}]{}^* e'$, *then* $\|C\| \models_{T_j} e \sim e'$;

2. if $P :: C \xrightarrow[\text{cs-extend}]{}^* C'$, *then* $P, \|C\| \models_{T_j} \bigwedge \|C'\|$.

The proof is by induction on the structure of the reductions. It is fairly simple and therefore it is omitted.

From Theorem 1 it readily follows the soundness of formula simplification, i.e. that $\Phi \xrightarrow[\text{simp}]{} \Phi'$ only if $\models_{T_j} (\Phi \leftrightarrow \Phi')$.

Theorem 2 (Termination). $CCR(X)$ *is terminating.*

Proof. Let $\langle C', e' \rangle \prec_{\text{ccr}} \langle C, e \rangle \doteq \langle \{e'\}, C' \rangle \prec^{cs} \langle \{e\}, C \rangle$ and $\langle P', C' \rangle \prec_{\text{cs-extend}} \langle P, C \rangle \doteq \langle P', C' \rangle \prec^{cs} \langle P, C \rangle$. Both relations are obviously well-founded since $\prec^{cs}$ is. We prove that all the **cs-extend**-reductions $P :: C \xrightarrow[\text{cs-extend}]{\Pi} C'$ are s.t. *(i)* $\langle P, C' \rangle \prec_{\text{cs-extend}} \langle P, C \rangle$ and *(ii)* all the maximal **cs-extend**-reductions of the form $P_1 :: C_1 \xrightarrow[\text{cs-extend}]{\Pi_1'} \cdots$ properly occurring in Π are s.t. $\langle P_1, C_1 \rangle \prec_{\text{cs-extend}} \langle P, C \rangle$.[9] A **cs-extend**-reduction $P :: C \xrightarrow[\text{cs-extend}]{\Pi} C'$ can only be the result of the application of the rules *cs-simp* or *augment*.

- Rule *cs-simp*. We have that $\Pi = \left[P :: C \xrightarrow[\text{cs-simp}]{\Pi_1} \cdots C' \right]$ for $n \geq 1$, i.e. $P ::$ $C \xrightarrow[\text{cs-simp}]{}^+ C'$.[10] Condition *(i)* then follows from (*decreasing.cs-simp*) by

---

[9] The case corresponding to the relation $\xrightarrow[\text{ccr}]{}$ is simple and therefore is omitted.

[10] Notice that $C' \neq C$ follows from the implicit condition $\langle P, C' \rangle \neq \langle P, C \rangle$—cf. footnote 4 in Section 3.2.

transitivity. Condition *(ii)* is vacuously true since Π_1 does not contain any cs-extend-reduction.

- Rule *augment*. We have that

$$\Pi = \left[C_1 :: Q \xrightarrow[\text{ccr}]{\Pi_1} \cdots \emptyset, \quad C_1 :: c\sigma \xrightarrow[\text{ccr}]{\Pi_2} \cdots c', \quad \{c'\} :: C_1 \xrightarrow[\text{cs-simp}]{\Pi_3} \cdots C' \right]$$

that is $C_1 :: p\sigma \xrightarrow[\text{ccr}]{}^* \text{true}$ for all $p \in Q$, $C_1 :: c\sigma \xrightarrow[\text{ccr}]{}^* c'$ (and therefore $c' \preceq^e c\sigma$), and $\{c'\} :: C_1 \xrightarrow[\text{cs-simp}]{}^+ C'$, where $(Q \to c) \in R$, $\langle P, C, c \rangle \mapsto \langle C_1, \sigma \rangle$, and $Q\sigma \npreceq^e \{c\sigma\}$. Since $\langle P, C, c \rangle \mapsto \langle C_1, \sigma \rangle$ by *(decreasing.select)* we have that

$$\{c\sigma\} \npreceq^e (P \cup \|C\|) \tag{7}$$
$$\langle P, C_1 \rangle \prec^{cs} \langle P, C \rangle \tag{8}$$

Since $\{c'\} :: C_1 \xrightarrow[\text{cs-simp}]{}^+ C'$ by *(decreasing.cs-simp)* and the transitivity of $\prec^{cs}$ we have

$$\langle \{c'\}, C' \rangle \prec^{cs} \langle \{c'\}, C_1 \rangle \tag{9}$$

and by *(cs-prec2)* we obtain:

$$\langle P \cup \{c'\}, C' \rangle \preccurlyeq^{cs} \langle P \cup \{c'\}, C_1 \rangle \tag{10}$$

Since $c' \preceq^e c\sigma$ then by (7) we have that $\{c'\} \npreceq^e (P \cup \|C\|)$ and by the *(sound.select)* we have that $\{c'\} \npreceq^e (P \cup \|C_1\|)$. This fact and *(cs-prec3)* allow us to turn (10) into $\langle P \cup \{c'\}, C' \rangle \preccurlyeq^{cs} \langle P, C_1 \rangle$. By applying *(cs-prec1)* we then get: $\langle P, C' \rangle \preccurlyeq^{cs} \langle P, C_1 \rangle$ which together with (8) allows us to conclude $\langle P, C' \rangle \preccurlyeq^{cs} \langle P, C \rangle$.

By looking at the available inference rules, it can be easily verified that all the maximal cs-extend-reductions properly occurring in Π_1 and Π_2 are of the form $\{q\} :: C_1 \xrightarrow[\text{cs-extend}]{\Pi'} \cdots$ with $q \preceq^e c\sigma$. Condition *(ii)* can thus be reduced to proving that $\langle \{q\}, C_1 \rangle \prec^{cs} \langle P, C \rangle$. This fact readily follows from *(decreasing.select)*.

The termination of formula simplification, i.e. of the $\xrightarrow[\text{simp}]{}$ relation, readily follows from Theorem 2.

5 Related Work

[14] describes the combination of a decision procedure for LA and a decision procedure for ground equality and the integration of the compound procedure with rewriting using an integration schema which is essentially that of Boyer & Moore. The presentation of the integration schema is informal and the termination of the resulting form of simplification is not discussed.

STEP [9] features a form of contextual rewriting integrated with a semi-decision procedure for first-order logic and a set of decision procedures extended to deal with non-ground constraints [3, 2]. However the interplay with rewriting is not addressed. On the other hand, the treatment of non-ground constraints is a significant extension which we have not yet investigated.

[8] describes a prover which incorporates several decision procedures following the approach described in [17]. A form of augmentation is implemented via the use of *matching rules*, however the interplay between the decision procedures and the simplifier is presented informally and termination is not ensured.

The work most closely related to ours is [12]. The paper presents an approach to the flexible integration of decision procedures into theorem provers by elaborating and extending the Boyer & Moore ideas. Termination of the resulting form of simplification is ensured but at the cost of loosing much of the efficiency of the Boyer & Moore schema.

The *Open Mechanized Reasoning Systems* project (OMRS for short) [10] provides a specification framework for supporting the activity of composing reasoning systems. Similarly to OMRS we use a rule-based approach to specifying reasoning modules. A rational reconstruction of the logic level of the Boyer & Moore's integration schema is reported in [5]. However the presentation is not abstracted from the specific decision procedure for LA and the control issues are not discussed.

6   Conclusions

We have presented a refined version of Constraint Contextual Rewriting which is guaranteed to terminate. The new version of CCR(X) retains the clear separation between rewriting and the decision procedure employed. Moreover the services provided by the decision procedure are specified abstractly (i.e., independently from the theory decided). This is in sharp contrast with the descriptions available in the literature in which the integration schemas are difficult to reuse and the fundamental properties of the resulting forms of simplification are not investigated.

In future work we plan to investigate further properties of CCR(X). In particular we plan to show that CCR(X) preserves the refutational completeness of many resolution-based proof procedures by generalizing the analogous result for contextual rewriting given in [20].

CCR is implemented in the system **RDL** (**R**ewrite and **D**ecision procedure **L**aboratory) which can be retrieved via the Constraint Contextual Rewriting home page at http://www.mrg.dist.unige.it/ccr/.

References

1. A. Armando and S. Ranise. Constraint Contextual Rewriting. In Ricardo Caferra and Gernot Salzer, editors, *Proc. of the 2nd Intl. Workshop on First Order Theorem Proving (FTP'98). Vienna, Austria, November 23-25, 1998*, pages 65–75, 1998.
2. N. S. Bjørner. *Integrating Decision Procedures for Temporal Verification*. PhD thesis, Computer Science Department, Stanford University, 1998.

3. N. S. Bjørner, M. E. Stickel, and T. E. Uribe. A practical integration of first-order reasoning and decision procedures. In William McCune, editor, *Proc. of the 14th International Conference on Automated deduction*, volume 1249 of *LNAI*, pages 101–115, Berlin, July 13–17 1997. Springer.

4. R.S. Boyer and J S. Moore. Integrating Decision Procedures into Heuristic Theorem Provers: A Case Study of Linear Arithmetic. *Machine Intelligence*, 11:83–124, 1988.

5. A. Coglio, F. Giunchiglia, P. Pecchiari, and C. L. Talcott. A Logic Level Specification of the NQTHM Simplification Process. Technical report, IRST, 1997.

6. David Cyrluk, Patrick Lincoln, and Natarajan Shankar. On Shostak's decision procedure for combinations of theories. In M. A. McRobbie and J. K. Slaney, editors, *Proc. of the 13th International Conference on Automated Deduction*, volume 1104 of *Lecture Notes in Artificial Intelligence*, pages 463–477. Springer-Verlag, 1996.

7. N. Dershowitz and J.P. Jouannaud. Rewriting systems. In *Handbook of Theoretical Computer Science*, pages 243–320. Elsevier Publishers, Amsterdam, 1990.

8. Dave Detlefs, Greg Nelson, and Jim Saxe. *Simplify: the ESC Prover*. COMPAQ, http://www.research.digital.com/SRC/esc/Esc.html.

9. Z. Manna *et. al.* STEP: The Stanford Temporal Prover. Technical Report CS-TR-94-1518, Stanford University, June 1994.

10. F. Giunchiglia, P. Pecchiari, and C. L. Talcott. Reasoning Theories: Towards an Architecture for Open Mechanized Reasoning Systems. Technical Report TR-9409-15, IRST, Nov. 1994.

11. J. Jaffar and J-L. Lassez. Constraint logic programming. In *Proceedings 14th ACM Symposium on Principles of Programming Languages*, pages 111–119, 1987.

12. Predrag Janičić, Alan Bundy, and Ian Green. A framework for the flexible integration of a class of decision procedures into theorem provers. In Harald Ganzinger, editor, *Proc. of the 16th International Conference on Automated Deduction (CADE-16)*, volume 1632 of *LNAI*, pages 127–141, Berlin, July 7–10, 1999. Springer-Verlag.

13. D. Kapur, D.R. Musser, and X. Nie. An Overview of the Tecton Proof System. *Theoretical Computer Science*, Vol. 133, October 1994.

14. D. Kapur and X. Nie. Reasoning about Numbers in Tecton. Technical report, Department of Computer Science, State University of New York, Albany, NY 12222, March 1994.

15. Matt Kaufmann and J S. Moore. Industrial strength theorem prover for a logic based on Common Lisp. *IEEE Transactions on Software Engineering*, 23(4):203–213, April 1997.

16. Jan Willem Klop. Term rewriting systems. In S. Abramsky, D. M. Gabbay, and T. S. E. Maibaum, editors, *Handbook of Logic in Computer Science*, volume 2, chapter 1, pages 1–117. Oxford University Press, Oxford, 1992.

17. G. Nelson and D.C. Oppen. Simplification by Cooperating Decision Procedures. Technical Report STAN-CS-78-652, Stanford Computer Science Department, April 1978.

18. G. Nelson and D.C. Oppen. Fast Decision Procedures Based on Congruence Closure. *Journal of the ACM*, 27(2):356–364, 1980.

19. R.E. Shostak. Deciding Combination of Theories. *Journal of the ACM*, 31(1):1–12, 1984.

20. H. Zhang. Contextual Rewriting in Automated Reasoning. *Fundamenta Informaticae*, 24(1/2):107–123, 1995.

21. H. Zhang and J. L. Remy. Contextual rewriting. In Jean-Pierre Jouannaud, editor, *Proc. of the 1st International Conference on Rewriting Techniques and Applications*, volume 202 of *LNCS*, pages 46–62, Dijon, France, May 1985. Springer.

Axioms vs. Rewrite Rules:
From Completeness to Cut Elimination

Gilles Dowek

INRIA-Rocquencourt, B.P. 105,
78153 Le Chesnay Cedex, France.
Gilles.Dowek@inria.fr, http://coq.inria.fr/~dowek

Abstract. Combining a standard proof search method, such as resolution or tableaux, and rewriting is a powerful way to cut off search space in automated theorem proving, but proving the completeness of such combined methods may be challenging. It may require in particular to prove cut elimination for an extended notion of proof that combines deductions and computations. This suggests new interactions between automated theorem proving and proof theory.

When we search for a proof of the proposition $2 + 2 = 4$, we can use the axioms of addition and equality to transform this proposition into $4 = 4$ and conclude with the reflexivity axiom. We could also use these axioms to transform this proposition into $2 + 2 = 2 + 2$ and conclude, but this proof is redundant with the first, that is in many ways better. Indeed, the axioms of addition are better used in one direction only: to compute values. In automated proof search systems, we often suppress axioms, such as the axioms of addition, and replace them by rewrite rules. This permits to cut off search space while keeping completeness.

The rules of addition apply to terms, but rules applying to propositions may also be considered. For instance, the axiom

$$\forall xy \ ((x \times y) = 0 \Leftrightarrow (x = 0 \lor y = 0))$$

can be replaced by the rule

$$x \times y = 0 \to x = 0 \lor y = 0$$

that rewrites an atomic proposition into a disjunction. Such rules are of special interest in set theory, e.g.

$$x \in \{y, z\} \to x = y \lor x = z$$

and in type theory [1,2], e.g.

$$\varepsilon(x \dot{\Rightarrow} y) \to \varepsilon(x) \Rightarrow \varepsilon(y)$$

When we orient an axiom, we want to cut off search space, but we do not want to loose completeness. In this note, we discuss the properties that the rewrite system must fulfill, so that orientation does not jeopardize completeness.

H. Kirchner and C. Ringeissen (Eds.): FroCoS 2000, LNAI 1794, pp. 62–72, 2000.

1   Orientation

1.1   From Replacement to Rewriting

We consider a set A and a binary relation $\to$ defined on A. We write $\to^*$ for its reflexive-transitive closure and $\equiv$ for its reflexive-symmetrical-transitive closure.

If t and t' are two elements of A, we have $t \equiv t'$ if and only if there is a sequence of terms $t_1, ..., t_n$ such that $t = t_1$, $t' = t_n$ and for each i, either $t_i \to t_{i+1}$ or $t_i \leftarrow t_{i+1}$.

For instance, if A is a set of terms built with a binary operator $+$ and a finite number of constants and $\to$ is the relation such that $t \to u$ if and only if u is obtained from t by replacing a subterm of the form $(x + y) + z$ by $x + (y + z)$. We can establish that

$$(a + (b + c)) + (d + e) \equiv (a + b) + ((c + d) + e)$$

with the sequence

$$((a + b) + c) + (d + e)$$

$$(a + (b + c)) + (d + e) \qquad\qquad (a + b) + ((c + d) + e)$$

$$(a + b) + (c + (d + e))$$

To establish that $t \equiv t'$, we search for such a sequence. We can apply the following *replacement* method: we start with the equality $t = t'$ and we derive more equalities with two rules. The first permits to derive $v = v'$ from $u = u'$ if $u = u' \to v = v'$ and the second permits to derive $v = v'$ from $u = u'$ if $u = u' \leftarrow v = v'$. When reach an equality of the form $u = u$, we are done.

Search is more efficient if we restrict to *rewriting sequences*, i.e. sequences of the form

$$t = t_1 \to ... \to t_n = u_1 \leftarrow ... \leftarrow u_p = t'$$

For instance

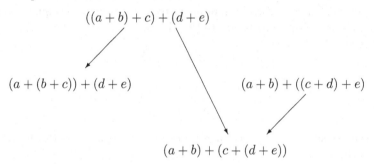

$$(a + (b + c)) + (d + e) \qquad\qquad\qquad (a + b) + ((c + d) + e)$$

$$a + ((b + c) + (d + e)) \qquad (a + b) + (c + (d + e))$$

$$a + (b + (c + (d + e)))$$

Indeed, to establish that $t \equiv t'$, we can restrict the replacement method, to the first rule and rewrite the equality $t = t'$ to an equality of the form $u = u$.

Such a restriction may be incomplete. Consider, for instance, the relation defined by $p \rightarrow' q$ and $p \rightarrow' r$. We have $q \leftarrow' p \rightarrow' r$ but there is no rewriting sequence relating q to r and the equality $q = r$ cannot be rewritten.

A relation is said to be *confluent* if whenever $t \rightarrow^* t_1$ and $t \rightarrow^* t_2$ there is a object u such that $t_1 \rightarrow^* u$ and $t_2 \rightarrow^* u$.

For instance, the relation $\rightarrow$ above is confluent, but the relation $\rightarrow'$ is not. It is easy to prove that when a relation is confluent, two objects are related if and only if they are related by a rewriting sequence and thus that rewriting is a complete search method.

1.2   From Paramodulation to Narrowing

Let us now turn to proof search in predicate calculus with equality. We assume that we have an equality predicate and the associated axioms. Paramodulation [12] is an extension of resolution that allows to replace equals by equals in clauses.

For instance, if we want to refute the clauses

$$(X + Y) + Z = X + (Y + Z)$$

$$P((a + (b + c)) + (d + e))$$

$$\neg P((a + b) + ((c + d) + e))$$

we use the associativity to rearrange brackets in the other clauses until we can resolve them.

Exactly as above, we can restrict the method and use of some equalities in one direction only. We suppress these equalities from the set of clauses to refute, we orient them as rewrite rules and we use them with a *narrowing* (or *oriented paramodulation*) rule that permits to unify a subterm of a clause with the left-hand side of a rewrite rule, and replace the instantiated subterm with the right-hand side of this rule (see, for instance, [10,8]). Notice that since clauses may contain variables, we need to use unification (and not matching as in rewriting).

The completeness of this method mostly rests on confluence of the rewrite system[1]: instead of using the rewrite rules in both directions to unfold and fold terms, we can use them in one direction only and meet on a common reduct.

1.3   Equational Resolution

Equational resolution [11,13] is another proof search method for predicate calculus with equality. In this method, like in narrowing, we suppress some equalities from the set of clauses to refute and we orient them as rewrite rules. We write

---

[1] Actually, although confluence plays the major rôle in completeness proofs, termination also is used. Whether or not completeness fails for non terminating systems in unknown to the author.

$t \to t'$ if t' is obtained by rewriting a subterm in t. We write $\to^*$ for the reflexive-transitive closure of this relation and $\equiv$ for its reflexive-symmetrical-transitive closure.

Morally, we identify equivalent propositions and we work on equivalence classes. Hence, we use an extended resolution rules where unification is replaced by equational unification [4,7]: a unifier of two-terms t and u is a substitution σ such that $\sigma t \equiv \sigma u$ (and not $\sigma t = \sigma u$). For instance the terms $(a + (b + c)) + (d + e)$ and $(a + b) + ((c + d) + e)$ are trivially unifiable because they are equivalent and hence with the clauses

$$P((a + (b + c)) + (d + e))$$

$$\neg P((a + b) + ((c + d) + e))$$

we can apply the resolution rule and get the empty clause.

To solve equational unification problems, we can use narrowing. But, in contrast with the previous method, the narrowing rule is applied to the unification problems and not to the clauses.

2 Completeness

We focus now on the completeness of equational resolution. We first recall a completeness proof of resolution and then discuss how it can be adapted.

2.1 Completeness of Resolution

Cut elimination A way to prove completeness of resolution is to prove that whenever a sequent $A_1, ..., A_p \vdash B_1, ..., B_q$ has a proof in sequent calculus (see, for instance,[5,6]), the empty clause can be derived from the clausal form of the propositions $A_1, ..., A_p, \neg B_1, ..., \neg B_q$.

Usually this proof proceeds in two steps: we first prove that if a sequent $\Gamma \vdash \Delta$ has a proof, then it has also a proof that does not use the cut rule of sequent calculus (cut elimination theorem). Then we prove, by induction over proof structure, that if $\Gamma \vdash \Delta$ has a cut free proof, then the empty clause can be derived from clausal form $cl(\Gamma, \neg\Delta)$ of the propositions $\Gamma, \neg\Delta$.

A variant A variant of this proof isolates the clausification step. A first lemma proves that if the sequent $\Gamma \vdash \Delta$ has a proof and $cl(\Gamma, \neg\Delta) = \{P_1, ..., P_n\}$, then the sequent $\overline{\forall}P_1...\overline{\forall}P_n \vdash$ also has a proof, where $\overline{\forall}P$ is the universal closure of the proposition P.

Then, by the cut elimination theorem, if the sequent $\overline{\forall}P_1...\overline{\forall}P_n \vdash$ has a proof, it has also a cut free proof. At last, we prove by induction over proof structure that if $\overline{\forall}P_1...\overline{\forall}P_n \vdash$ has a cut free proof then the empty clause can be derived from $\{P_1, ..., P_n\} = cl(\Gamma, \neg\Delta)$.

Herbrand theorem Instead of using the cut elimination theorem some authors prefer to use Herbrand theorem that is a variant of it.

According to Herbrand theorem, if $P_1, ..., P_n$ are quantifier free propositions then the sequent $\overline{\forall}P_1...\overline{\forall}P_n \vdash$ has a proof if and only if there are instances $\sigma_1^1 P_1, \ ..., \ \sigma_{k_1}^1 P_1, \ ..., \ \sigma_1^n P_n, \ ..., \ \sigma_{k_n}^n P_n$ of the propositions $P_1, ..., P_n$ such that the quantifier free proposition $\neg(\sigma_1^1 P_1 \wedge ... \wedge \sigma_{k_1}^1 P_1 \wedge ... \wedge \sigma_1^n P_n \wedge ... \wedge \sigma_{k_n}^n P_n)$ is tautologous.

As above, a first lemma proves that if the sequent $\Gamma \vdash \Delta$ has a proof and $cl(\Gamma, \neg\Delta) = \{P_1, ..., P_n\}$, then the sequent $\overline{\forall}P_1...\overline{\forall}P_n \vdash$ also has a proof. By Herbrand theorem, if the sequent $\overline{\forall}P_1...\overline{\forall}P_n \vdash$ has a proof, then there are instances $\sigma_1^1 P_1, \ ..., \ \sigma_{k_1}^1 P_1, \ ..., \ \sigma_1^n P_n, \ ..., \ \sigma_{k_n}^n P_n$ of the propositions $P_1, ..., P_n$ such that the quantifier free proposition $\neg(\sigma_1^1 P_1 \wedge ... \wedge \sigma_{k_1}^1 P_1 \wedge ... \wedge \sigma_1^n P_n \wedge ... \wedge \sigma_{k_n}^n P_n)$ is tautologous. At last, we prove that if the proposition $\neg(\sigma_1^1 P_1 \wedge ... \wedge \sigma_{k_1}^1 P_1 \wedge ... \wedge \sigma_1^n P_n \wedge ... \wedge \sigma_{k_n}^n P_n)$ is tautologous then the empty clause can be derived from $\{P_1, ..., P_n\} = cl(\Gamma, \neg\Delta)$.

2.2   Completeness of Equational Resolution

This proof can be adapted to equational resolution. First, the identification of equivalent propositions used in proof search can be used in sequent calculus also. This leads to the *sequent calculus modulo* [1]. For instance, the axiom rule

$$\frac{}{A \vdash A} \text{ axiom}$$

is transformed into the rule

$$\frac{}{A \vdash_{\equiv} B} \text{ axiom if } A \equiv B$$

and the left rule of disjunction

$$\frac{\Gamma \ A \vdash \Delta \quad \Gamma \ B \vdash \Delta}{\Gamma \ A \vee B \vdash \Delta} \vee\text{-left}$$

is transformed into the rule

$$\frac{\Gamma \ A \vdash_{\equiv} \Delta \quad \Gamma \ B \vdash_{\equiv} \Delta}{\Gamma \ C \vdash_{\equiv} \Delta} \vee\text{-left if } C \equiv A \vee B$$

In sequent calculus modulo the rules of addition and multiplication, we have a very short proof that the number 4 is even

$$\frac{\dfrac{\dfrac{}{4 = 4 \vdash_{\equiv} 4 = 2 \times 2} \text{ axiom}}{\forall x \ (x = x) \vdash_{\equiv} 4 = 2 \times 2} \vee\text{-left}}{\forall x \ (x = x) \vdash_{\equiv} \exists y \ (4 = 2 \times y)} \exists\text{-right}$$

while proving this proposition would be much more cumbersome in sequent calculus with the axioms of addition and multiplication.

Using sequent calculus modulo, we can prove the completeness of equational resolution. First, we prove the *equivalence lemma*: if the rewrite system encodes a theory $\mathcal{T}$ then the sequent $\Gamma \mathcal{T} \vdash \Delta$ has a proof in sequent calculus if and only if the sequent $\Gamma \vdash_{\equiv} \Delta$ has a proof in sequent calculus modulo. Then, the cut elimination theorem extends trivially to sequent calculus modulo when the equivalence is generated by rewrite rules applying to terms. At last, we prove by induction over proof structure, that if $\Gamma \vdash_{\equiv} \Delta$ has a cut free proof in sequent calculus modulo, then the empty clause can be derived from $cl(\Gamma, \neg \Delta)$ with the rules of equational resolution. The confluence of the rewrite system is only needed to prove that narrowing is a complete equational unification method.

2.3   Equational Resolution as Narrowing

Equational resolution can be seen as a formulation of narrowing. This suggests another completeness proof, reducing the completeness of equational resolution to that of narrowing.

Indeed, instead of resolving two clauses and narrowing the unification problem, as we do in equational unification, we could first narrow the clauses and apply the usual resolution rule to get the same result.

For instance, instead of resolving

$$P((a + (b + c)) + (d + e))$$

$$\neg P((a + b) + ((c + d) + e))$$

and narrowing the unification problem

$$(a + (b + c)) + (d + e) = (a + b) + ((c + d) + e)$$

we could take an option on these two clauses, and narrow them until they can be resolved.

So, narrowing a unification problem is just a way to narrow the clauses it comes from. Hence, equational resolution can be seen as a formulation of narrowing and thus as an implementation of paramodulation with the restriction that some equations should be used in one direction only. Hence, the completeness of equational resolution rests on confluence.

3   Rewriting Proposition

So far, we have considered methods where axioms are replaced by rewrite rules applying to terms. We have seen that the proof search methods obtained this way could be seen as restrictions of paramodulation and that their completeness rested on confluence. We consider now more powerful rewrite rules applying to propositions. For instance, the axiom

$$\forall xy \ (x \times y = 0 \Leftrightarrow (x = 0 \vee y = 0))$$

can be replaced by the rule

$$x \times y = 0 \rightarrow x = 0 \vee y = 0$$

that applies directly to propositions. For technical reasons, we restrict to rules, such as this one, whose left-hand side is an atomic proposition.

For example, with this rule, we can prove the proposition

$$\exists z \ (a \times a = z \Rightarrow a = z)$$

in sequent calculus modulo

$$\cfrac{\cfrac{\overline{a = 0 \vdash_{\equiv} a = 0} \ \text{axiom} \qquad \overline{a = 0 \vdash_{\equiv} a = 0} \ \text{axiom}}{a \times a = 0 \vdash_{\equiv} a = 0} \ \vee\text{-left}}{\cfrac{\vdash_{\equiv} a \times a = 0 \Rightarrow a = 0}{\vdash_{\equiv} \exists z \ (a \times a = z \Rightarrow a = z)} \ \exists\text{-right}} \ \Rightarrow\text{-right}$$

3.1   Resolution Modulo

In this case, equational resolution can be extended to *resolution modulo* [1]. In resolution modulo, like in equational resolution, the rewrite rules applying to terms are used to narrow unification problems, but, like in narrowing, the rules applying to propositions are used to narrow the clauses directly.

For instance, if we take the clausal form of the negation of the proposition

$$\exists z \ (a \times a = z \Rightarrow a = z)$$

i.e. the clauses

$$a \times a = Z$$

$$\neg a = Z$$

we can narrow the first clause with the rule

$$x \times y = 0 \rightarrow x = 0 \vee y = 0$$

yielding the clause

$$a = 0$$

Then, we resolve this clause with $\neg a = Z$ and get the empty clause.

3.2   Resolution with Axioms

An alternative is to keep the axiom

$$\forall xy \ (x \times y = 0 \Leftrightarrow (x = 0 \vee y = 0))$$

and to use resolution. In this case, we have to refute the clauses

$$\neg X \times Y = 0, X = 0, Y = 0$$

$$X \times Y = 0, \neg X = 0$$

$$X \times Y = 0, \neg Y = 0$$

$$a \times a = Z$$

$$\neg a = Z$$

We resolve the clause $a \times a = Z$ with the clause $\neg X \times Y = 0, X = 0, Y = 0$ yielding the clause $a = 0$. Then, we resolve this clause with $\neg a = Z$ and get the empty clause.

An above, resolution modulo can be seen as a restriction of resolution, where some clauses can be used in one direction only. As above we could think that resolution modulo is complete as soon as the rewrite system is confluent. Unfortunately this is not the case.

3.3   A Counter Example to Completeness

The axiom

$$A \Leftrightarrow (B \wedge \neg A)$$

can be transformed into a rewrite rule (Crabbé's rule)

$$A \to B \wedge \neg A$$

Resolution modulo cannot prove the proposition $\neg B$, because from the clause B, neither the resolution rule nor the narrowing rule can be applied.

But, surprisingly, with the axiom $A \Leftrightarrow (B \wedge \neg A)$, we can prove the proposition $\neg B$. Indeed, the clausal form of the proposition $A \Rightarrow (B \wedge \neg A)$ yields the clauses

$$\neg A, B \quad \neg A$$

the clausal form of the proposition $(B \wedge \neg A) \Rightarrow A$ yields

$$A, \neg B$$

and the clausal form of the negation of the proposition $\neg B$ yields

$$B$$

From B and $A, \neg B$, we derive A and then from this clause and $\neg A$ we get the empty clause. Hence, a resolution proof with axioms cannot always be transformed into a resolution modulo proof. Resolution modulo cannot be seen as a restriction of resolution where some clauses can be used in one direction only and completeness may be lost even if the rewrite system is confluent.

Notice that, in the resolution proof, clauses coming from both propositions $A \Rightarrow (B \wedge \neg A)$ and $(B \wedge \neg A) \Rightarrow A$ are used. Hence, although the rewrite system is confluent, we need both to unfold A to $B \wedge \neg A$ and to fold $B \wedge \neg A$ to A. Moreover, when we resolve B with $\neg B, A$ we fold $B \wedge \neg A$ to A although we have proved the proposition B, but not the proposition $\neg A$ yet. This *partial folding* that resolution allows but resolution modulo disallows is the reason why some resolution proofs cannot be transformed into resolution modulo proofs.

Hence, orientation cuts off search space more dramatically when we have rules rewriting propositions that when we have only rules rewriting terms. It can cut off search space so dramatically that completeness may be lost.

3.4 Completeness of Resolution Modulo

Since resolution can prove the proposition $\neg B$ with the axiom $A \Leftrightarrow (B \wedge \neg A)$, the sequent $A \Leftrightarrow (B \wedge \neg A) \vdash \neg B$ can be proved in sequent calculus and hence the sequent $\vdash \neg B$ can be proved in sequent calculus modulo. For instance, it has the proof

$$
\cfrac{
  \cfrac{
    \cfrac{B \vdash B \text{ axiom} \quad
      \cfrac{
        \cfrac{
          \cfrac{
            \cfrac{
              \cfrac{\overline{B, A, B \vdash A} \text{ weak. + ax.}}{B, \neg A, A, B \vdash} \text{ ¬-left}
            }{A, A, B \vdash} \text{ ∧-left}
          }{A, B \vdash} \text{ contraction-left}
        }{B \vdash \neg A} \text{ ¬-right}
      }
    }{B \vdash A} \text{ ∧-right}
    \quad
    \cfrac{
      \cfrac{
        \cfrac{
          \cfrac{\overline{B, A, B \vdash A} \text{ weak. + ax.}}{B, \neg A, A, B \vdash} \text{ ¬-left}
        }{A, A, B \vdash} \text{ ∧-left}
      }{A, B \vdash} \text{ contraction-left}
    }{}
  }{B \vdash} \text{ cut}
}{\vdash \neg B} \text{ ¬-right}
$$

The proposition $\neg B$ has a proof in sequent calculus modulo but not in resolution modulo and thus resolution modulo is incomplete.

The completeness theorem of resolution modulo does not prove that each time a proposition has a proof in sequent calculus modulo, it has a proof in resolution modulo (because this is false), but it proves, using techniques similar to those developed in section 2.1 and 2.2, that if a proposition has a cut free proof in sequent calculus modulo, it has a proof in resolution modulo.

The proposition $\neg B$ has proof in sequent calculus modulo, but no cut free proof and sequent calculus modulo the rule $A \rightarrow B \wedge \neg A$ does not have the cut elimination property.

Modulo some rewrite systems such as

$$x \times y = 0 \rightarrow x = 0 \vee y = 0$$

or

$$A \rightarrow B \wedge A$$

sequent calculus modulo has the cut elimination property and thus resolution modulo is complete. Hence modulo these rewrite systems there is no occurrence of the phenomenon we have observed above, where a resolution proof could not be transformed into a resolution modulo proof because it used partial folding. Still the translation of resolution proofs to resolutions modulo proofs is non trivial because it involves cut elimination.

In [3] we present cut elimination proofs for large classes of rewrite systems (including various presentations of type theory, all confluent and terminating quantifier free rewrite systems, ...) and we conjecture that cut elimination holds for all terminating and confluent rewrite system, although we know that such a conjecture cannot be proved in type theory or in large fragments of set theory because it implies their consistency. Cut elimination also holds for sequent calculus modulo some non terminating rewrite systems such as

$$A \to B \land A$$

Notice that when resolution modulo is complete (for instance for type theory), this completeness result cannot be proved in the theory itself (because it implies its consistency), while confluence usually can. So, it is not surprising that tools more powerful than confluence, such as cut elimination, are required.

3.5   Towards a New Kind of Completion

Completion [9] transforms a rewrite system into a confluent one. We can imagine a similar process that transforms a rewrite system into one modulo which sequent calculus has the cut elimination property.

For instance, the rule $A \to B \land \neg A$ is equivalent to the axiom $A \Leftrightarrow (B \land \neg A)$ whose clausal form is

$$A, \neg B \quad \quad B, \neg A \quad \quad \neg A$$

like that of the axioms $A \Leftrightarrow \bot$ and $B \Leftrightarrow A$ that are equivalent to the rewrite system $A \to \bot$, $B \to A$ modulo which sequent calculus has the cut elimination property and resolution is complete.

So we could imagine to transform the rewrite system

$$A \to (B \land \neg A)$$

into

$$A \to \bot$$

$$B \to A$$

or even into

$$A \to \bot$$

$$B \to \bot$$

in order to recover completeness.

Acknowledgements. Thanks to Th. Hardin, C. Kirchner, Ch. Lynch and B. Werner for many helpful discussions on this subject.

References

1. G. Dowek, Th. Hardin, and C. Kirchner, Theorem proving modulo, *Rapport de Recherche INRIA* 3400 (1998).
2. G. Dowek, Th. Hardin, and C. Kirchner, HOL-lambda-sigma: an intentional first-order expression of higher-order logic, *Rewriting Techniques and applications*, P. Narendran and M. Rusinowitch (Eds.), Lecture Notes in Computer Science 1631, Springer-Verlag (1999), pp. 317-331.
3. G. Dowek and B. Werner. Proof normalization modulo. *Types for proofs and programs* 98, T. Altenkirch, W. Naraschewski, B. Rues (Eds.), Lecture Notes in Computer Science 1657, Springer-Verlag (1999), pp. 62-77. Rapport de Recherche 3542, INRIA (1998)
4. M.J. Fay, First-order unification in an equational theory, *Fourth Workshop on Automated Deduction* (1979), pp. 161-167.
5. J. Gallier, Logic in computer science, *Harper and Row* (1986).
6. J.Y. Girard, Y. Lafont, and P. Taylor. Types and proofs, *Cambridge University Press* (1989).
7. J.-M. Hullot, Canonical forms and unification, W. Bibel and R. Kowalski (Eds.) *Conference on Automated Deduction*, Lecture Notes in Computer Science 87, Springer-Verlag (1980), pp. 318-334.
8. J. Hsiang and M. Rusinowitch, Proving refutational completeness of theorem proving strategies: the transfinite semantic tree method, Journal of the ACM, 38, 3 (1991) pp. 559-587.
9. D. E. Knuth and P. B. Bendix. Simple word problems in universal algebras. In J. Leech, editor, *Computational Problems in Abstract Algebra*, pages 263–297. Pergamon Press, Oxford, 1970.
10. G. Peterson, A Technique for establishing completeness results in theorem proving with equality, *Siam J. Comput.*, 12, 1 (1983) pp. 82-100.
11. G. Plotkin, Building-in equational theories *Machine Intelligence*, 7 (1972), pp. 73–90
12. G.A. Robinson and L. Wos, Paramodulation and theorem proving in first-order theories with equality, *Machine Intelligence*, 4, B. Meltzer and D. Michie (Eds.), American Elsevier (1969) pp. 135-150.
13. M. Stickel, Automated deduction by theory resolution, *Journal of Automated Reasoning*, 4, 1 (1985), pp. 285-289.

Normal Forms and Proofs in Combined Modal and Temporal Logics

U. Hustadt[1], C. Dixon[1], R. A. Schmidt[2], and M. Fisher[1]

[1] Centre for Agent Research and Development,
Manchester Metropolitan University,
Chester Street, Manchester M1 5GD, United Kingdom
{U.Hustadt,M.Fisher,C.Dixon}@doc.mmu.ac.uk
[2] Department of Computer Science, University of Manchester,
Oxford Road, Manchester M13 9PL, United Kingdom
schmidt@cs.man.ac.uk

Abstract. In this paper we present a framework for the combination of modal and temporal logic. This framework allows us to combine different normal forms, in particular, a separated normal form for temporal logic and a first-order clausal form for modal logics. The calculus of the framework consists of temporal resolution rules and standard first-order resolution rules.

We show that the calculus provides a sound, complete, and terminating inference systems for arbitrary combinations of subsystems of multi-modal S5 with linear, temporal logic.

1 Introduction

For a number of years, temporal and modal logics have been applied outside pure logic in areas such as formal methods, theoretical computer science and artificial intelligence. A variety of sophisticated methods for reasoning with these logics have been developed, and in many cases applied to real-world problems.

With the advent of more complex applications, a *combination* of modal and temporal logics is increasingly required, particularly in areas such as security in distributed systems (Halpern 11), specifying multi-agent system (Jennings 14; Wooldridge and Jennings 20), temporal databases (Finger 7), and accident analysis (Johnson 15). Further motivation for the importance of combinations of modal logics can be found in (Blackburn and de Rijke 3).

In all these cases, combinations of multi-modal and temporal logics are used to capture the detailed behaviour of the application domain. While many of the basic properties of such combinations are well understood (Baader and Ohlbach 1; Fagin et al. 6; Gabbay 8; Wolter 18), very little work has been carried out on proof methods for such logics. Wooldridge, Dixon, and Fisher (19) present a tableaux-based calculus for the combination of discrete linear temporal logic with the modal logics KD45 (characterising belief) and S5 (characterising knowledge). Dixon, Fisher, and Wooldridge (5) present a resolution-based calculus for the combination of discrete linear temporal logic with the modal logic S5. A

H. Kirchner and C. Ringeissen (Eds.): FroCoS 2000, LNAI 1794, pp. 73–87, 2000.

combination of calendar logic for specifying everyday temporal notions with a variety of other modal logics has been considered in (Ohlbach 16; Ohlbach and Gabbay 17).

Our aim in this paper is to present an approach that is general enough to capture a wide range of combinations of temporal and modal logics, but still provides viable means for effective theorem proving. The following aspects of our approach are novel:

- The approach covers the combination of discrete, linear, temporal logic with extensions of multi-modal K_m by any combination of the axiom schemata 4, 5, B, D, and T. This extends the results presented in (Dixon et al. 5; Wooldridge et al. 19).
- Instead of combining two calculi operating according to the same underlying principles, like for example two tableaux-based calculi, we combine two different approaches to theorem-proving in modal and temporal logics, namely the translation approach for modal logics (using first-order resolution) and the SNF approach for temporal logics (using modal resolution).
- The particular translation we use has only recently been proposed by de Nivelle (4) and can be seen as a special case of the T-encoding introduced by Ohlbach (16). It allows for conceptually simple decision procedures for extensions of K4 by ordered resolution without any reliance on loop checking or similar techniques.

2   Combinations of Temporal and Modal Logics

In this section, we give the syntax and semantics of a class of combinations of a discrete linear temporal logic with normal multi-modal logics.

Syntax

Let P be a set of propositional variables, A a set of agents, and M a set of modalities. Then the tuple $\Sigma = (\mathsf{P}, \mathsf{A}, \mathsf{M})$ is a *signature* of MTL. If $m \in \mathsf{M}$ and $a \in \mathsf{A}$, then the ordered pair (m, a) is a *modal parameter*. The set of *well-formed formulae* of MTL over a signature Σ is inductively defined as follows: (i) true and false are formulae, (ii) every propositional variable is a formula, (iii) if φ and ψ are formulae, then $\neg\varphi$, $\varphi \vee \psi$, $\varphi \wedge \psi$, $\varphi \Rightarrow \psi$, and $\varphi \Leftrightarrow \psi$ are formulae, (iv) if φ and ψ are formulae, then $\bigcirc\varphi$, $\Diamond\varphi$, $\Box\varphi$, $\varphi\mathcal{U}\psi$, and $\varphi\mathcal{W}\psi$ are formulae, (v) if φ is a formula, κ is a modal parameter, then $[\kappa]\,\varphi$ is a formula.

We also use true and false to denote the empty conjunction and disjunction, respectively. A literal is either p or $\neg p$ where p is a propositional variable. A *modal literal* is either $[\kappa]\,L$ or $\neg\,[\kappa]\,L$ where κ is a modal parameter and L is a literal. For any (modal) literal L we denote by $\sim\!L$ the negation normal form of $\neg L$.

With each modal parameter we can associate a set of axiom schemata defining its properties. We assume that the axiom schemata K and Nec hold for every modal operator. In this paper we allow in addition any combination of the axiom schemata 4, 5, B, D, and T.

Semantics

Let S be a set of states. A *timeline* is an infinite, linear, discrete sequence of states indexed by the natural numbers. A *point* is an ordered pair (t, k) where t is a timeline and k is a natural number, a so-called *temporal index*. P denotes the set of all points. A *valuation* ι is a mapping from P to a subset of P. An *interpretation* $\mathcal{M}$ is a tuple $(\mathcal{T}, t_0, \mathcal{R}, \iota)$ where $\mathcal{T}$ is a set of timelines with a distinguished timeline $t_0 \in \mathcal{T}$, $\mathcal{R}$ is a collection of binary relations on P containing for every modal parameter κ a relation R_κ, and ι is a valuation.

We define a binary relation $\models$ between a formula φ and a pair $\mathcal{M}$ and (t, k) where $\mathcal{M}$ is an interpretation and (t, k) is a point as follows.

$\mathcal{M}, (t, k) \models \mathsf{true}$

$\mathcal{M}, (t, k) \not\models \mathsf{false}$

$\mathcal{M}, (t, k) \models \mathsf{start}$ iff $t = t_0$ and $k = 0$

$\mathcal{M}, (t, k) \models p$ iff $p \in \iota((t, k))$

$\mathcal{M}, (t, k) \models \neg\varphi$ iff $\mathcal{M}, (t, k) \not\models \varphi$

$\mathcal{M}, (t, k) \models \varphi \wedge \psi$ iff $\mathcal{M}, (t, k) \models \varphi$ and $\mathcal{M}, (t, k) \models \psi$

$\mathcal{M}, (t, k) \models \varphi \Rightarrow \psi$ iff $\mathcal{M}, (t, k) \not\models \varphi$ or $\mathcal{M}, (t, k) \models \psi$

$\mathcal{M}, (t, k) \models \varphi \vee \psi$ iff $\mathcal{M}, (t, k) \models \varphi$ or $\mathcal{M}, (t, k) \models \psi$

$\mathcal{M}, (t, k) \models \bigcirc\varphi$ iff $\mathcal{M}, (t, k{+}1) \models \varphi$

$\mathcal{M}, (t, k) \models \Box\varphi$ iff for all $n \in \mathbb{N}$, $n \geq k$ implies $\mathcal{M}, (t, n) \models \varphi$

$\mathcal{M}, (t, k) \models \Diamond\varphi$ iff there exists $n \in \mathbb{N}$ such that $n \geq k$ and $\mathcal{M}, (t, n) \models \varphi$

$\mathcal{M}, (t, k) \models \varphi\,\mathcal{U}\psi$ iff there exists $n \in \mathbb{N}$ such that $n \geq k$, $\mathcal{M}, (t, n) \models \psi$, and
                             for all $m \in \mathbb{N}$, $k \leq m < n$ implies $\mathcal{M}, (t, m) \models \varphi$

$\mathcal{M}, (t, k) \models \varphi\,\mathcal{W}\psi$ iff $\mathcal{M}, (t, k) \models \varphi\,\mathcal{U}\psi$ or $\mathcal{M}, (t, k) \models \Box\varphi$

$\mathcal{M}, (t, k) \models [\kappa]\,\varphi$ iff for all $t' \in \mathcal{T}$ and for all $k' \in \mathbb{N}$, $((t, k), (t', k')) \in R_\kappa$
                             implies $\mathcal{M}, (t', k') \models \varphi$

If $\mathcal{M}, w \models \varphi$ then we say φ is *true* or *holds* at w in $\mathcal{M}$. An interpretation $\mathcal{M}$ *satisfies* a formula φ iff φ holds at $(t_0, 0)$ and it *satisfies* a set N of formula iff for every formula $\psi \in N$, $\mathcal{M}$ satisfies ψ. In this case $\mathcal{M}$ is a *model* for φ and N, respectively.

Let $\mathcal{M}$ be an interpretation and let T be a relation on points such that $((t, k), (t', k')) \in T$ iff $t = t'$ and $k' = k{+}1$. Let v, w be points in $\mathcal{M}$. Then w is *reachable* from v iff $(v, w) \in (T \cup \bigcup_\kappa R_\kappa)^*$. An interpretation $\mathcal{M}$ such that every point in $\mathcal{M}$ is reachable from $(t_0, 0)$ is a *connected interpretation*.

Note that the semantics of MTL is a notational variation of the standard (Kripke) semantics of a multi-modal logics with points corresponding to worlds. For every modal parameter κ we have a relation R_κ on worlds. In addition, there is a *temporal relation* relation T on worlds defined as the union of a family of disjoint, discrete, linear orders on the set of worlds. The semantics of $\bigcirc\varphi$ is given in terms of T while the semantics of the remaining temporal operators is given in terms of the reflexive, transitive closure T^* of T. In case we have associated additional axiom schemata to a modal operator $[\kappa]$, the relation R_κ has to satisfy the well-known corresponding properties, that is, transitivity for 4, euclideanness for 5, symmetry for B, seriality for D, and reflexivity for T.

3   A Normal Form for **MTL** Formulae

Dixon et al. (5) have shown that every well-formed formulae of MTL can be transformed to a set of *SNF$_K$ clauses* in a satisfiability equivalence preserving way. The use of SNF$_K$ clauses eases the presentation of a resolution calculus for MTL as well as the soundness, completeness and termination proof.

The transformation Π_K to SNF$_K$ clauses uses a renaming technique where particular subformulae are replaced by new propositions. To ease the presentation of SNF$_K$ clauses we use the universal modality $\square^*$ as an auxiliary modal operator. The modal operator $\square^*$ has the following important property.

Theorem 1 (Goranko and Passy 10). *Let $\mathcal{M}$ be an interpretation, let v, w be points in $\mathcal{M}$, and let φ be a formula such that $\mathcal{M}, v \models \square^*\varphi$. Then $\mathcal{M}, w \models \varphi$ iff w is reachable from v.*

In connected interpretations also the following stronger result holds.

Theorem 2. *Let $\mathcal{M}$ be a connected interpretation and φ be a well-formed formula of* MTL. *Then $\mathcal{M}, (t_0, 0) \models \square^*\varphi$ iff $\mathcal{M}, w \models \varphi$ for every point w in $\mathcal{M}$.*

SNF$_K$ clauses have the following form

(initial clause)	$\mathsf{start} \Rightarrow \bigvee_{i=1}^n L_i$
(global clause)	$\square^*(\bigwedge_{j=1}^m K_j \Rightarrow \bigcirc(\bigvee_{i=1}^n L_i))$
(sometime clause)	$\square^*(\bigwedge_{j=1}^m K_j \Rightarrow \Diamond L)$
(literal clause)	$\square^*(\mathsf{true} \Rightarrow \bigvee_{i=1}^n L_i)$
(modal clause)	$\square^*(\mathsf{true} \Rightarrow L_1 \vee M_1)$

where K_j, L_i, and L (with $1 \le j \le m$ and $1 \le i \le n$) are literals and M_1 is a modal literal.

We only present the part of Π_K dealing with formulae of the form $\square\varphi$ and $\varphi \mathcal{W} \psi$ which is important for the understanding of the temporal resolution rule and the example derivation presented later. For a complete description of Π_K see Dixon et al. (5).

$$\{A \Rightarrow \square\varphi\} \rightarrow \begin{cases} A \Rightarrow B \\ B \Rightarrow \bigcirc B \\ B \Rightarrow \varphi \end{cases} \qquad \text{if } \varphi \text{ is a literal and } B \text{ is new}$$

$$\{A \Rightarrow \varphi \mathcal{W} \psi\} \rightarrow \begin{cases} A \Rightarrow \varphi \vee \psi \\ A \Rightarrow B \vee \psi \\ B \Rightarrow \bigcirc(\varphi \vee \psi) \\ B \Rightarrow \bigcirc(B \vee \psi) \end{cases} \text{if } \varphi \text{ and } \psi \text{ are literals and } B \text{ is new}$$

Theorem 3 (Dixon et al. 5). *Let φ be a formula of* MTL. *Then φ is satisfiable if and only if $\Pi_K(\varphi)$ is satisfiable in a connected interpretation.*

4 Translation of SNF$_K$ Clauses into SNF$_r$ Clauses

In the approach of Dixon et al. (5) the calculus for MTL consists of a set of special resolution inference rules for SNF$_K$ clauses. Broadly, these rules can be divided into two classes: those dealing with SNF$_K$ clauses containing temporal operators and those dealing with SNF$_K$ clauses containing modal operators. Inference rules in the later class also take care that the calculus is complete if we have associated the axiom schemata of S5 with all modal operators.

Instead of using additional resolution rules for modal literals we use the translation approach to modal theorem proving. That is, we translate occurrences of modal literals into first-order logic, in particular, we do so in a way that preserves satisfiability and makes the use of first-order resolution possible. However, the temporal connectives are not translated and we will include additional inference rules for them in our calculus. Intuitively, the proposed translation makes the underlying relations R_κ on points explicit as well as the quantificational effect of the modal operator $[\kappa]$ explicit in our language, but still leaves the relations and quantificational effect of the temporal operators implicit.

The translation function π_r on literals, and conjunctions and disjunctions of literals is defined as follows.

$$\pi_r(\text{true}, x) = \text{true} \qquad\qquad \pi_r(\Diamond\varphi, x) = \Diamond\pi_r(\varphi, x)$$
$$\pi_r(\text{false}, x) = \text{false} \qquad\qquad \pi_r(\bigcirc\varphi, x) = \bigcirc\pi_r(\varphi, x)$$
$$\pi_r(p, x) = q_p(x) \qquad\qquad \pi_r([\kappa]\,L, x) = \forall y\,(\neg r_\kappa(x, y) \vee \pi_r(L, y))$$
$$\pi_r(\neg p, x) = \neg q_p(x) \qquad\qquad \pi_r(\neg\,[\kappa]\,L, x) = r_\kappa(x, f(x)) \wedge \pi_r(\sim L, f(x))$$
$$\pi_r(\varphi \star \psi, x) = \pi_r(\varphi, x) \star \pi_r(\psi, x) \quad \text{for } \star \in \{\wedge, \vee, \Rightarrow\}$$

p is a propositional variable, q_p is a unary predicate symbol uniquely associated with p, L is a literal, and f is a Skolem function uniquely associated with an occurrence of $[\kappa]\,L$. In addition, following de Nivelle (4) the mapping τ_r on modal literals is defined by

$$\tau_r([\kappa]\,L, x) = q_{[\kappa]\,L}(x) \qquad\qquad \tau_r(\neg\,[\kappa]\,L, x) = \neg q_{[\kappa]\,L}(x)$$

where $q_{[\kappa]\,L}$ is a new predicate symbol uniquely associated with $[\kappa]\,L$.

The translation Π_r on SNF$_K$ clauses is defined in the following way:

$$\Pi_r(\text{start} \Rightarrow \textstyle\bigvee_{i=1}^n L_i) = \{\pi_r(\text{true} \Rightarrow \textstyle\bigvee_{i=1}^n L_i, \text{now})\}$$
$$\Pi_r(\Box^*(\textstyle\bigwedge_{j=1}^m K_j \Rightarrow \bigcirc(\textstyle\bigvee_{i=1}^n L_i))) = \{\forall x\,\pi_r(\textstyle\bigwedge_{j=1}^m K_j \Rightarrow \bigcirc(\textstyle\bigvee_{i=1}^n L_i), x)\}$$
$$\Pi_r(\Box^*(\textstyle\bigwedge_{j=1}^m L_j \Rightarrow \Diamond L)) = \{\forall x\,\pi_r(\textstyle\bigwedge_{j=1}^m K_j \Rightarrow \Diamond L, x)\}$$
$$\Pi_r(\Box^*(\text{true} \Rightarrow \textstyle\bigvee_{i=1}^n L_i)) = \{\forall x\,\pi_r(\text{true} \Rightarrow \textstyle\bigvee_{i=1}^n L_i, x)\}$$
$$\Pi_r(\Box^*(\text{true} \Rightarrow L_1 \vee M_1)) = \left\{\begin{array}{l}\forall x\,(\text{true} \Rightarrow \pi_r(L_1, x) \vee \tau_r(M_1, x)) \\ \forall x\,(\text{true} \Rightarrow {\sim}\tau_r(M_1, x) \vee \pi_r(M_1, x))\end{array}\right\}$$

The translation of a set N of SNF$_K$ clauses is given by

$$\Pi_r(N) = \bigcup_{C \in N} \Pi_r(C).$$

Table 1. Translation of axiom schemata

4	Transitivity	$\forall x, y \,(\text{true} \Rightarrow \neg q_{[\kappa]\,L}(x) \vee \neg r_\kappa(x, y) \vee q_{[\kappa]\,L}(y))$
5	Euclideanness	$\forall x, y \,(\text{true} \Rightarrow \neg q_{[\kappa]\,L}(y) \vee \neg r_\kappa(x, y) \vee q_{[\kappa]\,L}(x))$
B	Symmetry	$\forall x, y \,(\text{true} \Rightarrow \neg q_{[\kappa]\,L}(y) \vee \neg r_\kappa(x, y) \vee \pi_r(L, x))$
D	Seriality	$\forall x \,(\text{true} \Rightarrow r_\kappa(x, f(x)))$
T	Reflexivity	$\forall x \,(\text{true} \Rightarrow r_\kappa(x, x))$

The formulae obtained by applying Π_r to SNF_K clauses will be called SNF_r clauses. The target language of Π_r can be viewed as a fragment of first-order logic allowing only unary and binary predicate symbols extended by the the temporal operators $\bigcirc$ and $\Diamond$ or as a fragment of first-order temporal logic with the same restriction on predicate symbols and temporal operators. However, the semantics of the target language does not coincide with either of these as we will see below. In Section 5 we present a syntactic characterisation of the class of SNF_r clauses. The universal quantifiers in a SNF_r clause are usually omitted in our presentation. Any free variable in a SNF_r clause is assumed to be implicitly universally quantified.

Again, following de Nivelle (4), depending on the additional properties of a modal operator $[\kappa]$ SNF_r clauses from Table 1 are added to the set of SNF_r clauses for every predicate symbol $q_{[\kappa]\,L}$ introduced by Π_r. The semantics of SNF_r clauses is given by temporal interpretations. A *temporal interpretation* is a tuple $(\mathcal{M}_r, \mathcal{I})$ where $\mathcal{M}_r$ is a tuple $(\mathcal{T}, t_0, \iota)$ such that $\mathcal{T}$ is a set of timelines with a distinguished timeline $t_0 \in \mathcal{T}$, ι is a morphism mapping n-ary predicate symbols to n-ary relations on $\mathcal{P}$, and $\mathcal{I}$ is a *interpretation function* mapping the constant now to $(t_0, 0)$, every variable symbol x to an element of $\mathcal{P}$, and every unary Skolem function f to a morphism $\mathcal{I}(f) : \mathcal{P} \to \mathcal{P}$. The function $\mathcal{I}$ is extended to a function $\cdot^{\mathcal{I}}$ mapping terms to $\mathcal{P}$ in the standard way, that is, $t^{\mathcal{I}} = \mathcal{I}(t)$ if t is a variable or constant, and $f(t_1, \dots, t_n)^{\mathcal{I}} = \mathcal{I}(f)(t_1^{\mathcal{I}}, \dots, t_n^{\mathcal{I}})$, otherwise.

Let $\mathcal{I}$ be an interpretation function. By $\mathcal{I}[x/w]$, where x is a variable and w is a point, we denote a interpretation function $\mathcal{I}'$ such that $\mathcal{I}'(y) = \mathcal{I}(y)$ for any symbol y distinct from x, and $\mathcal{I}'(x) = w$. If $x_1, \dots, x_n$ are distinct variables and $w_1, \dots, w_n$ are points, then $\mathcal{I}[x_1/w_1, \dots, x_n/w_n]$ denotes $\mathcal{I}[x_1/w_1] \dots [x_n/w_n]$. If $w = (t, k)$ is a point and $n \in \mathbb{N}$, then w^{+n} denotes the point $(t, k + n)$. If f is a mapping from points to points, then f^{+n} denotes a function defined by $f^{+n}(w) = f(w)^{+n}$ for every $w \in \mathcal{P}$, that is, for every $w \in \mathcal{P}$, if $f(w) = (t, k)$, then $f^{+n}(w) = (t, k + n)$. By $\mathcal{I}^{+n}$ we denote a interpretation function defined by $\mathcal{I}^{+n}(s) = \mathcal{I}(s)^{+n}$ for every symbol s in the domain of $\mathcal{I}$.

$(\mathcal{M}_r, \mathcal{I}) \models \text{true}$

$(\mathcal{M}_r, \mathcal{I}) \not\models \text{false}$

$(\mathcal{M}_r, \mathcal{I}) \models p(t_1, \dots, t_n)$ iff $(t_1^{\mathcal{I}}, \dots, t_n^{\mathcal{I}}) \in \iota(p)$

$(\mathcal{M}_r, \mathcal{I}) \models \neg\varphi$ iff $(\mathcal{M}_r, \mathcal{I}) \not\models \varphi$

$(\mathcal{M}_r, \mathcal{I}) \models \varphi \wedge \psi$ iff $(\mathcal{M}_r, \mathcal{I}) \models \varphi$ and $(\mathcal{M}_r, \mathcal{I}) \models \psi$

$$(\mathcal{M}_r, \mathcal{I}) \models \varphi \vee \psi \qquad \text{iff } (\mathcal{M}_r, \mathcal{I}) \models \varphi \text{ or } (\mathcal{M}_r, \mathcal{I}) \models \psi$$

$$(\mathcal{M}_r, \mathcal{I}) \models \varphi \Rightarrow \psi \qquad \text{iff } (\mathcal{M}_r, \mathcal{I}) \not\models \varphi \text{ or } (\mathcal{M}_r, \mathcal{I}) \models \psi$$

$$(\mathcal{M}_r, \mathcal{I}) \models \bigcirc\varphi \qquad \text{iff } (\mathcal{M}_r, \mathcal{I}^{+1}) \models \varphi$$

$$(\mathcal{M}_r, \mathcal{I}) \models \Diamond\varphi \qquad \text{iff there exists } n \in \mathbb{N} \text{ such that } (\mathcal{M}_r, \mathcal{I}^{+n}) \models \varphi$$

$$(\mathcal{M}_r, \mathcal{I}) \models \forall x\, \varphi \qquad \text{iff for every } w \in \mathcal{P}, (\mathcal{M}_r, \mathcal{I}[x/w]) \models \varphi.$$

If $(\mathcal{M}_r, \mathcal{I}) \models \varphi$, then $(\mathcal{M}_r, \mathcal{I})$ *satisfies* φ and φ is *satisfiable*. Although the syntax of SNF_r clauses resembles that of first-order temporal logic, the semantics is different. Unlike in first-order temporal logic variables are not interpreted as elements of domains attached to points, but are interpreted as points. Likewise constants and function symbols are not interpreted as morphisms on domains but as morphisms on points. In fact, the semantics is still based on the same building blocks: timelines, points, relations on points and mappings between points and symbols of our language (or vice versa).

However, note that for formulae not containing any occurrences of a $\bigcirc$ and $\Diamond$ operators, temporal interpretations act like standard first-order interpretations.

Theorem 4. *Let N be a set of SNF_K clauses. Then N is satisfiable if and only if $\Pi_r(N)$ is satisfiable.*

Proof. For an arbitrary connected model $\mathcal{M}$ for N we are able to construct a temporal interpretation $(\mathcal{M}_r, \mathcal{I})$ for $\Pi_r(N)$ and vice versa. The proof that $\mathcal{M}$ and $(\mathcal{M}_r, \mathcal{I})$ satisfy N and $\Pi_r(N)$, respectively, proceeds by induction on the structure of formula in N and $\Pi_r(N)$.

5 A Calculus for MTL

We call a literal L *shallow* if all its argument terms are either variables or constants, otherwise it is *deep*. If C is a disjunction or conjunction of shallow, unary literals, then by $C[x]$ we indicate that all literals in C have a common variable argument term x, and $C[x/y]$ is obtained by replacing every occurrence of x in C by y. Similarly, if L is a monadic literal, we write $L[t]$ to indicate that the term t is the argument of L.

A *clause* is a formula of the form $P \Rightarrow \varphi$ where P is a conjunction of literals and φ is a disjunction of literals, a formula of the form $\Diamond L$, or a formula $\bigcirc C$ where C is disjunction of literals. The *empty clause* is false or true $\Rightarrow$ false. We regard the logical connectives $\wedge$ and $\vee$ in clauses to be associative, commutative, and idempotent. Equality of clauses is taken to be equality modulo variable renaming.

A clause C is a *simple clause* if and only if all literals in C are shallow, unary, and share the same argument term. A conjunction (disjunction) C is a *simple conjunction (disjunction)* iff C is a conjunction (disjunction) of shallow, unary literals that share the same argument term. $\sim P$ denotes the negation normal form of a conjunction P.

A clause C is a *temporal clause* if and only if it either has the form $P[t] \Rightarrow \bigcirc D[t]$ or $P[x] \Rightarrow \Diamond L[x]$ where $P[x]$ is a simple conjunction, $D[t]$ is a simple clause and t is either a variable or the constant now. A clause C is a *modal clause* iff it either has the form $P[x] \Rightarrow r_\kappa(x, f(x))$, $P[x] \Rightarrow D[f(x)]$, or $P[x] \Rightarrow \neg r_\kappa(x, y) \vee D[y]$ where P is a simple conjunction and $D[f(x)]$ is a clause of unary, deep literals with common argument $f(x)$. For a simple or modal clause C we do not distinguish between $P \Rightarrow D$ and $true \Rightarrow \sim P \vee D$.

For every ground atom A, let the complexity measure $c(A)$ be the multiset of arguments of A. We compare complexity measures by the multiset extension $\succ_m$ of the strict subterm ordering. The ordering is lifted from ground to non-ground expressions as follows: $A \succ' B$ if and only if $c(A\sigma) \succ_m c(B\sigma)$, for all ground instances $A\sigma$ and $B\sigma$ of atoms A and B. The ordering $\succ'$ on atoms can be lifted to literals by associating with every positive literal A the multiset $\{A\}$ and with every negative literal $\neg A$ the multiset $\{A, A\}$, and comparing these by the multiset extension of $\succ'$. We denote the resulting ordering by $\succ$. The ordering $\succ$ is an admissible ordering in the sense of Bachmair and Ganzinger (2). Thus, the following two inference rules provide a sound and complete calculus for first-order logic in clausal form:

$$\text{Res} \qquad \frac{true \Rightarrow C \vee A \qquad true \Rightarrow D \vee \neg B}{true \Rightarrow (C \vee D)\sigma}$$

where (i) $C \vee A$ and $D \vee \neg B$ are simple or modal clauses, (ii) σ is the most general unifier of A and B, (iii) $A\sigma$ is strictly $\succ$-maximal with respect to $C\sigma$, and (iv) $\neg B\sigma$ is $\succ$-maximal with respect to $D\sigma$. As usual we assume that premises of resolution inference steps are variable disjoint.

$$\text{Fac} \qquad \frac{true \Rightarrow C \vee L_1 \vee L_2}{true \Rightarrow (C \vee L_1)\sigma}$$

where (i) $C \vee L_1 \vee L_2$ is a simple or modal clause, (ii) σ is the most general unifier of L_1 and L_2, and (iii) $L_1\sigma$ is $\succ$-maximal with respect to $C\sigma$.

Lemma 10 will show that the factoring inference rule is not required for the completeness of our calculus if we assume that the logical connective $\vee$ is idempotent.

The remaining inference rules are similar to those of the calculus for linear temporal logic presented in Dixon et al. (5). An inference by *step resolution* takes one of the following forms

$$\text{SRes1} \qquad \frac{P \Rightarrow \bigcirc(C \vee L_1) \qquad Q \Rightarrow \bigcirc(D \vee L_2)}{(P \wedge Q)\sigma \Rightarrow \bigcirc(C \vee D)\sigma}$$

$$\text{SRes2} \qquad \frac{true \Rightarrow C \vee L_1 \qquad Q \Rightarrow \bigcirc(D \vee L_2)}{Q\sigma \Rightarrow \bigcirc(C \vee D)\sigma}$$

$$\text{SRes3} \qquad \frac{P \Rightarrow \bigcirc false}{true \Rightarrow \sim P}$$

where (i) P and Q are simple conjunctions, (ii) $C \vee L_1$ and $D \vee L_2$ are simple clauses, and (iii) σ is the most general unifier of L_1 and $\sim L_2$.

The following *merge rule* allows the formation of the conjunction of temporal clauses.

$$P_0[x_0] \Rightarrow \bigcirc C_0[x_0]$$

$$\ldots$$

Merge

$$\dfrac{P_n[x_n] \Rightarrow \bigcirc C_n[x_n]}{\bigwedge_{i=0}^{n} P_i[y] \Rightarrow \bigcirc \bigwedge_{i=0}^{n} C_i[y]}$$

where (i) each P_i, $1 \leq i \leq n$, is a simple conjunction, (ii) each C_i, $1 \leq i \leq n$ is a simple clause, (iii) y is a new variable. The conclusion of an inference step by the merge rule is a *merged temporal clause*. The only purpose of this rule is to ease the presentation of the following temporal resolution rule.

An inference by *temporal resolution* takes the following form

$$P_0[x_0] \Rightarrow \bigcirc G_0[x_0]$$

$$\ldots$$

TRes

$$\dfrac{\begin{array}{c} P_n[x_n] \Rightarrow \bigcirc G_n[x_n] \\ A(y) \Rightarrow \Diamond L[y] \end{array}}{A(y) \Rightarrow (\bigwedge_{i=0}^{n} \neg P_i[y]) \mathcal{W} L(y)}$$

where (i) each $P_i[x_i] \Rightarrow \bigcirc G_i[x_i]$, $1 \leq i \leq n$, is a merged temporal clause, (ii) for all i, $0 \leq i \leq n$, $\forall x_i\, G_i[x_i] \Rightarrow \neg L[x_i]$ and $\forall x_i\, G_i[x_i] \Rightarrow \bigvee_{j=0}^{n} P_j[x_j/x_i]$ are provable.

The conclusion $A(y) \Rightarrow (\bigwedge_{i=0,\ldots,n} \neg P_i[y]) \mathcal{W} L(y)$ of an inference by temporal resolution has to be transformed into normal form. Thus, we obtain

(1) $\text{true} \Rightarrow \neg A(y) \vee L(y) \vee {\sim} P_i(y)$

(2) $\text{true} \Rightarrow \neg A(y) \vee L(y) \vee q_L^w(y)$

(3) $q_L^w(y) \Rightarrow \bigcirc(L(y) \vee {\sim} P_i(y))$

(4) $q_L^w(y) \Rightarrow \bigcirc(L(y) \vee q_L^w(y))$,

where q_L^w is a new unary predicate symbol uniquely associated with L.

The calculus $\mathsf{C_{MTL}}$ consists of the inference rules Res, Fac, SRes1, SRes2, SRes3, Merge, and TRes. It is possible to replace Merge and TRes by a single inference rule which uses ordinary temporal clauses as premises and forms the merged clauses only in an intermediate step to compute the conclusion of an application of the temporal resolution rule. Thus, in our consideration in Section 6 and 7 we will not explicitly mention the Merge inference rule and merged temporal clauses.

6   Soundness of $\mathsf{C_{MTL}}$

Lemma 5. *Let* $\text{true} \Rightarrow C$ *be a clause and let* σ *be a substitution. If* $\text{true} \Rightarrow C$ *is satisfiable, then* $\text{true} \Rightarrow C\sigma$ *is satisfiable.*

Theorem 6 (Soundness). *Let* φ *be a well-formed formula of* MTL. *If a refutation of* $\Pi_r(\Pi_K(\varphi))$ *in* $\mathsf{C_{MTL}}$ *exists then* φ *is unsatisfiable.*

Proof. We show that for every instance of an inference rule of $\mathsf{C_{MTL}}$ that the satisfiability of the premises implies the satisfiability of the conclusion. In the case of Res and Fac this is straightforward since temporal interpretations act like first-order interpretations and we know that Res and Fac are sound with respect to first-order logic. The proof for the remaining inference rules can be found in (Hustadt et al. 12). □

7   Termination of $\mathsf{C_{MTL}}$

The termination proof for our calculus will take advantage of the following observations.

1. SNF_r clauses can be divided into three disjoint classes: *temporal clauses*, *modal clauses*, and *simple clauses*. It is straightforward to check that if N is a set of SNF_K clauses then all clauses in $\Pi_r(N)$ including the clauses we add for one of the axiom schemata 4, 5, B, D, and T belong to exactly one of these classes.
   Note that simple and modal clauses are standard first-order clauses.
2. The inference rules of our calculus can be divided into two classes: Res and Fac are the standard inference rules for ordered resolution and only modal clauses and simple clauses are premises in inference steps by these rules. The conclusion of such an inference step will again be a clause belonging to one of these two classes as we show in Lemma 9 and Lemma 10 below.
   SRes1, SRes2, SRes3, and TRes are variants of the inference rules of the resolution calculus for linear temporal logic presented in Dixon et al. (5). Only temporal and simple clauses can be premises of inference steps by these rules. The conclusion of such an inference step will consist of clauses which again belong to one these classes as is shown in Lemma 11 and Lemma 12 below.

Thus, the clauses under consideration and the calculus enjoy a certain modularity. Interaction between the two classes of inference rules and the class of temporal and modal clauses are only possible via the class of simple clauses. Given a finite signature, the classes of simple, modal, and temporal clauses are finitely bounded. Termination of any derivation from SNF_r is a direct consequence of the closure properties of the inference rules mentioned above.

Lemma 7. *Let φ be a well-formed formula of* MTL. *Every clause in $\Pi_r(\Pi_K(\varphi))$ is either a simple, a temporal, or a modal clause.*

Lemma 8. *Given a finite signature Σ, the classes of simple, modal and temporal clauses over Σ are finitely bounded.*

Proof. Note that the only terms which can occur in these clauses are either variables, the constant now, or terms of the from $f(t)$ where t is either a variable or a constant. Furthermore, no clause has more than two variables. Given that we can show that the length of clauses is linear in the size of the signature, limiting the number of non-variant clauses to an exponential number in the size of the signature.

The proof shows that SNF_r clauses have a linear length in the size of the signature. That gives us a single-exponential space (and time) bound for our decision procedure.

Due to side condition (i) of the inference rules Res and Fac, temporal clauses cannot be premises of inference steps by these rules. Simple clauses and modal clauses are special cases of *DL-clauses* (Hustadt and Schmidt 13). The following two lemmata follow directly from the corresponding result for DL-clauses.

Lemma 9. *Let $C_1 \vee A$ and $C_2 \vee \neg B$ be SNF_r clauses and let $C = (C_1 \vee C_2)\sigma$ be an ordered resolvent of these clauses. Then C is either a modal clause or a simple clause.*

Lemma 10. *Let $C_1 = D_1 \vee L_1 \vee L_2$ be a SNF_r clause and let $C = (D_1 \vee L_1)\sigma$ be a factor of C_1. Then C is either a modal clause or a simple clause and an application of the factoring rule simply amounts to the removal of duplicate literals in C_1.*

By a case analysis of all possible inference steps by SRes1, SRes2, SRes3, and TRes on SNF_r clauses we obtain the following two lemmata.

Lemma 11. *Let C_1 and C_2 be SNF_r clauses and let C be the conclusion of inference steps by SRes1, SRes2, or SRes3 from C_1 and C_2. Then C is a temporal clause or a simple clause.*

Lemma 12. *Let $C_1, \ldots, C_n$ be SNF_r clauses and let C be one of the clauses resulting from the transformation of the conclusion of an application of TRes to $C_1, \ldots, C_n$. Then C is a simple or a temporal clause.*

We are now in the position to state the main theorem of this section.

Theorem 13 (Termination). *Let φ be a well-formed formula of MTL. Any derivation from $\Pi_r(\Pi_K(\varphi))$ in $\mathsf{C_{MTL}}$ terminates.*

Proof. By induction on the length of the derivation from $\Pi_r(\Pi_K(\varphi))$ we can show that any clause occurring in the derivation is either a simple, modal, or temporal clause. Lemma 7 proves the base case that every clauses in $\Pi_r(\Pi_K(\varphi))$ satisfies this property. Lemmata 9, 10, 11, and 12 establish the induction step of the proof.

The signature of clauses in $\Pi_r(\Pi_K(\varphi))$ is obviously finite. By Lemma 8 the classes of simple, modal, and temporal clauses based on a finite signature is finitely bounded. Thus, after a finitely bounded number of inference step we will have derived the empty clause or no new clauses will be added to the set of clauses. In both cases the derivation terminates.

8   Completeness of $\mathsf{C_{MTL}}$

The proof of completeness proceeds as follows. We describe a canonical construction of a *behaviour graph* and *reduced behaviour graph* for a given set N

of SNF$_r$ clauses. In Theorem 14 we show that N is unsatisfiable if and only if its reduced behaviour graph is empty. Theorem 16 shows that if the reduced behaviour graph for N is empty, then we are able to derive a contradiction using C$_{\mathsf{MTL}}$. Theorem 14 and 16 together imply that for any unsatisfiable set N of SNF$_r$ clauses we can derive a contradiction. Thus, C$_{\mathsf{MTL}}$ is complete. Details of the construction of behaviour graphs, reduced behaviour graphs, and the proof for the results of this section can be found in (Hustadt et al. 12).

Theorem 14. *Let N be a set of SNF$_r$ clauses. Then N is unsatisfiable if and only if its reduced behaviour graph is empty.*

Proof. The constructions are similar to those in Dixon et al. (5), except that we have explicit nodes and edges for the modal dimension for our logic, which were not necessary in the case that we have only the modal logic S5.

Lemma 15. *Let N be a set of SNF$_r$ clauses. If the unreduced behaviour graph for N is empty, then we can derive a contradiction from N using only the inference rules Res, SRes1, SRes2, and SRes3 of C$_{\mathsf{MTL}}$.*

Proof. If the unreduced behaviour graph is empty, then any node we have constructed originally, has been deleted because one of the simple, modal, or temporal clauses of the form $P[x] \Rightarrow \bigcirc C[x]$ is not true at n_s. Thus, we can use the inference rules Res, SRes1, SRes2, and SRes3 to derive a contradiction.

Theorem 16. *Let N be a set of SNF$_r$ clauses. If the reduced behaviour graph for N is empty, then we can derive a contradiction from N by C$_{\mathsf{MTL}}$.*

Proof. Let N be an unsatisfiable set of SNF$_r$ rules. The proof is by induction on the number of nodes in the behaviour graph of N. If the unreduced behaviour graph is empty, then by Lemma 15 we can obtain a refutation using the inference rules Res, SRes1, SRes2, and SRes3.

Suppose that the unreduced behaviour graph G is non-empty. By Theorem 14 the reduced behaviour graph must be empty, so each node in G can be deleted by reduction rules similar to those in (Dixon et al. 5). The deletion of these nodes are shown to correspond to applications of step resolution and temporal resolution along the lines of Dixon et al. (5).

The completeness theorem now follows from Theorems 3, 4, and 16.

Theorem 17 (Completeness). *Let φ be a well-formed formula of MTL. If φ is unsatisfiable, then there exists a refutation of $\Pi_r(\Pi_K(\varphi))$ by C$_{\mathsf{MTL}}$.*

9   Example Refutation

We show that $[K]\bigcirc p \wedge \Box\,[K](p \Rightarrow \bigcirc p) \Rightarrow \bigcirc\Box p$ is valid if $[K]$ is a T modality. This is done by proving the unsatisfiability of

$$\varphi = [K]\bigcirc p \wedge \Box\,[K](p \Rightarrow \bigcirc p) \wedge \bigcirc\Diamond\neg p$$

$\Pi_r(\Pi_K(\varphi))$ is equal to the following set of clauses:

(5) true $\Rightarrow q_0(\text{now})$
(6) true $\Rightarrow \neg q_0(x) \vee q_{[K]\bigcirc p}(x)$
(7) true $\Rightarrow \neg q_0(x) \vee q_1(x)$
(8) true $\Rightarrow \neg q_0(x) \vee q_2(x)$
(9) true $\Rightarrow \neg q_{[K]\bigcirc p}(x) \vee \neg r_K(x, y) \vee q_3(y)$
(10) $q_3(z) \Rightarrow \bigcirc q_p(z)$
(11) true $\Rightarrow \neg q_1(x) \vee q_{[K](p\Rightarrow\bigcirc p)}(x)$
(12) $q_1(x) \Rightarrow \bigcirc q_1(x)$
(13) true $\Rightarrow \neg q_{[K](p\Rightarrow\bigcirc p)}(x) \vee \neg r_K(x, y) \vee q_4(y)$
(14) true $\Rightarrow \neg q_4(x) \vee \neg q_p(x) \vee q_5(x)$
(15) $q_5(x) \Rightarrow \bigcirc q_p(x)$
(16) $q_2(x) \Rightarrow \bigcirc q_6(x)$
(17) $q_6(x) \Rightarrow \Diamond \neg q_p(x)$
(18) $r_K(x, x)$

The derivation proceeds as follows:

$[(18)1,(13)2,\text{Res}]$	(19)	true $\Rightarrow \neg q_{[K](p\Rightarrow\bigcirc p)}(x) \vee q_4(x)$
$[(11)1,(12)2,\text{SRes2}]$	(20)	$q_1(x) \Rightarrow \bigcirc q_{[K](p\Rightarrow\bigcirc p)}(x)$
$[(19)1,(20)2,\text{SRes2}]$	(21)	$q_1(x) \Rightarrow \bigcirc q_4(x)$
$[(14)1,(21)2,\text{SRes2}]$	(22)	$q_1(x) \Rightarrow \bigcirc(\neg q_p(x) \vee q_5(x))$
$[(15)2,(22)2,\text{SRes1}]$	(23)	$q_1(x) \wedge q_5(x) \Rightarrow \bigcirc q_5(x)$
$[(12),(15),(23),\text{Merge}]$	(24)	$q_1(x) \wedge q_5(x) \Rightarrow \bigcirc(q_1(x) \wedge q_5(x) \wedge q_p(x))$

Intuitively clause (24) says that once q_1 and q_5 hold at a point x, q_1, q_5, and q_p will hold at any temporal successor of x. Thus, once we reach x we will not be able to satisfy $\Diamond\neg q_p(x)$. This gives rise to an application of the temporal resolution rule to (17) and (24). We obtain the following four clauses from the conclusion of this inference step.

$[(17),(24),\text{TRes}]$	(25)	true $\Rightarrow \neg q_6(x) \vee \neg q_p(x) \vee \neg q_1(x) \vee \neg q_5(x)$
	(26)	true $\Rightarrow \neg q_6(x) \vee \neg q_p(x) \vee q_7(x)$
	(27)	$q_7(x) \Rightarrow \bigcirc(\neg q_p(x) \vee \neg q_1(x) \vee \neg q_5(x))$
	(28)	$q_7(x) \Rightarrow \bigcirc(\neg q_p(x) \vee q_7(x))$

In the following only clause (25) will be relevant. We show now that q_1, q_2 and q_3 cannot be true at the same point x.

$[(14)3,(25)4,\text{Res}]$	(29)	true $\Rightarrow \neg q_6(x) \vee \neg q_p(x) \vee \neg q_1(x) \vee \neg q_4(x)$
$[(19)2,(29)4,\text{Res}]$	(30)	true $\Rightarrow \neg q_6(x) \vee \neg q_p(x) \vee \neg q_1(x) \vee \neg q_{[K](p\Rightarrow\bigcirc p)}(x)$
$[(11)2,(30)4,\text{Res}]$	(31)	true $\Rightarrow \neg q_6(x) \vee \neg q_p(x) \vee \neg q_1(x)$
$[(12)2,(31)3,\text{SRes2}]$	(32)	$q_1(x) \Rightarrow \bigcirc(\neg q_6(x) \vee \neg q_p(x))$
$[(16)2,(32)3,\text{SRes1}]$	(33)	$q_1(x) \wedge q_2(x) \Rightarrow \bigcirc\neg q_p(x)$
$[(10)2,(33)3,\text{SRes1}]$	(34)	$q_1(x) \wedge q_2(x) \wedge q_3(x) \Rightarrow \bigcirc\text{false}$
$[(34),\text{SRes3}]$	(35)	true $\Rightarrow \neg q_1(x) \vee \neg q_2(x) \vee \neg q_3(x)$

The remainder of the refutation is straightforward. Based on the clauses (5) to (9) and the reflexivity of r_K, it is easy to see that q_1, q_2, and q_3 are true at the point now which contradicts clause (35).

[(18)1,(9)2,Res]	(36)	true $\Rightarrow \neg q_{[K]}\bigcirc_p(x) \vee q_3(x)$
[(36)2,(35)3,Res]	(37)	true $\Rightarrow \neg q_1(x) \vee \neg q_2(x) \vee \neg q_{[K]}\bigcirc_p(x)$
[(6)2,(37)3,Res]	(38)	true $\Rightarrow \neg q_1(x) \vee \neg q_2(x) \vee \neg q_0(x)$
[(7)2,(38)1,Res]	(39)	true $\Rightarrow \neg q_2(x) \vee \neg q_0(x)$
[(8)2,(39)1,Res]	(40)	true $\Rightarrow \neg q_0(x)$
[(5)1,(40)1,Res]	(41)	true $\Rightarrow$ false

10 Conclusion

We have presented a framework for the combination of modal and temporal logics consisting of (i) a normal form transformation of formulae of the combined logics into sets of SNF_K clauses, (ii) a translation of modal subformula in SNF_K clauses into a first-order language, and (iii) a calculus $\mathsf{C_{MTL}}$ for the combined logic which can be divided into standard resolution inference rules for first-order logic and modified resolution inference rules for discrete linear temporal logic.

The calculus $\mathsf{C_{MTL}}$ provides a decision procedure for combinations of subsystems of multi-modal S5 with linear, temporal logic.

Note that instead of modifying the inference rules for discrete linear temporal logic we could have retained them in their original form and added *bridging rules* between the two logics. We have shown that the only clauses which can be premises of the first-order inference rules as well as of the temporal inference rules are simple clauses. So, assume that Π_r leaves any SNF_K clauses with occurrences of the temporal connective $\bigcirc$ and $\Diamond$ unchanged. Furthermore, let δ be the homomorphic extension of a function that maps atoms $q_p(x)$ to q_p. Then an alternative calculus to $\mathsf{C_{MTL}}$ consists of Res, Fac, the original step resolution rules and temporal inference rule by Dixon et al. (5), and the two bridging rules

$$br_{pl} \quad \frac{\text{true} \Rightarrow C}{\text{true} \Rightarrow \delta(C)} \qquad\qquad br_{fol} \quad \frac{\text{true} \Rightarrow \delta(C)}{\text{true} \Rightarrow C}$$

where true $\Rightarrow C$ is a simple clause. This again stresses the importance of the observation that simple clauses control the interaction between the two calculi involved. The bridging rules allow for the translation of simple clauses during the derivation, thus providing an interface between the two calculi we have combined. This is approach is closely related to the work by Ghidini and Serafini (9).

Although we have only considered the basic modal logic K and its extensions by the axiom schemata 4, 5, B, D, and T, we are confident that soundness, completeness, and termination can be guaranteed for a much wider range of modal logics.

An important extension of the combinations of modal logics we are currently investigating is the addition of interactions between modal and temporal logics. In the presence of interactions the modularity of the calculus and the modularity of our proofs, in particular the proof of termination, can no longer be preserved.

References

Baader, F. and Ohlbach, H. J. (1995). A multi-dimensional terminological knowledge representation language. *Journal of Applied Non-Classical Logics*, 2:153–197.

Bachmair, L. and Ganzinger, H. (1997). A theory of resolution. Research report MPI-I-97-2-005, Max-Planck-Institut für Informatik, Saarbrücken, Germany. To appear in J. A. Robinson and A. Voronkov, editors, *Handbook of Automated Reasoning*.

Blackburn, P. and de Rijke, M. (1997). Why combine logics? *Studia Logica*, 59:5–27.

de Nivelle, H. (1999). Translation of S4 into GF and 2VAR. Manuscript.

Dixon, C., Fisher, M., and Wooldridge, M. (1998). Resolution for temporal logics of knowledge. *Journal of Logic and Computaton*, 8(3):345–372.

Fagin, R., Halpern, J. Y., Moses, Y., and Vardi, M. Y. (1996). *Reasoning About Knowledge*. MIT Press.

Finger, M. (1994). Notes on several methods for combining temporal logic systems. Presented at ESSLLI'94.

Gabbay, D. M. (1996). Fibred semantics and the weaving of logics. Part 1. Modal and intuitionistic logics. *Journal of Symbolic Logic*, 61(4):1057–1120.

Ghidini, C. and Serafini, L. (1998). Distributed first order logics. In Gabbay, D. M. and de Rijke, M., editors, *Proc. FroCoS'98*. To appear.

Goranko, V. and Passy, S. (1992). Using the universal modality: Gains and questions. *Journal of Logic and Computation*, 2(1):5–30.

Halpern, J. Y. (1987). Using reasoning about knowledge to analyse distributed systems. *Annual Review of Computer Science*, 2.

Hustadt, U., Dixon, C., Schmidt, R., and Fisher, M. (2000). Normal forms and proofs in combined modal and temporal logics. Extended version of this paper, available at http://www.card.mmu.ac.uk/U.Hustadt/publications/HDSF2000b.ps.gz.

Hustadt, U. and Schmidt, R. A. (2000). Issues of decidability for description logics in the framework of resolution. In Caferra, R. and Salzer, G., editors, *Automated Deduction in Classical and Non-Classical Logics*, volume 1761 of *LNAI*, pages 192–206. Springer.

Jennings, N. R. (1999). Agent-based computing: Promise and perils. In Dean, T., editor, *Proc. IJCAI'99*, pages 1429–1436. Morgan Kaufmann.

Johnson, C. W. (1994). The formal analysis of human-computer interaction during accidents investigations. In *People and Computers IX*, pages 285–300. Cambridge University Press.

Ohlbach, H. J. (1998). Combining Hilbert style and semantic reasoning in a resolution framework. In Kirchner, C. and Kirchner, H., editors, *Proc. CADE-15*, volume 1421 of *LNAI*, pages 205–219. Springer.

Ohlbach, H. J. and Gabbay, D. M. (1998). Calendar logic. *Journal of Applied Non-Classical Logics*, 8(4).

Wolter, F. and Zakharyaschev, M. (1998). Satisfiability problem in description logics with modal operators. In Cohn, A. G., Schubert, L. K., and Shapiro, S. C., editors, *Proc. KR'98*, pages 512–523. Morgan Kaufmann.

Wooldridge, M., Dixon, C., and Fisher, M. (1998). A tableau-based proof method for temporal logics of knowledge and belief. *Journal of Applied Non-Classical Logics*, 8(3):225–258.

Wooldridge, M. and Jennings, N. R. (1995). Intelligent agents: Theory and practice. *The Knowledge Engineering Review*, 10(2):115–152.

Structured Sequent Calculi for Combining Intuitionistic and Classical First-Order Logic*

Paqui Lucio

Dpto de L.S.I., Facultad de Informática,
Paseo Manuel de Lardizabal, 1, Apdo 649,
20080-San Sebastián, SPAIN.
Tel: +34 (9)43 015049, Fax: +34 (9)43 219306,
jiplucap@si.ehu.es.

Abstract. We define a sound and complete logic, called $\mathcal{FO}^{\supset}$, which extends classical first-order predicate logic with intuitionistic implication.

As expected, to allow the interpretation of intuitionistic implication, the semantics of $\mathcal{FO}^{\supset}$ is based on structures over a partially ordered set of worlds. In these structures, classical quantifiers and connectives (in particular, implication) are interpreted within one (involved) world. Consequently, the *forcing relation* between worlds and formulas, becomes *non-monotonic* with respect to the ordering on worlds. We study the effect of this lack of monotonicity in order to define the satisfaction relation and the *logical consequence relation* which it induces.

With regard to proof systems for $\mathcal{FO}^{\supset}$, we follow Gentzen's approach of *sequent calculi* (*cf.* [8]). However, to deal with the two different implications simultaneously, the sequent notion needs to be more *structured* than the traditional one. Specifically, in our approach, the antecedent is structured as a sequence of sets of formulas. We study how inference rules preserve soundness, defining a structured notion of logical consequence. Then, we give some general sufficient conditions for the *completeness* of this kind of sequent calculi and also provide a sound calculus which satisfies these conditions. By means of these two steps, the completeness of $\mathcal{FO}^{\supset}$ is proved in full detail. The proof follows Hintikka's set approach (*cf.* [11]), however, we define a more general procedure, called *back-saturation*, to saturate a set with respect to a sequence of sets.

1 Introduction

Combining different logical systems has become very common in several areas of computer science. In this paper, we introduce a logical system, called $\mathcal{FO}^{\supset}$, obtained by using a combination of classical and intuitionistic logic. Our original motivation for defining this logic was to provide logical foundations for *logic programming* (LP) languages which combine classical ($\rightarrow$) and intuitionistic ($\supset$) implication. A well-founded LP language should be entirely supported by an

---

* This work has been partially supported by project TIC98-0949-C02-02.

underlying logic. This means that the declarative semantics of the LP language is based on the logical model-theory, and at the same time, the operational semantics is supported by the logical proof-theory; *cf.* [15] for technical details. Sequent calculi provide a very natural and direct way to formalise the operational semantics of LP languages. It is well known that standard LP (Horn clauses programming) is well-founded in classical first-order predicate logic. However, as mentioned in [10]:

> "If the two implications are considered altogether, the resulting semantics differs from that of both intuitionistic and classical logic."

Then, for the extension of LP languages with intuitionistic implication, our aim is to obtain complete sequent calculi for a sound extension of first-order predicate logic with intuitionistic implication. We want to make it clear that the concern of this paper is the logic $\mathcal{FO}^{\supset}$ itself, but not its application to LP. In fact, the latter is the main subject of [1].

The problem in combining Hilbert axiomatizations of classical and intuitionistic logic is shown in [2]:

> "It cannot be attained by simply extending the union of both axiomatizations by so called interaction axioms"

In fact, whenever both kinds of implicative connectives are mixed, some equivalences and axiom schemas are lost, and they collapse into classical logic. What they do to achieve a Hilbert-style axiomatization (and also a tableaux method) is to restrict some axiom schemas (and tableaux rules) to a special kind of formulas called persistent. The same restriction is used in the natural deduction system introduced in [12] which also combines classical and intuitionistic logic. Our proposal enables greater flexibility for developing deductions, by introducing structure in the sequents, since this structure makes unnecessary the above persistence-restriction. We will return to this matter in the last section, after $\mathcal{FO}^{\supset}$ has been presented.

$\mathcal{FO}^{\supset}$ syntax is first-order without any restriction, hence, we deal with variables, functions, predicates, connectives and quantifiers. For semantical differentiation of both implications, it suffices to consider standard Kripke structures, over a partially ordered set of worlds. Worlds have an associated classical first-order structure for the semantical interpretation of terms over its universe ([20,21]). An intuitionistic implication $\varphi \supset \psi$ is satisfied in a world w if every world greater than w satisfying φ also satisfies ψ. By contrast, classical connectives and quantifiers are interpreted within one (involved) world. In particular, a world w satisfies $\varphi \rightarrow \psi$ if either w does not satisfy φ or satisfies ψ. As a consequence, the forcing relation between worlds and formulas becomes non-monotonic with respect to the ordering on worlds. This lack of monotonicity is taken into account to define the *satisfaction* and *logical consequence* relations. In particular, a Kripke model satisfies a formula whenever all its minimal worlds force it. This satisfaction relation avoids the collapse of both implications. Moreover, it induces (in the usual way) a logical consequence relation for $\mathcal{FO}^{\supset}$. With regard to sequent calculi for $\mathcal{FO}^{\supset}$ the central point is the nature of sequents.

Gentzen's original notion (*cf.* [8]) considers a sequent to be a pair (Γ, Φ) where the *antecedent* Γ and the *consequent* Φ are finite sequences of formulas. It is usual (in classical logic) to consider sets, instead of sequences, because this prevents some extra inference rules for duplication of formulas (contraction rule), interchanges of formulas (interchange rule), and so on. In intuitionistic logic, the consequent usually consists of a single formula. To deal with classical and intuitionistic implications within the same logic, the antecedent requires more structure, in order to avoid the collapse of both implications. We introduce this idea by means of two simple examples. Let us consider the three propositional symbols p, q, r. It is obvious that $p \to q$ is semantically weaker than $p \supset q$. Therefore, we should not allow the derivation of the sequent with $p \to q$ as antecedent and $p \supset q$ as consequent. Roughly speaking, $p \to q$ is a "one world sentence", whereas $p \supset q$ is "about all greater worlds". This meaning suggests that the "$\supset$-in-the-right" rule, to split $p \supset q$ for putting p on the left and leaving q on the right, must be used with care about which worlds $p \to q$ and p are speaking about. In fact, the antecedent of the next sequent in our derivation tree has the sequence (but not the set) $\langle p \to q; p \rangle$; obviously with q as consequent. This sequent should not be derivable, since $p \to q$ is saying nothing about greater worlds satisfying p. Moreover, to derive, for example, the valid (classical) sentence $(p \to q) \to ((q \to r) \to (p \to r))$, it becomes useful to relate $p \to q$, $q \to r$ and p to the same world in the antecedent sequence, taking r as the consequent. Thus sequences of sets arise as antecedents. Accordingly, in this paper, a sequent consists of a pair (Δ, χ) where the antecedent Δ is a (finite) sequence of (finite) sets of formulas and the consequent χ is a single formula. Hence, we say *structured sequent calculus* to emphasise its nature. For the study of structured deductive systems in a more general setting, see [5].

The rest of the paper is organised as follows: in Section 2 we give preliminary details about syntax and notation; in Section 3 we establish the model theory and the necessary semantical notions and properties; in Section 4 we present the soundness and completeness results for structured sequent calculi; finally, in Section 5, we summarise conclusions and related work.

2   Preliminaries

We consider *signatures* Σ consisting of countable (pairwise disjoint) sets VS_Σ of variable symbols, FS_Σ of function symbols, and PS_Σ of predicate symbols, with some specific arity for each function and predicate symbol. Function symbols of arity 0 are called *constant symbols*. We denote by $Term_\Sigma$ the set of all well-formed first-order Σ-terms, inductively defined by:

- A variable $x \in VS_\Sigma$ is a Σ-term
- If $f \in FS_\Sigma$ is n-ary and $t_1, \ldots, t_n \in Term_\Sigma$, then $f(t_1, \ldots, t_n)$ is a Σ-term.

The set $Form_\Sigma$ of all well-formed Σ-formulas contains the atomic formulas and the recursive composition of formulas by means of intuitionistic implication and classical connectives and quantifiers. For convenience, we consider the following atomic formulas:

– $p(t_1, \ldots, t_n)$, for n-ary $p \in PS_\Sigma$ and $t_1, \ldots, t_n \in Term_\Sigma$
– F (falsehood).

In addition, we consider the classical connectives: negation ($\neg$) and implication ($\rightarrow$); the intuitionistic implication ($\supset$); and the classical universal quantifier $\forall$. The remaining connectives and quantifiers, i.e. conjunction ($\wedge$), disjunction ($\vee$), existential quantification ($\exists$), and intuitionistic negation ($\sim$), can be defined as abbreviations: $\varphi \wedge \psi$ for $\neg(\varphi \rightarrow \neg\psi)$, $\varphi \vee \psi$ for $\neg\varphi \rightarrow \psi$, $\exists x\varphi$ for $\neg\forall x\neg\varphi$, and $\sim \varphi$ for $\varphi \supset F$. Even the intuitionistic universal quantifier, namely $\check{\forall}$, can be expressed in $\mathcal{FO}^\supset$ by the formula $(\neg F) \supset \forall x\varphi$ as definition of $\check{\forall}x\varphi$.

Terms without variable symbols are called *closed terms*. By means of quantifiers, formulas have free and bound variables. We denote by $free(\varphi)$ the set of all free variables of the formula φ. We call Σ-*sentence* a Σ-formula without free variables.

The uppercase Greek letters Γ and Φ (possibly with sub- and superscripts) will be used as metavariables for *sets* of formulas, whereas Δ, Δ', Δ'', ... are reserved to be metavariables for *sequences of sets* of formulas.

We denote by $cons(t)$ the set of all constant symbols appearing in the term t. A very simple structural induction extends $cons$ to formulas, sets of formulas and sequences of sets of formulas.

Structures (or models) are Kripke models of first-order intuitionistic logic ([20,21]), that is, we consider worlds with universes and function symbols interpretations, as well as predicate symbols interpretations. Variables assignments and substitution notation are especially awkward for dealing with Kripke models. For that reason, to define the semantics of quantifiers, we follow the approach in [19] of *naming* all elements of the universe:

Definition 1. Given a set of elements (or universe) A, a signature Σ can be extended to the signature Σ_A, by adding a new constant symbol $\hat{a}$ for each element $a \in A$. ∎

In order to simplify substitution notation we write $\varphi(\bar{t})$ for the *simultaneous substitution instance* of the formula $\varphi(\bar{x})$ with terms $\bar{t}$ for variables $\bar{x}$. The notation $\varphi(\bar{x})$ is a meta-expression, which does not mean that $\bar{x}$ is the exhaustive list of free variables of φ, nor that all of them occur as free variables in φ.

3   Non-monotonic Forcing and Logical Consequence

In this section we firstly establish the semantical structures for $\mathcal{FO}^\supset$, and then define the *forcing, satisfaction and logical consequence* relations. At the same time, we point out some interesting meta-logical properties, such as, for example, the monotonicity of terms evaluation and the substitution lemma, which are usual and useful in most of the well-known logics.

Semantical structures for $\mathcal{FO}^\supset$ are standard first-order Kripke structures with a partially ordered set of worlds ([20,21]). Worlds have an associated classical first-order structure over a universe. We consider that the signature of every world includes names for individuals of its universe. We abbreviate by Σ_w the signature Σ_{A_w}, which extends Σ with a name $\widehat{a}$ for each $a \in A_w$ (see Definition 1).

Definition 2. A *Kripke Σ-structure* is a triple $\mathcal{K} = (W(\mathcal{K}), \preceq, \langle \mathcal{A}_w \rangle_{w \in W(\mathcal{K})})$ where

(a) $(W(\mathcal{K}), \preceq)$ is a partially ordered set (of worlds)
(b) Each $\mathcal{A}_w$ is a first-order Σ_w-structure defined by:
   - A non-empty universe A_w
   - A function $f^{\mathcal{A}_w} : (A_w)^n \to A_w$ for each n-ary $f \in FS_{\Sigma_w}$
   - A set At_w of atomic Σ-formulas of the form $p(\widehat{a_1}, \ldots, \widehat{a_n})$
   such that for any pair of worlds $v \preceq w$:
   1. $A_v \subseteq A_w$,
   2. $At_v \subseteq At_w$, and
   3. $f^{\mathcal{A}_w}(\overline{a}) = f^{\mathcal{A}_v}(\overline{a})$, for all n-ary $f \in FS_{\Sigma_v}$ and all $\overline{a} \in (A_v)^n$. ∎

These Kripke structures allow us to interpret Σ_w-terms in worlds in a monotonic way.

Definition 3. Let w be a world of a Kripke Σ-structure and let $t \in Term_{\Sigma_w}$, its interpretation $t^{\mathcal{A}_w}$ (briefly t^w) is inductively defined as follows:

- $\widehat{a}^{\mathcal{A}_w} = a$ for any $a \in A_w$
- $(f(t_1, \ldots, t_n))^{\mathcal{A}_w} = f^{\mathcal{A}_w}(t_1^{\mathcal{A}_w}, \ldots, t_n^{\mathcal{A}_w})$. ∎

Proposition 4. *For any $t \in Term_{\Sigma_v}$, any Kripke-structure $\mathcal{K}$ and any pair of worlds $v, w \in W(\mathcal{K})$ such that $v \preceq w$: $t^w = t^v \in A_v \subseteq A_w$.*

Proof: The usual proof by structural induction on t is suitable. ∎

The satisfaction of sentences in worlds is handled by the following *forcing relation*:

Definition 5. Letting $\mathcal{K}$ be a Kripke Σ-structure, the binary *forcing relation* $\Vdash$ between worlds in $W(\mathcal{K})$ and Σ-formulas is inductively defined as follows:

$w \nVdash F$
$w \Vdash p(t_1, \ldots, t_n)$ iff $p(\widehat{t_1^w}, \ldots, \widehat{t_n^w}) \in At_w$
$w \Vdash \neg\varphi$ iff $w \nVdash \varphi$
$w \Vdash \varphi \to \psi$ iff $w \nVdash \varphi$ or $w \Vdash \psi$
$w \Vdash \varphi \supset \psi$ iff for all $v \in W(\mathcal{K})$ such that $v \succeq w$: if $v \Vdash \varphi$ then $v \Vdash \psi$
$w \Vdash \forall x \varphi(x)$ iff $w \Vdash \varphi(\widehat{a})$ for all $a \in A_w$.

For a set of formulas Φ, $w \Vdash \Phi$ means that $w \Vdash \varphi$ for all $\varphi \in \Phi$. ∎

It is obvious that $\Vdash$ is relative to $\mathcal{K}$, hence we will write $\Vdash_{\mathcal{K}}$ whenever the simplified notation $\Vdash$ may be ambiguous.

$\mathcal{FO}^{\supset}$ satisfies the following meta-logical property:

Lemma 6. (Substitution Lemma) *For any Kripke Σ-structure $\mathcal{K}$, any $w \in W(\mathcal{K})$, any closed Σ-terms t_1, t_2 and any Σ-formula φ with $free(\varphi) \subseteq \{x\}$:*

$$\text{If } t_1^w = t_2^w \text{ and } w \Vdash \varphi(t_1), \text{ then } w \Vdash \varphi(t_2).$$

Proof: A very easy structural induction on φ. ∎

By Definition 2, it is obvious that the forcing relation $\Vdash$ behaves monotonically for atomic sentences, therefore we say that $\mathcal{FO}^{\supset}$ is *atomically monotonic*. However, in contrast with (pure) intuitionistic logic, monotonicity can not be extended to arbitrary sentences. This happens because $\mathcal{FO}^{\supset}$ gives a non-intuitionistic semantics to negation, classical implication and universal quantification. The following example illustrates this behaviour.

Example 7. Consider the following Kripke structure $\mathcal{K}$ with $W(\mathcal{K}) = \{u, v, w\}$ such that $u \preceq v \preceq w$, $\mathcal{A}_u$ with universe $\{a\}$, $\mathcal{A}_v$ with universe $\{a, b\}$, $\mathcal{A}_w$ with universe $\{a, b, c\}$, $At_u = \emptyset$, $At_v = \{r(\widehat{a}), r(\widehat{b})\}$, and $At_w = \{r(\widehat{a}), r(\widehat{b})\}$. In such a model, the world u forces $\neg r(\widehat{a})$ and also $r(\widehat{a}) \to q(\widehat{a})$, but v does not force either. Moreover, v forces $\forall x r(x)$, but w does not. ∎

Now, our aim is to define a satisfaction relation between Kripke structures and sentences, from which a logical consequence relation between sets of sentences and sentences may be induced in the usual way.

Definition 8. Let $\approx$ be a satisfaction relation between Σ-structures and Σ-sentences, then the induced logical consequence relation, denoted by the same symbol $\approx$, is defined as follows.

For every set of Σ-sentences $\Gamma \cup \{\varphi\}$:
$\Gamma \approx \varphi$ iff for all Kripke Σ-structure $\mathcal{K}$: $\mathcal{K} \approx \Gamma \implies \mathcal{K} \approx \varphi$. ∎

It should be noted at the outset that we can not define the satisfaction relation as it is commonly done in well-known logics with possible worlds semantics, like intuitionistic logic (*cf.* [20,21]) and modal logics (*cf.* [4]). The reason is that, under this usual notion (recalled in Definition 9), both implications collapse into the intuitionistic one.

Definition 9. A Kripke structure $\mathcal{K}$ *satisfies* (or is a *model* of) a sentence φ if and only if $w \Vdash \varphi$ for all $w \in W(\mathcal{K})$. We call this relation *global satisfaction* and it is denoted by $\mathcal{K} \models_G \varphi$. ∎

It is easy to observe that, for logics with monotonic forcing relation (e.g. intuitionistic logic), the fact to be forced in all worlds is equivalent to the fact to be forced in all minimal worlds. Hence, for this kind of logics, the logical consequence induced (in the sense of Definition 8) from global satisfaction ($\models_G$) is equivalent to the one induced from the following *minimal worlds satisfaction relation*:

Definition 10.

(a) We say that a world $w \in W(\mathcal{K})$ is *minimal* if and only if there does not exist $v \in W(\mathcal{K})$ such that $v \prec w$.

(b) A Kripke Σ-structure $\mathcal{K}$ *satisfies* (or is a *model* of) a Σ-sentence φ ($\mathcal{K} \models \varphi$) if and only if $w \Vdash \varphi$ for each minimal world $w \in W(\mathcal{K})$.

(c) For sets of sentences: $\mathcal{K} \models \Gamma$ if and only if $\mathcal{K} \models \varphi$ for every $\varphi \in \Gamma$.   ∎

We adopt, for $\mathcal{FO}^{\supset}$, the satisfaction relation $\models$ of Definition 10(b) and the logical consequence relation (also denoted by $\models$) induced from it, in the sense of Definition 8.

As a matter of fact, $\models$ has a direct (instead of induced) equivalent definition, which is based on the *local* (in contrast with global) point of view. We are going to define the local relation $\models_L$ and then we prove that it is equivalent to $\models$ and stronger than $\models_G$.

Definition 11. $\Gamma \models_L \varphi$ if and only if for every Kripke structure $\mathcal{K}$ and for every world $w \in W(\mathcal{K})$, if $w \Vdash \Gamma$ then $w \Vdash \varphi$.   ∎

Proposition 12.

(i) $\models$ and $\models_L$ are equivalent logical consequence relations.

(ii) $\models_L$ is stronger than $\models_G$.

(iii) If the forcing relation is monotonic, then $\models_L$ and $\models_G$ are equivalent logical consequence relations.

Proof:

(i) It is trivial that $\models_L$ is stronger than $\models$. To prove the converse, let us suppose that there exist $\mathcal{K}$ and $w \in W(\mathcal{K})$ such that $w \Vdash \Gamma$ and $w \nVdash \varphi$. Then, we can define

$$\mathcal{K}^w = (\{v | v \in W(\mathcal{K}), v \succeq w\}, \preceq, \langle \mathcal{A}_v \rangle_{v \in W(\mathcal{K}), v \succeq w})$$

$\mathcal{K}^w$ is a Kripke structure with a unique minimal world w such that $w \Vdash \Gamma$ and $w \nVdash \varphi$. Therefore, $\mathcal{K}^w \Vdash \Gamma$ and $\mathcal{K}^w \nVdash \varphi$.

(ii) is trivial.

(iii) Suppose that $\Gamma \models_G \varphi$ and consider any $\mathcal{K}$ and $w \in W(\mathcal{K})$ such that $w \Vdash_{\mathcal{K}} \Gamma$. Now, consider $\mathcal{K}^w$ as above. By forcing monotonicity $w' \Vdash_{\mathcal{K}^w} \Gamma$ holds for all $w' \in W(\mathcal{K}^w)$. Then, $w' \Vdash_{\mathcal{K}^w} \varphi$ holds for all $w' \in W(\mathcal{K}^w)$. So, $w \Vdash_{\mathcal{K}} \varphi$.   ∎

Overall, the $\mathcal{FO}^{\supset}$ logical consequence $\models$ has been induced from the minimal world satisfaction relation of Definition 10(b), however, the equivalent formulation given by Definition 11 and Proposition 12(i) can be considered when convenient.

It is easy to see that $\models$ is monotonic with respect to set inclusion:

Fact 13. If $\Gamma \subseteq \Gamma'$ and $\Gamma \models \varphi$, then $\Gamma' \models \varphi$.   ∎

4 Structured Sequent Calculi and Completeness

As explained in Section 1, we consider structured sequents. Specifically, a sequent consists of a pair (Δ, χ) (written as $\Delta \mapsto \chi$) where the antecedent Δ is a *(finite) sequence of (finite) sets of formulas* and the consequent χ is a single formula. In order to simplify sequent notation we reserve Γ and Φ (possibly with sub- and superscripts) for *sets* of formulas, and Δ, Δ', Δ'', ... for *sequences of sets* of formulas; the semicolon sign $(;)$ will represent the infix operation for concatenation of sequences; $\Gamma \cup \{\varphi\}$ will be abbreviated by Γ, φ; and a set Γ is identified with the sequence consisting of this unique set. Thus, the semicolon sign $(;)$ is used to split a sequence into its components (sets of formulas) and the comma sign $(,)$ splits sets of formulas into its elements. For instance, $\Delta; \Gamma, \varphi; \Delta'$ denotes the sequence beginning with the sequence of sets Δ, followed by the set $\Gamma \cup \{\varphi\}$, and ending with the sequence of sets Δ'. With an abuse of notation we will write $\varphi \in \Delta$ to mean that φ belongs to some of the sets of formulas in Δ.

In this section we firstly define the notion of proof in a structured sequent calculus. Secondly, we explain what is required for soundness of this kind of calculi. For this purpose a structured consequence notion is defined. Then, we give general sufficient conditions for completeness, which have a dual advantage. On the one hand, different calculi for different purposes could be obtained from the general conditions. On the other hand, these conditions make the completeness proof easier. Lastly, we provide a sound and complete structured sequent calculus for $\mathcal{FO}^{\supset}$.

Definition 14. A *proof* of a sequent $\Delta \mapsto \chi$ in a calculus C (or *C-proof*) is a finite tree, constructed using inference rules from C, whose root is the sequent $\Delta \mapsto \chi$ and whose leaves are axioms (or initial sequents) in the calculus C. When there exists a *C-proof* of the sequent $\Delta \mapsto \chi$, we write $\Delta \vdash_C \chi$. Moreover, $\vdash_C$ is extended to (finite) sequences of infinite sets as follows. For Δ which is a sequence $\Gamma_0; \ldots; \Gamma_n$ of (possibly infinite) sets of Σ-formulas and φ a Σ-formula, $\Delta \vdash_C \varphi$ holds if and only if there exists a sequence Δ' of finite sets $\Gamma'_0; \ldots; \Gamma'_n$ such that $\Gamma'_i \subseteq \Gamma_i$ for each $i \in \{1..n\}$ and there exists a C-proof of the sequent $\Delta' \mapsto \varphi$. ∎

Therefore, $\vdash_C$ is the *derivability relation* induced by the calculus C. Notice that every set in the antecedent of a sequent must be finite, whereas in the derivability relation, sets can be infinite. A trivial consequence of definition 14 is the following fact, which provides the *thinning rule* as meta-rule:

Fact 15. If $\Delta; \Gamma; \Delta' \vdash_C \chi$ and $\Gamma \subseteq \Gamma'$, then $\Delta; \Gamma'; \Delta' \vdash_C \chi$. ∎

Now, we define a relation $\models^*$ between sequences of sets of formulas and formulas, such that its restriction to a single set of formulas coincides with the logical consequence relation $\models$.

Definition 16. Let Δ be a sequence $\Gamma_0; \Gamma_1; \ldots; \Gamma_n$ of sets of Σ-sentences and χ a Σ-sentence, then we say that $\Delta \models^* \chi$ if and only if for all Kripke Σ-structure

$\mathcal{K}$ and for all sequence of worlds $u_0, u_1, \ldots u_n \in W(\mathcal{K})$ such that $u_{i-1} \preceq u_i$ (for $i = 1, \ldots, n$):

$$u_i \Vdash \Gamma_i \text{ (for all } i = 0, \ldots, n) \implies u_n \Vdash \chi. \qquad \blacksquare$$

Reflecting on the direct (local) definition of the logical consequence relation (Definition 11 and Proposition 12), it is obvious that:

Proposition 17. *For every set of Σ-sentences $\Phi \cup \{\chi\}$: $\Phi \models^* \chi$ iff $\Phi \models \chi$.* $\blacksquare$

The soundness of a calculus C is warranted whenever each proof rule of C preserves $\models^*$. In other words, let us consider the following schema of proof rule:

$$\frac{\Delta_1 \Mapsto \chi_1 \quad \ldots \quad \Delta_n \Mapsto \chi_n}{\Delta \Mapsto \chi}$$

then, because of finiteness of sets in a C-proof and by induction on its length, it suffices that each proof rule of C (with the above scheme) satisfies:

$$\Delta_i \models^* \chi_i \text{ (for all } i = 1, \ldots, n) \implies \Delta \models^* \chi.$$

Notice that the monotonicity of $\models$ with respect to set inclusion (Fact 13) can be generalized to $\models^*$ in the following sense:

Fact 18. If $\Gamma \subseteq \Gamma'$ and $\Delta; \Gamma; \Delta' \models^* \varphi$, then $\Delta; \Gamma'; \Delta' \models^* \varphi$. $\blacksquare$

However, $\models^*$ is non-monotonic with respect to sub-sequence relations, in general. For instance, $\Delta; \Gamma \models^* \varphi$ could not hold, although $\Delta \models^* \varphi$ holds.

As a first step to achieve completeness, we provide general sufficient conditions for $\vdash_C$, which are formulated on the basis of Hintikka sets (*cf.* [11]). We have to saturate sets which are involved in a sequence. Therefore, we must take into account previous sets in the sequence. For this purpose, we introduce the notion of *back-saturated set* with respect to a sequence of sets Δ. Because of technical reasons, back-saturation is also made with respect to a set of auxiliary constant symbols.

Definition 19. Let Δ be a sequence of sets of Σ-sentences and let AC be a countable set of new auxiliary constants. We say that a set Γ of $\Sigma \cup AC$-sentences is *back-saturated* with respect to (AC, Δ) if and only if it satisfies the following conditions:

1. $F \notin \Gamma$
2. if A is atomic and $A \in \Delta; \Gamma$, then $\neg A \notin \Gamma$
3. $\varphi \rightarrow \psi \in \Gamma \implies \neg\varphi \in \Gamma$ or $\psi \in \Gamma$
4. $\varphi \supset \psi \in \Delta; \Gamma \implies \neg\varphi \in \Gamma$ or $\psi \in \Gamma$
5. $\forall x \varphi(x) \in \Gamma \implies \varphi(t) \in \Gamma$ for all closed $t \in Term_{\Sigma \cup AC}$
6. $\neg(\varphi \rightarrow \psi) \in \Gamma \implies \varphi \in \Gamma$ and $\neg\psi \in \Gamma$
7. $\neg\forall x \varphi(x) \in \Gamma \implies \neg\varphi(c) \in \Gamma$ for some $c \in AC$.

The first two conditions constitute the so-called *atomic coherence property* of Γ.

$\blacksquare$

Now, we will show that, for calculi satisfying the conditions in Figure 1, any set of sentences, in the antecedent of a sequent, can be back-saturated preserving non-derivability. This is the crucial lemma for completeness.

Lemma 20. *Let $\Delta; \Gamma; \Delta'$ be a sequence of sets of Σ-sentences, χ a Σ-sentence, and AC be a countable set of new auxiliary constants. If $\Delta; \Gamma; \Delta' \nvdash_C \chi$ and $\vdash_C$ satisfies the conditions of Figure 1, then there exists a set Γ^* such that:*

(i) Γ^* is back-saturated with respect to (AC, Δ),
(ii) $\Gamma \subseteq \Gamma^*$ and
(iii) $\Delta; \Gamma^*; \Delta' \nvdash_C \chi$.

Proof: We will build Γ^*, starting with Γ, by iteration of a procedure which adds sentences. At each iteration step $k \in I\!N$ a set Γ_k of sentences is built. To do that we enumerate the set AC by $\{c_0, c_1, \ldots, c_n, \ldots..\}$; the set of all closed $\Sigma \cup AC$-terms by $\{t_0, t_1, \ldots, t_n, \ldots..\}$; and the set of all $\Sigma \cup AC$-sentences by $\{\gamma_0, \gamma_1, \ldots, \gamma_n, \ldots..\}$. The procedure initializes $\Gamma_0 := \Gamma$. Then, for any $k \geq 1$ it obtains Γ_k from Γ_{k-1} in the following way.

(a) If k is odd, it takes the least pair $(i, j) \in I\!N^2$ (in lexicographic order) such that $\gamma_i \in \Gamma_{k-1}$ have the form $\forall x \varphi(x)$ and $\varphi(t_j) \notin \Gamma_{k-1}$. Then, it makes $\Gamma_k := \Gamma_{k-1} \cup \{\varphi(t_j)\}$.

(b) If k is even, it takes the least $i \in I\!N$ such that γ_i has not been treated yet (in the step k) and one of the following two facts holds:

(b1) $\gamma_i \in \Delta; \Gamma_{k-1}$ and it has the form $\varphi \supset \psi$
(b2) $\gamma_i \in \Gamma_{k-1}$ and its form is either $\varphi \to \psi$ or $\neg(\varphi \to \psi)$ or $\neg \forall x \varphi(x)$.
Then, it obtains Γ_k, depending on the case, as follows:
(b1) If $\Delta; \Gamma_{k-1}, \neg \varphi; \Delta' \nvdash_C \chi$ then $\Gamma_k := \Gamma_{k-1} \cup \{\neg \varphi\}$ else $\Gamma_k := \Gamma_{k-1} \cup \{\psi\}$
(b2) γ_i is $\varphi \to \psi$: If $\Delta; \Gamma_{k-1}, \neg \varphi; \Delta' \nvdash_C \chi$ then $\Gamma_k := \Gamma_{k-1} \cup \{\neg \varphi\}$
else $\Gamma_k := \Gamma_{k-1} \cup \{\psi\}$
γ_i is $\neg(\varphi \to \psi)$: $\Gamma_k := \Gamma_{k-1} \cup \{\varphi, \neg \psi\}$
γ_i is $\neg \forall x \varphi(x)$:   It takes the least j such that $c_j \in AC \setminus cons(\Delta; \Gamma_{k-1}; \Delta'; \chi)$
and then it makes $\Gamma_k := \Gamma_{k-1} \cup \{\neg \varphi(c_j)\}$.

If Γ_k is back-saturated with respect to (AC, Δ), the procedure stops at this step k for which $\Gamma^* = \Gamma_k$. Otherwise we approximate

$$\Gamma^* = \bigcup_{k \in \mathcal{N}} \Gamma_k.$$

By construction $\Gamma \subseteq \Gamma^*$. To finish the proof, we must justify that $\Delta; \Gamma^*; \Delta' \nvdash_C \chi$ and Γ^* is back-saturated with respect to (AC, Δ).

We will show that $\Delta; \Gamma_k; \Delta' \nvdash_C \chi$, for all $k \in I\!N$ (by induction on k). For $k = 0$ the assertion holds by hypothesis. Now, consider any $k \in I\!N$, $\Delta; \Gamma_{k-1}; \Delta' \nvdash_C \chi$ holds by the induction hypothesis. For each possible case, in the extension of Γ_{k-1} to Γ_k, there are some conditions in Figure 1 which ensure that $\Delta; \Gamma_k; \Delta' \nvdash_C \chi$. The case (a) works because of condition 4, (b1) because of condition 3 and the case (b2), because of conditions 2, 6, and 8.

Finally, the atomic coherence of Γ^* results from $\Delta; \Gamma^*; \Delta' \nvdash_C \chi$ because of conditions 1, 5 and 10. The remaining back-saturation properties of Γ^* hold by construction.   ∎

Now, we will prove that conditions of Figure 1 are, in fact, sufficient for completeness.

---

1. $A \in \Delta \Longrightarrow \Delta \vdash_C A$
2. $\Delta; \Gamma, \neg\varphi; \Delta' \vdash_C \chi$ and $\Delta; \Gamma, \psi; \Delta' \vdash_C \chi \Longrightarrow \Delta; \Gamma, \varphi \to \psi; \Delta' \vdash_C \chi$
3. $\Delta; \Gamma; \Delta'; \Gamma', \neg\varphi; \Delta'' \vdash_C \chi$ and $\Delta; \Gamma; \Delta'; \Gamma', \psi; \Delta'' \vdash_C \chi \Longrightarrow$
   $\Delta; \Gamma, \varphi \supset \psi; \Delta'; \Gamma'; \Delta'' \vdash_C \chi$
4. $\Delta; \Gamma, \varphi(t); \Delta' \vdash_C \chi \Longrightarrow \Delta; \Gamma, \forall x \varphi(x); \Delta' \vdash_C \chi$
5. $\Delta; \Gamma \vdash_C A \Longrightarrow \Delta; \Gamma, \neg A; \Delta' \vdash_C \chi$
6. $\Delta; \Gamma, \varphi, \neg\psi; \Delta' \vdash_C \chi \Longrightarrow \Delta; \Gamma, \neg(\varphi \to \psi); \Delta' \vdash_C \chi$
7. $\Delta; \Gamma; \varphi \vdash_C \psi \Longrightarrow \Delta; \Gamma, \neg(\varphi \supset \psi); \Delta' \vdash_C \chi$
8. $\Delta; \Gamma, \neg\varphi(c); \Delta' \vdash_C \chi \Longrightarrow \Delta; \Gamma, \neg\forall x \varphi(x); \Delta' \vdash_C \chi$
9. $\Delta; \Gamma, \neg\varphi \vdash_C F \Longrightarrow \Delta; \Gamma \vdash_C \varphi$
10. $\Delta; \Gamma, F; \Delta' \vdash_C \chi$

where A is atomic, $t \in Term_{\Sigma \cup AC}$ and $c \in AC \setminus cons(\Delta; \Gamma, \varphi(x); \Delta')$.

Fig. 1. Sufficient conditions for completeness

Lemma 21. *If the derivability relation $\vdash_C$ of a calculus C satisfies the conditions of Figure 1, then for any set of Σ-sentences $\Phi \cup \{\chi\}$: $\Phi \models \chi \Longrightarrow \Phi \vdash_C \chi$*

Proof: Suppose that $\Phi \nvdash_C \chi$. We will prove that $\Phi \nvDash \chi$ by showing the existence of a counter-model $\mathcal{K}$ with worlds in the set $I\!N^*$ of sequences of natural numbers, ordered by $u \leq v$ if and only if u is an initial segment of v. [1] This structure $\mathcal{K}$ will be such that $\epsilon \Vdash \Phi$ and $\epsilon \nVdash \chi$. Let us consider the existence of a countable family of disjoint sets of new auxiliary constants $\langle AC_i \rangle_{i \in I\!N}$. With each sequence $s \in I\!N^*$ we associate the set $AC_{\#s}$ and the signature

$$\Sigma_s \equiv \Sigma \cup \bigcup_{n \leq \#s} AC_n.$$

Now, we inductively associate a set Γ_s of Σ-formulas with each $s = \langle n_0, n_1, \ldots, n_k \rangle \in I\!N^*$. Let Δ_s denote the sequence $\Gamma_\epsilon; \Gamma_{\langle n_0 \rangle}; \Gamma_{\langle n_0, n_1 \rangle}; \ldots; \Gamma_{\langle n_0, n_1, \ldots, n_k \rangle}$ and $\Delta_{<s}$ the sequence $\Gamma_\epsilon; \ldots; \Gamma_{\langle n_0, n_1, \ldots, n_{k-1} \rangle}$.

The collection $\{\Gamma_s | s \in I\!N^*\}$ is defined to satisfy that each Γ_s is back-saturated with respect to $(AC_{\#s}, \Delta_{<s})$, $\Phi \cup \{\neg\chi\} \subseteq \Gamma_\epsilon$, and $\Delta_s \nvdash_C F$.

---

[1] ϵ will denote the empty sequence, $\cdot$ is the infix function for adding a natural number to a sequence and $\#$ is the length function over sequences.

As basis step, we define Γ_ϵ as the back-saturated set with respect to AC_0 and the empty sequence of sets such that $\Phi \cup \{\neg\chi\} \subseteq \Gamma_\epsilon$ and $\Gamma_\epsilon \nvdash_C F$. Γ_ϵ exists because $\Phi, \neg\chi \nvdash_C F$ holds by condition 9. As inductive step, we define $\Gamma_{s \cdot j}$ for $s \in I\!\!N^*$ and $j \in I\!\!N$, provided that $\Delta_s \nvdash_C F$. In order to do this, we consider an enumeration $\{\gamma_0, \gamma_1, \ldots, \gamma_n, \ldots\ldots\}$ of all sentences of Γ_s of the form $\neg(\varphi_1 \supset \varphi_2)$. Then, let γ_j be $\neg(\varphi \supset \psi)$. By conditions 7 and 9, we have that $\Delta_{<s}; \Gamma_s; \varphi, \neg\psi \nvdash_C F$. So, there exists $\Gamma_{s \cdot j} \supseteq \{\varphi, \neg\psi\}$ back-saturated with respect to $(AC_{\#s+1}, \Delta_s)$ such that $\Delta_s; \Gamma_{s \cdot j} \nvdash_C F$.

Now, by means of $\{\Gamma_s \mid s \in I\!\!N^*\}$, we can define $\mathcal{K} = (I\!\!N^*, \leq, \langle \mathcal{A}_s \rangle_{s \in I\!\!N^*})$ as follows:

- $A_s = \{t \mid t \in Term_{\Sigma_s} \text{ and } t \text{ is closed}\}$
- $f^{\mathcal{A}_s}(t_1, \ldots, t_n) = f(t_1, \ldots, t_n)$
- $At_s = \{p(\widehat{t_1}, \ldots, \widehat{t_n}) \mid p(t_1, \ldots, t_n) \in \Delta_s\}$.

It is easy to see that $\mathcal{K}$ is a Kripke structure and also that $\widehat{t}^s = t = t^s$ holds for any $s \in I\!\!N^*$. To finish the proof we have to check that $\epsilon \Vdash \Phi$ and $\epsilon \nVdash \chi$. Hence, since $\Phi \cup \{\neg\chi\} \subseteq \Gamma_\epsilon$, it suffices to show that $\eta \in \Gamma_s \implies s \Vdash \eta$ holds for any $s \in I\!\!N^*$ and any Σ-sentence η. We prove this fact by induction on η, using Definition 19, since each Γ_s is back-saturated with respect to $(AC_{\#s}, \Delta_{<s})$:

- $p(t_1, \ldots, t_n) \in \Gamma_s \implies p(\widehat{t_1}, \ldots, \widehat{t_n}) \in At_s \implies s \Vdash p(t_1, \ldots, t_n)$
- $\neg\varphi$ requires induction on φ:
  - $\neg p(t_1, \ldots, t_n) \in \Gamma_s \implies p(t_1, \ldots, t_n) \notin \Delta_s \implies p(\widehat{t_1}, \ldots, \widehat{t_n}) \notin At_s$
    $\implies s \nVdash p(t_1, \ldots, t_n) \implies s \Vdash \neg p(t_1, \ldots, t_n)$
  - $\neg(\varphi \to \psi) \in \Gamma_s \implies \varphi \in \Gamma_s$ or $\neg\psi \in \Gamma_s \implies s \Vdash \varphi$ or $s \nVdash \psi \implies$
    $s \Vdash \varphi \to \psi$
  - $\neg(\varphi \supset \psi) \in \Gamma_s \implies \varphi, \neg\psi \in \Gamma_{s \cdot j}$ for some $j \in I\!\!N \implies s \cdot j \Vdash \varphi$ and
    $s \cdot j \nVdash \psi \implies s \nVdash \varphi \supset \psi \implies s \Vdash \neg(\varphi \supset \psi)$
  - $\neg \forall x \varphi(x) \in \Gamma_s \implies \neg\varphi(c) \in \Gamma_s$ for some $c \in AC_{\#s} \implies s \Vdash \neg\varphi(c)$ for
    some $c \in AC_{\#s} \implies$ (by $c^s = \widehat{c}^s$ and Lemma 6) $s \Vdash \neg\varphi(\widehat{c})$ for some
    $c \in AC_{\#s} \implies s \Vdash \neg\forall x \varphi(x)$
- $\varphi \to \psi \in \Gamma_s \implies \neg\varphi \in \Gamma_s$ or $\psi \in \Gamma_s \implies s \Vdash \neg\varphi$ or $s \Vdash \psi \implies s \Vdash \varphi \to \psi$
- $\varphi \supset \psi \in \Gamma_s \implies \varphi \supset \psi \in \Delta_w$ for all $w \geq s \implies \neg\varphi \in \Gamma_w$ or $\psi \in \Gamma_w$ for all
  $w \geq s \implies w \Vdash \neg\varphi$ or $w \Vdash \psi$ for all $w \geq s \implies s \Vdash \varphi \supset \psi$
- $\forall x \varphi(x) \in \Gamma_s \implies \varphi(t) \in \Gamma_s$ for all closed $t \in Term_{\Sigma_s} \implies s \Vdash \varphi(t)$ for all
  closed $t \in Term_{\Sigma_s} \implies$ (by $t^s = \widehat{t}^s$ and Lemma 6) $s \Vdash \varphi(\widehat{t})$ for all $t \in A_s$
  $\implies s \Vdash \forall x \varphi(x)$. ∎

The conditions of Figure 1 could be rewritten as inference rules, changing $\vdash_C$ by $\Rightarrow$. By Fact 15 and Lemma 21, this is a sound and complete calculus for $\mathcal{FO}^\supset$, which is cut-free, but it is not very natural. In Figure 2, we give a more natural calculus, with two inference rules for introducing each connective and quantifier on the left and on the right, respectively. Notice that the differences between the rules for both implications reflect that $\supset$ can be used for any "ulterior" set in the sequence, whereas $\to$ only is valid for the set containing it. We also would like to remark that by looking at the antecedent as a set of formulas, the rules in Figure 2 become well-known inference rules in both classical and intuitionistic sequent calculi. In particular, viewed in this manner, the rules for both implications collapse into a single pair of rules.

The structural rules $(Init)$, (Abs), (Cas) and (RaA) respectively allow us to built initial sequents, to put any consequent in the place of falsehood, to reason by distinction of the two cases given by the *law of the excluded middle* (classical negation), and to make *reductio ad absurdum* reasoning. Notice also that by a combination of $(\neg L)$ and (Cas) this calculus provides a derived cut-rule (see Figure 3). Besides, it is easy to check that, using the abbreviations for the rest of connectives and quantifiers $(\wedge, \vee, \exists, \sim)$, the expected inference rules can be derived for them.

Structural Rules

$(Init)\ \Delta \Rightarrow A$ if $A \in \Delta$ and A is atomic(incl. F) $(Abs)\ \dfrac{\Delta \Rightarrow F}{\Delta \Rightarrow \chi}$

$(Cas)\ \dfrac{\Delta; \Gamma, \varphi; \Delta' \Rightarrow \chi \quad \Delta; \Gamma, \neg\varphi; \Delta' \Rightarrow \chi}{\Delta; \Gamma; \Delta' \Rightarrow \chi}$ $(RaA)\ \dfrac{\Delta; \Gamma, \neg\varphi \Rightarrow F}{\Delta; \Gamma \Rightarrow \varphi}$

Connective Rules

$(\neg L)\ \dfrac{\Delta; \Gamma \Rightarrow \varphi}{\Delta; \Gamma, \neg\varphi; \Delta \Rightarrow \chi}$ $(R\neg)\ \dfrac{\Delta; \Gamma, \varphi \Rightarrow F}{\Delta; \Gamma \Rightarrow \neg\varphi}$

$(\to L)\ \dfrac{\Delta; \Gamma \Rightarrow \varphi \quad \Delta; \Gamma, \psi; \Delta' \Rightarrow \chi}{\Delta; \Gamma, \varphi \to \psi; \Delta' \Rightarrow \chi}$ $(R \to)\ \dfrac{\Delta; \Gamma, \varphi \Rightarrow \psi}{\Delta; \Gamma \Rightarrow \varphi \to \psi}$

$(\supset L)\ \dfrac{\Delta; \Gamma; \Delta'; \Gamma' \Rightarrow \varphi \quad \Delta; \Gamma; \Delta'; \Gamma', \psi; \Delta'' \Rightarrow \chi}{\Delta; \Gamma, \varphi \supset \psi; \Delta'; \Gamma'; \Delta'' \Rightarrow \chi}$ $(R \supset)\ \dfrac{\Delta; \{\varphi\} \Rightarrow \psi}{\Delta \Rightarrow \varphi \supset \psi}$

Quantifier Rules

$(\forall L)\ \dfrac{\Delta; \Gamma, \varphi(t); \Delta' \Rightarrow \chi}{\Delta; \Gamma, \forall x\varphi; \Delta' \Rightarrow \chi}$ $(R\forall)\ \dfrac{\Delta \Rightarrow \varphi(c)}{\Delta \Rightarrow \forall x\varphi}$

(t is a term in $(\forall L)$, and c is a new constant which does not appear in the lower sequent of $(R\forall)$).

Fig. 2. A sound and complete calculus for $\mathcal{FO}^{\supset}$

Now, we will show that the set of inference rules in the Figure 2 constitutes a sound and complete sequent calculus for $\mathcal{FO}^{\supset}$.

Firstly, the soundness of this calculus can be proved easily but requires a long proof. The proof consists in checking that each inference rule is sound with respect to (or preserves) the relation $\models^*$ (Definition 16). As an example, we show that $(\to L)$ is sound. Let Δ be some sequence of sets $\Gamma_0; \ldots; \Gamma_n$ and let Δ' be another sequence $\Phi_0; \ldots; \Phi_m$. Suppose that there exists a Kripke structure $\mathcal{K}$ and a sequence of worlds in $W(\mathcal{K})$: $u_0 \preceq \ldots \preceq u_n \preceq w \preceq v_0 \preceq \ldots \preceq v_m$ such that:

- $u_i \Vdash \Gamma_i$ for all $i = 1, \ldots, n$
- $w \Vdash \Gamma \cup \{\varphi \to \psi\}$

$-\ v_j \Vdash \Phi_j$ for all $j = 1, \ldots, m$
$-\ v_m \not\Vdash \chi$

Since $w \Vdash \varphi \to \psi$, then $w \not\Vdash \varphi$ or $w \Vdash \psi$. The first case means that $\Delta; \Gamma \not\models^* \varphi$ and in the second case it turns out that $\Delta; \Gamma, \psi; \Delta' \not\models^* \psi$. Therefore, if $\Delta; \Gamma \models^* \varphi$ and $\Delta; \Gamma, \psi; \Delta' \models^* \chi$, then $\Delta; \Gamma, \varphi \to \psi; \Delta' \models^* \chi$.

Secondly, we will prove completeness by checking that the derivability relation induced by the calculus of Figure 2 satisfies the conditions of Figure 1. In order to do it easier, we introduce the derived inference rules of Figure 3. These rules provide cut, contraposition (in the last set of the antecedent), derivation of assumptions (of the last set of the antecedent), and contradiction respectively.

$$(Cut) \ \frac{\Delta; \Gamma \Rightarrow \varphi \quad \Delta; \Gamma, \varphi; \Delta' \Rightarrow \chi}{\Delta; \Gamma; \Delta' \Rightarrow \chi} \qquad (Ctp) \ \frac{\Delta; \Gamma, \neg\chi \Rightarrow \varphi}{\Delta; \Gamma, \neg\varphi \Rightarrow \chi}$$

$$(Ass) \ \Delta; \Gamma, \varphi \Rightarrow \varphi \qquad\qquad (Ctd) \ \Delta; \Gamma, \varphi, \neg\varphi; \Delta' \Rightarrow \chi$$

Fig. 3. Derived inference rules

(Cut) is derived by using $(\neg L)$ and (Cas), and (Ctp) by $(\neg L)$ and (RaA). (Ass) can be derived by induction on φ. In the basic case, it suffices to use $(Init)$. For the induction step it is enough to use the induction hypothesis together with the corresponding rules $(R\odot)$ and $(\odot L)$ in each case of binary connective or quantifier $\odot$. Finally, (Ctd) comes from $(\neg L)$ and (Ass).

Theorem 22. (Completeness) *The calculus of Figure 2 is complete for $\mathcal{FO}^\supset$.*

Proof: By Lemma 21, it is enough to check that the calculus satisfies the conditions of Figure 1. Some of the conditions are directly obtained from an inference rule. This is the case for condition 1 by $(Init)$, for 4 by $(\forall L)$, for 5 by $(\neg L)$ and for 9 by (RaA). Conditions 7 and 10 are also very easy to check; condition 7 holds by rules $(R \supset)$ and $(\neg L)$ and 10 by $(Init)$ and (Abs).
From now on, the thinning meta-rule (see Fact 15) is often implicitly assumed. For both conditions 2 and 3 we use the same scheme of proof. We firstly apply (Cas) with φ and $\neg\varphi$. Then, the sequent with φ is a premise of the condition. For the other sequent we use (Cut) with $\neg\varphi$ as cut-formula. We obtain the other premise of the condition and also a sequent which is easily derived by $(R\neg)$ and $(\to L)$. The resulting leaves are (Ass) and (Ctd) sequents.
A proof for condition 6 can be similarly obtained beginning with two consecutive applications of (Cut) with cut-formulas φ and $\neg\psi$ on the set Γ.
For condition 8, let us suppose that $\Delta; \Gamma, \neg\varphi(c); \Delta' \Rightarrow \chi$ is a derivable sequent. That is, we are assuming that it is composed of finite sets. With an abuse of notation, we use the same names for the finite sets in the sequent as for its (possibly infinite) extensions in the derivability condition 8. We apply finitely many times $(R \to)$ and $(R \supset)$ to put, one by one, the formulas of Δ' in the consequent. In this way we obtain a sequent of the form $\Delta; \Gamma, \neg\varphi(c) \Rightarrow \eta(\Delta', \chi)$,

where $\eta(\Delta', \chi)$ is the above explained implicative formula combining both implications. Now, by (Ctp), $(R\forall)$ (c does not appear anywhere in the sequent), and again using (Ctp) we have a proof of the sequent (S_1) $\Delta; \Gamma, \neg\forall x\varphi(x) \Mapsto \eta(\Delta', \chi)$. Further, by systematic application of $(\to L)$ and $(\supset L)$ to $\eta(\Delta', \chi)$, we prove the sequent (S_2) $\Delta; \Gamma, \neg\forall x\varphi(x), \eta(\Delta', \chi); \Delta' \Mapsto \chi$. Applying (Cut) to (S_1) and (S_2) we derive the sequent $\Delta; \Gamma, \neg\forall x\varphi(x); \Delta' \Mapsto \chi$. By thinning meta-rule (Fact 15) the corresponding derivability relation is obtained. ∎

5   Conclusions and Related Work

We have shown that structured sequents enable the sequent calculi approach as proof method for a logic combining classical and intuitionistic implication. We have given semantical foundations for a logic which achieves such combination in the (unrestricted) first-order case. Some design aspects of the logic $\mathcal{FO}^{\supset}$ have been influenced and inspired by its original motivation in the area of logic programming. In particular, antecedents consisting of sequences of sets arose as a tool to formalize the operational semantics of LP languages combining classical and intuitionistic implication (see [10,1]). We have improved these structured sequents to combine both whole logics. We have provided general sufficient conditions for completeness of sequent calculi dealing with this kind of sequents, and also a sound and complete calculus. The completeness proof is based on a procedure for saturating sets with respect to sequences of sets.
The well-known modal logic $S4$ (introduced in [14]) is closely related to intuitionistic logic, by a translation of intuitionistic formulas into $S4$ formulas ([3,4]). Similarly, $\mathcal{FO}^{\supset}$ can be translated into $S4$. The $S4$ connective $\square$ allows one to translate an intuitionistic implication $\varphi \supset \psi$ into $\square(\varphi \to \psi)$. For atomic monotonicity, atomic formulas $p(\bar{t})$ also have to be translated into $\square p(\bar{t})$. This transformational approach enables logical (model-theoretic) foundations for LP languages combining both implications ([9]), and also provides a useful connection between the *semantical aspects* $\mathcal{FO}^{\supset}$ and $S4$. On the contrary, this connection is not so useful for relating its *proof-theoretical aspects*. Structured sequents calculi provide a new *deduction style* allowing proofs which can not be translated to $S4$. Roughly speaking, structured sequent calculi are more flexible than traditional $S4$ proof methods. In other words, a probable sequent has more (essentially different) proofs in $\mathcal{FO}^{\supset}$ than its translation has in $S4$. For instance, consider the following $S4$-rule:

$$(R\square) \frac{\Phi^{\#} \Mapsto \chi}{\Phi \Mapsto \square\chi} \quad \text{where } \Phi^{\#} = \{\square\varphi \mid \square\varphi \in \Phi\} \text{ [2]}$$

The indispensable non-$\square$-formulas in the antecedent of a sequent with a $\square$-consequent must be used before one application of the $(R\square)$ rule removes them. The semantical reason is that $\square$-formulas are properties of "every greater world" and the rule removes the "one world" assumptions. In $\mathcal{FO}^{\supset}$, the structure of the antecedent allows us to preserve all its "information", even in deduction steps dealing with a "every greater world" consequent. A remarkable consequence is

---

[2] $S4$ sequents are symmetric, but this is not relevant for our discussion.

that $\mathcal{FO}^\supset$ is more suitable (than $S4$) for goal-directed proofs. A *goal-directed (or uniform) proof* ([16,17]) essentially is a sequent calculus proof obtained by applying (at each step) the ($\odot R$) rule, where $\odot$ is the top-level logical symbol of the consequent (or goal). When the goal is an atom A, the legal step (so-called "backchaining") essentially consists on of applying ($\rightarrow L$) to some formula $\varphi \rightarrow A$ in the antecedent. Goal-directed proofs have become an important proof-theoretical foundation for LP languages and goal-directed proof systems have been developed for fragments of first order logic ([18]), intuitionistic logic ([7,16]), intermediate logics lying in between intuitionistic and classical logic ([6]), and many other fragments of (higher-order, modal, etc.) logics. A more detailed discussion of this topic is outside the scope of this paper. [1] deals with a fragment of $\mathcal{FO}^\supset$ which is a logic programming language satisfying the existence of a goal-directed proof for every provable sequent. Let us illustrate the goal-oriented ability of $\mathcal{FO}^\supset$ by means of a simple example. Consider the sequent $p, p \rightarrow q \Mapsto p \supset q$. It has a very easy goal-directed proof, whereas its translation to $S4$ modal logic, $\Box p, \Box p \rightarrow \Box q \Mapsto \Box(\Box p \rightarrow \Box q)$, does not have a goal-directed proof. For the former, by ($R \supset$), we obtain $p, p \rightarrow q \; ; p \Mapsto q$. Now, by applying ($\rightarrow L$) ("backchaining") to $p \rightarrow q$, two initial sequents $p \Mapsto p$ and $p, q; p \Mapsto q$ are obtained. For the latter, a goal-directed proof must firstly apply ($R\Box$), hence it obtains the non-provable sequent $\Box p \Mapsto \Box p \rightarrow \Box q$. There is a (non goal-directed) $S4$-proof which begins with the ($\rightarrow L$) to obtain the initial sequent $\Box p \Mapsto \Box p$ and also the sequent $\Box p, \Box q \Mapsto \Box(\Box p \rightarrow \Box q)$ which now can be proved by ($R\Box$).

Fariñas and Herzig ([2]) have investigated the combination of classical and intuitionistic logic, in the propositional case. They provide a Hilbert-style axiomatization and also a tableaux method. Actually, a completeness proof for the *propositional subset* of $\mathcal{FO}^\supset$ could be obtained by deriving the axiomatization of the propositional logic introduced in [2]. A similar work is [12] where a natural deduction system is given. A common characteristic of both systems, [2] and [12], is that some deduction steps depend on the *persistence* of formulas. From the semantical point of view, persistent formulas are the "every greater world" ones in the sense mentioned above. In $S4$ only $\Box$-formulas are persistent, whereas the combination of classical and intuitionistic connectives leads to an inductive characterization of persistence. In [2,12] persistent goals require the elimination of non-persistent premises, like the ($R\Box$)-rule does in $S4$. Hence their deductive styles are similar and quite far from the (structured sequents based) $\mathcal{FO}^\supset$ style, especially with regard to goal-directed proofs. Let us consider again the above example and the tableaux method introduced in [2], in order to obtain a closed tableau for the sentence $\neg((p \land (p \rightarrow q)) \rightarrow (p \supset q))$. After some simple steps, we have a linear tableau containing the three nodes: p, $p \rightarrow q$, and $\neg(p \supset q)$. From the goal-directed point of view, the first two play the role of premises and the last one is the goal. Therefore, to build a "goal-directed tableau" we must enlarge the unique branch with p and $\neg q$ and, simultaneously, we must eliminate all premises, since they are not persistent. Hence, this tableau will not close. A closed tableau could be obtained and to do this it suffices to use the premise $p \rightarrow q$ before the goal.

References

1. Arruabarrena, R., Lucio, P., and Navarro M., A Strong Logic Programming View for Static Embedded Implications, In: W. Thomas (ed.) *Proc. of the Second Intern. Conf. on Foundations of Software Science and Computation Structures, FOSSACS'99*, Lecture Notes in Computer Science, 1578:56-72, Springer-Verlag, (1999).
2. Fariñas, L. and Herzig, A., Combining classical and intuitionistic logic, or: intuitionistic implication as a conditional, In: Baader, F. and Schulz, K. U. (eds.), *Frontiers in Combining Systems (Proc. Int. Workshop FroCos'96)*, Applied Logic Series, Kluwer Academic Publisher, 93-102, (1996).
3. Fitting, M., *Proof Methods for Modal and Intuitionistic Logics*, D. Reidel Publishing Company, (1983).
4. Fitting, M., Basic Modal Logic, in: Gabbay, D. M., Hogger, C. J. and Robinson, J. A. (eds.) *Handbook of Logic in Artificial Intelligence and Logic Programming*, Vol. 1, 365-448, Oxford University Press, (1993).
5. Gabbay, D. M., *Labelled Deductive Systems*, Vol. I, Oxford University Press, (1996).
6. Gabbay, D. M. and Olivetti, N., Goal-directed algorithmic proof theory, Technical Report, Imperial College of Science, Technology and Medicine, (1998).
7. Gabbay, D. M. and Reyle, U., N-Prolog: an Extension of Prolog with Hypothetical Implications, *Journal of Logic Programming*, 4:319-355, (1984).
8. Gentzen, G., Investigation into logical deduction. In: M. E. Szabo (ed.), *The Collected Papers of Gerhard Gentzen*, North-Holland, 66-131, (1969).
9. Giordano, L., and Martelli, A., Structuring logic programs: A modal approach, *Journal of Logic Programming*, 21:59-94, (1994).
10. Giordano, L., Martelli, A., and Rossi, G., Extending Horn Clause Logic with Implications Goals, *Theoretical Computer Science*, 95:43-74, (1992).
11. Hintikka, K. J. J., Form and content in quantification theory, *Acta Philosophica Fennica*, 8:7-55, (1955).
12. Humberstone, L., Interval semantics for tense logic: some remarks, *Journal of Philosophical Logic*, 8:171-196, (1979).
13. Kripke, S., Semantical analysis of intuitionistic logic I, In: Crossley, J. and Dummett, M. (eds.), *Formal Systems and Recursive Functions.* North-Holland, 92-129, (1963).
14. Lewis, C. and Langford, C., *Symbolic Logic*, Dover, (1959, second edition), (1932, first edition).
15. Meseguer, J., Multiparadigm Logic Programming, In: *Proceedings of ALP'92*, L.N.C.S. 632. Springer-Verlag, 158-200, (1992).
16. Miller, D., Abstraction in Logic Programs. In: Odifreddi, P. (ed.), *Logic and Computer Science*, Academic Press, 329-359, (1990).
17. Miller, D., Nadathur, G., Pfenning, F. and Scedrov, A., Uniform proofs as a foundation for logic programming, *Annals of Pure and Applied Logic*, 51:125-157, (1991).
18. Nadathur, G., Uniform Provability in Classical Logic, Technical Report, Dept. of Computer Science, Univ. of Chicago, TR-96-09, (1996).
19. van Dalen, D., *Logic and Structure.* Springer-Verlag, (1980).
20. van Dalen, D., and Troelstra, *Constructivism in Mathematics: An Introduction*, Vol. 1, Elsevier Science, North-Holland, (1988).
21. van Dalen, D., and Troelstra, *Constructivism in Mathematics: An Introduction*, Vol. 2, Elsevier Science, North-Holland, (1988).

Handling Differential Equations with Constraints for Decision Support

Jorge Cruz and Pedro Barahona

Dep. de Informática, Universidade Nova de Lisboa, 2825 Monte de Caparica, Portugal
{jc,pb}@di.fct.unl.pt

Abstract. The behaviour of many systems is naturally modelled by a set of ordinary differential equations (ODEs) which are parametric. Since decisions are often based on relations over these parameters it is important to know them with sufficient precision to make those decisions safe. This is in principle an adequate field to use interval domains for the parameters, and constraint propagation to obtain safe bounds for them. Although complex, the use of interval constraints with ODEs is receiving increasing interest. However, the usual consistency maintenance techniques (box- and local hull-consistency) for interval domains are often insufficient to cope with parametric ODEs. In this paper we propose a stronger consistency requirement, global hull-consistency, and an algorithm to compute it. To speed up this computation we developed an incremental approach to refine as needed the precision of ODEs trajectories. Our methodology is illustrated with an example of decision support in a medical problem (diagnosis of diabetes).

1. Introduction

Model-based decision support relies on an explicit representation of some system (in a domain of interest). The dynamics of such system is often described by relating the rates at which the system variables change with the current values of these variables. Such models are naturally represented as a set of ordinary differential equations (ODEs) which are parametric, as they include parameters whose value is not known exactly.

The classification of the system behaviour (for instance, whether it oscillates) usually depends on some relations between the system parameters. When decisions have to be made according to the behaviour of the system, determining the value of the parameters becomes of course an important problem. Nevertheless, to know the exact values of the parameters is often not required, being sufficient to determine them with enough precision to make safe decisions.

Such determination is not an easy task, in general. Sometimes it is possible to obtain an analytical solution for the ODEs. In such systems, the model may be tuned by collecting data in a number of experiments, and adjusting the model parameters to fit the predicted to the observed values. However, many systems have no analytical solution, and the above values must be predicted by numeric simulation. This is a typical generate and test process: exact values for the parameters must be generated and the results of its simulation tested against the observations.

H. Kirchner and C. Ringeissen (Eds.): FroCoS 2000, LNAI 1794, pp.105-120.

This procedure has two main drawbacks. Firstly, approximation errors are inevitable and a mismatch between predictions and observations can be explained either because the values of the parameters are wrong, or because of approximation errors (or a combination of both types of errors). Secondly, even if these latter errors can be discarded, systems are often highly non linear, and small errors in the value of the parameters may cause important changes in the predicted results (chaotic systems are an extreme example of this non-linearity). Moreover, in a decision support context, safe decisions may only be made if all values that match the experimental data are considered, and this poses the extra difficulty of generating and testing all the (potentially infinite) hypotheses.

These difficulties with numerical simulation may, in principle, be circumvented by adopting interval domains for the parameters, and using constraint propagation to obtain safe bounds for them. Despite its complexity, the use of interval constraints with ODEs is recently receiving increasing interest [1], and was proposed for the representation of deep models of biomedical systems [2].

The success of this approach to handle parametric ODEs greatly rests on the ability of constraint propagation to narrow the uncertainty about the parameters (until a safe decision can be made). As with other domains, constraint propagation of intervals maintains some form of consistency of the constraint set. Typically, two types of consistency are considered: box-consistency and local hull-consistency [3]. Neither of them is complete; more importantly, in the context of decision support, neither seems to be able to sufficiently narrow the parameters in parametric ODEs so that decisions may be made safe.

To overcome this problem, we propose in this paper a new type of consistency, global hull-consistency, that combines the global nature of box-consistency with the simultaneous instantiation of all parameters of local hull-consistency. The paper is organised as follows. In section 2 we present a motivating example from the medical domain (diagnosis of diabetes), where the above considerations are exemplified. Section 3 overviews the main features of interval-based approaches to handle ODEs. Section 4 presents our definition of global hull-consistency, and justifies it in the context of decision support. This algorithm relies on a number of trajectory computations with sufficient precision, and section 5, presents an incremental validated approach to deal with ODEs, which speeds up the required computations. Section 6 presents and discusses the results of applying our approach on the diabetes example. Finally section 7 presents the main conclusions and discusses future research still needed.

2. A Differential Model for Diagnosing Diabetes

Diabetes mellitus is a disease that prevents the body from metabolising glucose due to an insufficient supply of insulin. A glucose tolerance test (GTT) is frequently used for diagnosing diabetes. In this test, the patient arrives at the hospital after an overnight fast and ingests a large dose of glucose. During the next hours, several blood tests are made and from the evolution of the glucose concentration values (and possibly insulin values as well) a diagnosis is made by the physicians.

Ackerman and al [4], proposed a well-known model for the description of the blood glucose regulatory system during a GTT. The model is the following parametric ordinary differential equation:

$$\frac{dg}{dt} = -p_1 g - l \qquad\qquad \frac{dh}{dt} = -p_3 h + r \qquad\qquad (1)$$

where   g is the deviation of the glucose blood concentration from its fasting level;
h is the deviation of the insulin blood concentration from its fasting level;
p_1, p_2, p_3 and p_4 are positive parameters

When the patient arrives after an overnight fast the concentrations of glucose and insulin are in the fasting level so g and h are zero. Let t_0 be the instant immediately after the ingestion of a large dose of glucose (g_0). According to the model, the evolution of glucose and insulin blood concentrations is described by the trajectory of the above differential equation with the initial values:

$$g(t_0) = g_0 \text{ and } h(t_0) = 0.$$

The parameter values, which vary from person to person, determine the evolution of the trajectory over time. Figure 1 shows two simulations of the evolution of the glucose blood concentration for a patient that arrived with a glucose fasting level concentration of 70 mg glucose / 100 ml blood. Immediately after the ingestion of an initial dose of glucose, the glucose concentration was 150 mg glucose / 100 ml blood $(g_0 = 150-70 = 80)$. The different trajectories obtained for the two simulations are due to the different values assumed for the parameters.

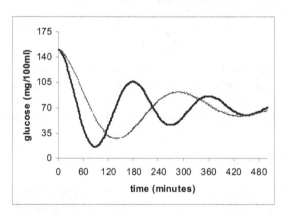

Fig. 1. Two simulations of the evolution of the blood glucose concentration. In case 1 (thick line), typical normal values were assumed (p_1=0.0044, p_2=0.04, p_3=0.0044, p_4=0.03). In case 2 (thin line), the values of the parameters p_2 and p_4 were reduced (p_2=0.03, p_4=0.015).

Independently from the specific values defined for the parameters, the general behaviour of the glucose trajectory (and insulin trajectory as well) oscillates around, and eventually converges to, the fasting concentration level. This trajectory is periodic with a period T given (in minutes) by:

$$T = \frac{2\pi}{\sqrt{p_1 p_3 + p_2 p_4}} \qquad\qquad (2)$$

A criterion used for diagnosing diabetes is based on the period T (and hence on the above relation on the parameters p_i), which is increased in diabetic patients. It is generally accepted that a value for T higher than 4 hours is an indicator of diabetes, otherwise normalcy is concluded.

Summarising, the decision problem (for the diagnosis of diabetes) is based on:

i) a parametric ordinary differential equation to model the glucose/insulin concentrations;

ii) the decision is based on a condition (T < 240) that can be obtained from the parameter values;

iii) the exact parameter values are unknown, but some ranges may be accepted;

iv) within these ranges different decisions may be taken (the patient is diabetic or not)

v) some points of the trajectory are known (from the blood tests).

In practice, for a particular patient, it is necessary to narrow the range of the parameters so as to safely evaluate the diagnostic condition. Notice that this differential equation has an analytical solution (otherwise it would not be so widely used in practice) and is used in this paper for illustration purposes. Nevertheless our approach may be used with any models, even with no analytical solutions.

3. Ordinary Differential Equations and Initial Value Problems

An Ordinary Differential Equation (ODE) defines the derivative of **u** (a vector) with respect to variable t by means of a function of t and **u**[5]:

$$\frac{d\mathbf{u}}{dt} = f(t, \tag{3}$$

Equations with higher order derivatives on the same variable (t) can be transformed into an ODE by including new variables (one for each higher order derivative). Given an ODE and a value for **u** at a given t_0, the initial value problem (IVP) aims at determining the value of **u** for other values of t (**u**(t) is called the trajectory).

3.1. Traditional Approaches

Classical numerical approaches to this problem try to approximate the trajectory **u**(t) at some discrete points of t (t_1, t_2, ..., t_n). This is achieved by considering each t_i in an increasing sequence and using the approximated values of the trajectory in the previous points to calculate the approximated value of the trajectory in the next point. Each of the n steps of this approximation process is based on a traditional numerical method of approximation (Taylor, Runge Kutta, etc) where the error term is truncated (Equation 4 show the Taylor method of order k).

$$\mathbf{u}(t_1) = \mathbf{u}(t_0) + \sum_{i=1}^{i=k-1} \frac{(t_1 - t_0)^i}{i!} f^{(i-1)}(t_0, \mathbf{u}(t_0)) + \frac{(t_1 - t_0)^k}{k!} f^{(k-1)}(\xi) \tag{4}$$

The fundamental problem with these approaches is the error introduced at every step of the procedure due to the truncation of the error term (the last term of Eq. 4) and to the floating point arithmetic round-off. Even if these errors are small in each step, the cumulative effect over the n steps usually makes the approximated trajectory to depart significantly from the correct trajectory.

In contrast, validated methods are able to verify the existence of a unique solution to the problem and to produce guaranteed error bounds for the true trajectory.

3.2. Validated Interval Arithmetic Approaches

The first validated approaches to ODEs originated from the interval arithmetic framework introduced by Moore [6]. The key idea is the use of interval arithmetic to calculate each approximation step explicitly taking into account the error term within appropriate interval bounds. Truncation and round-off errors are thus encapsulated within bounds of uncertainty around the true trajectory.

In validated approaches each step (between two consecutive points t_i and t_j) consists of two phases. In a first phase, it is necessary to validate the existence of a unique solution and to calculate an a priori enclosure of the true trajectory between the two points (this is required to bind the error term in the next phase). Usually this phase uses the Picard operator (see Section 5). In the second phase a tighter enclosure of the trajectory at point t_j is obtained through interval arithmetic over a chosen numerical approximation step (as the one presented in Eq. 4) with the error term bounded as a result of the a priori enclosure found in the previous phase.

The relative unpopularity of the direct application of interval methods to ODE problems derives from the additional computational work required and, in many early approaches, to the insufficient tightness produced on the trajectory bounds. Nevertheless, some widely used software was developed using the above ideas, namely Lohner's AWA program [7] and the Nedialkov's VNODE package [8].

To overcome the above difficulties, research has been carried out to take advantage of the efficiency and soundness of constraint techniques.

3.3. Validated Interval Constraint Approaches

The application of the interval constraints framework (introduced by Cleary [9]) for validated ODE solving was suggested by Older [10] and Hickey [11]. The idea is to represent a problem by a constraint network whose nodes are the trajectory values on the discrete points of t, and imposing constraints on them according to the approximation step (as in the interval arithmetic approaches). The goal is not only to improve efficiency (compared to the interval arithmetic approaches) but also to improve in terms of the generality of a constraint approach. Not only can IVPs be solved, but also any additional information (e.g. final or intermediate trajectory values) may propagate throughout the constraint network, preserving validated approximations on each trajectory point.

These constraint approaches are theoretically equivalent to the previous interval approaches (requiring as well a first phase -for error bounding- and a second phase

-for a tighter enclosure- in each approximation step) but benefit from the propagation technology developed for the constraint framework.

Early proposals [10][11], assumed an a priori enclosure for the whole trajectory, requiring only the second phase of the approximation step. More recently, Deville et al [1] defined a general approach where consistency techniques handle the main problems of the interval methods. For example, the wrapping effect (the error resulting from propagating at each step an entire box that encloses a smaller region) is reduced by piecewise interval extension (instead of entire box extension), expressed as unconstrained optimisation (a well known problem in the constraint community).

4. Differential Equations in the Context of Decision Support

If ODEs are considered in the context of decision support, the determination of the trajectory is no longer the only (main) goal, and this has to be considered together with other constraints of the decision model. Moreover, most models include parameters within the ODE definition, possibly common to other constraints.

Within this context, the whole ODE is just another constraint of the model relating several variables that can be either trajectory point values or parameters. The main goal of the constraint system is to reduce the bounds of all model variables to eliminate values which can be proved inconsistent with the constraint set.

The main difficulty with parametric ODEs is that parameter uncertainty can be quickly propagated and increased along the whole trajectory. Note that even perfect approximation methods may not solve this uncertainty (of course, less precise methods will give worst results). Most ODEs present such behaviour (unless they are contracting) and in the extreme case of chaotic ODEs any uncertainty, however small, is quickly magnified, preventing any reasonable, long-term, trajectory calculation.

Several partial consistency definitions and algorithms have been proposed to prune the domains of real variables in constraint systems (Collavizza et al [3] characterised the general properties of the filtering achieved by such types of partial consistency). However parametric ODEs introduce extra difficulties which challenge the quality of the filtering achieved by these different consistency types.

4.1. Partial Consistencies over Parametric ODEs

The two main partial consistencies used for solving non linear constraints over real variables are hull-consistency [12] (or 2B-consistency) and box-consistency [13]. These are approximations of arc-consistency, widely used in finite domains. Arc-consistency eliminates a value from a variable domain if it has no support, i.e. if no compatible value exists in the domain of another variable sharing the same constraint. In continuous (i.e. infinite) domains this kind of enumeration is no longer possible and both hull and box-consistency assume that the domains of the variables are convex, so they simply aim at tightening their outer bounds. If the correct domains are disjunctive they both admit inconsistencies within the outer limits and some kind of higher level algorithm is needed to prune further the solution set (to eliminate wrong solutions).

The key idea behind hull-consistency is to check arc-consistency only at the bounds of the variable domains. This means that if a variable is instantiated with the value of one of its bounds then, for each constraint on that variable, there must be a consistent instantiation of the other constraint variables. As it is very difficult to check this property for complex constraints, the algorithms that impose hull-consistency decompose the original constraint system into a set of basic primitive constraints, with the addition of extra variables, for which this property can be easily checked locally.

The major drawback of this decomposition approach is the worsening of the locality problem. The existence of intervals for the variables that satisfy (locally) each of the constraints does not necessarily mean that they satisfy simultaneously all of them. When a complex constraint is subdivided into primitive constraints this will only worsen this problem due to the addition of new variables and the increased complexity of the constraint network.

In parametric ODEs, this local hull-consistency achieves, in general, poor filtering. On the one hand the potential benefit from using more complex constraints at each step is largely eliminated by the increase of complexity on the constraint caused by the decomposition of these constraints into basic ones. On the other hand, if more intermediate points are considered, the constraint network becomes more complex and worsens the locality problems.

Box-consistency checks the consistency of each bound of the domain of each variable with the domains of the others. The motivation of this approach is that after substituting all but one variables by their domains, the problem of binding the remaining variable can be tackled by numerical methods, in particular a combination of interval Newton iterates and bisection [13]. The main advantage of this approach is that each constraint can be manipulated as a whole, not requiring the decomposition into primitive constraints, thus preventing the amplification of the locality problem.

Nevertheless, box-consistency is still inappropriate for dealing with parametric ODEs. The reason is that if there are several uncertain parameters, the algorithm to enforce box-consistency will try to tighten the bounds of each parameter substituting the other parameters by their domains. Hence, it is still necessary to work with a parametric ODE (though with one less parameter). Since for these parametric ODEs a wide range of trajectories is usually possible, it is rare to obtain inconsistencies required for domain reduction (see Section 6 for an example).

What seems to be needed to deal with parametric ODE problems is the simultaneous instantiation of all parameters (similarly to the consistency checking in the hull-consistency approach) and manipulation of the ODE constraint network as a single global constraint (similarly to the manipulation of each constraint as a whole in box-consistency).

4.2. Global Hull-Consistency over Parametric ODEs

In order to achieve a good filtering for the domains of the parameters appearing in an ODE, we propose an algorithm to guarantee hull-consistency on the parametric ODE as a single global constraint. The algorithm is based on two simple assumptions (derived from the properties of parametric ODEs): i) to conclude consistency it is necessary to find a solution; ii) to conclude inconsistency it is enough to fail a consistency check with uncertain domains.

The high level functioning of the algorithm is sketched in pseudo-code in Figure 2. The input is the set of the initial parameter domains. The output is a boolean denoting whether the problem is consistent and, if consistent, the narrowing of the parameter domains according to the global hull-consistency. To obtain these results, and while no inconsistency has been detected, each parameter domain ($D_i = [D_i^-..D_i^+]$) is processed by specialised functions (shrink_left and shrink_right) that compute the leftmost (lb_i) and rightmost (rb_i) value of each parameter domain that belong to some instantiated solution, thus imposing hull-consistency on each bound. Inconsistency is detected by these functions and interrupts the while loop.

```
function global_hull_consistency( in {D₁,...,Dₙ}, out {D₁',...,Dₙ'}): boolean
    consistent ← true
    i ← 1
    while i ≤ n and consistent do
        if shrink_left(i, {D₁',..., D_{i-1}', D_i , D_{i+1} ...,Dₙ},lb_i) then
            rD_i ← [lb_i..D_i^+]
            if shrink_right(i, {D₁',..., D_{i-1}', rD_i , D_{i+1} ...,Dₙ},rb_i)
                then D_i' ← [lb_i..rb_i]
                else consistent ← false
        else consistent ← false
        i ← i+1
    endwhile
    global_hull_consistency ← consistent
```

Fig. 2. The global hull-consistency algorithm.

Note that this procedure is similar to the narrowing method used by Newton [13] to guarantee box-consistency, although it requires no fix point algorithm. The reason is that, after narrowing each variable domain, each one of the new bounds appear in one solution and so, whatever filtering is done in the remaining variable domains, it will never delete the values that supported those bounds.

Function shrink_left is described in Fig. 3 (a symmetric approach is used for shrinking the right bound). The algorithm first tries to find an instantiated solution s_i for parameter i. If it fails, there is no possible solution within the range of D_i. If it succeeds it recursively tries to find one leftmost value within the sub-range $[D_i^-..s_i[$ (upper bounded by the solution value already found). If there is a new leftmost value in this sub-range then it is also the leftmost of the original range, otherwise the already found solution value is the value returned by the function.

```
function shrink_left( in i, in {D₁,...,Dₙ}, out lb_i): boolean
    if solution(i,{D₁,...,Dₙ},s_i) then
        D_i' ← [D_i^-..s_i[
        if shrink_left(i,{D₁,..., D_{i-1}, D_i', D_{i+1} ...,Dₙ},b_i)
            then lb_i ← b_i
            else lb_i ← s_i
        shrink_left ← true
    else shrink_left ← false
```

Fig. 3. The algorithm used to narrow the left bound of a variable domain.

Notice that the above algorithm always converges because in each recursion step the range considered to look for a leftmost value is narrowed and there is a machine limitation for the representation of real numbers, or fails when there are no solutions. Moreover, if it succeeds, then a solution has been found (with the value lb_i for the parameter i) guaranteeing the full consistency of the returned value. Therefore, to guarantee that lb_i is the leftmost possible value for the parameter i, it is only necessary to assure that function *solution* does not miss any possible solution, i.e. it will only fail in case of inconsistency.

Function *solution*, shown in Fig. 4, meets this condition and is in accordance with the two assumptions presented in the beginning of this section. It first tries to prove inconsistency by checking the consistency with the actual parameter ranges. If an inconsistency is detected than it fails immediately, otherwise it tries to guess a possible solution by instantiating all domains with their mid values and checking consistency again. If a solution is found then the value of the parameter i is returned, otherwise it is necessary to continue the search. The whole search space will be split in two and the search for a solution will be performed recursively in each subdivision (of course, once a solution is found, the algorithm stops). In order to split the search space, one parameter is chosen (the one with the widest domain) and its domain is divided into two halves.

```
function solution( in i, in {D₁,...,Dₙ}, out sᵢ): boolean
    if consistency_check({D₁,...,Dₙ})
    then for j = 1 to n do Dⱼ' ← (Dⱼ⁻+Dⱼ⁺)/2
            if consistency_check({D₁',...,Dₙ'})
            then sᵢ ← Dᵢ'
                 solution ← true
            else Dₖ ← widest_domain({D₁,...,Dₙ})
                 lDₖ ← [Dₖ⁻..(Dₖ⁻+Dₖ⁺)/2]
                 if solution (i,{D₁,..., lDₖ,...,Dₙ},s)
                 then sᵢ ← s
                      solution ← true
                 else rDₖ ← [(Dₖ⁻+Dₖ⁺)/2..Dₖ⁺]
                      if solution (i,{D₁,..., rDₖ,...,Dₙ},s)
                      then sᵢ ← s
                           solution ← true
                      else solution ← false
    else solution ← false
```

Fig. 4. The algorithm for finding a particular solution.

Again the above algorithm always converges (because in each recursion step the search space is reduced and there is a machine limitation for the representation of real numbers) or fails (if there are no solutions). However the number of consistency checks can be huge (in the worst case it corresponds to analyse every possible instantiation in the search space).

For practical reasons it is better to use a coarser approximation of this global hull-consistency enforcing algorithm. As such a minimum width (w) is defined as being sufficient to consider a parameter "instantiated". With this relaxation the algorithm will not split further a search space if all the variable domain widths are less than w.

Although this algorithm was implemented to deal with parametric ODEs (because none of the current consistency approaches were able to narrow the domains - see Section 6) it is still adequate for any other problem where narrowing the variable domains is a key issue.

5. An Incremental Constraint Approach for Parametric ODEs

The previous section has shown that dealing with parametric ODEs within a decision support context requires the simultaneous instantiation of all the parameters. The implications are, in general, a very large number of consistency checks.

Existing validated approaches were developed with the purpose of solving an ODE problem by minimising the uncertainty around the true trajectory. In order to do this, as seen in Section 3, the approximated trajectory is calculated in a number of intermediate points t_i, between t_0 and t_n, the initial and final points of the trajectory. If the purpose is to minimise the uncertainty around the true trajectory, a large number of intermediate points should be considered.

Of course there is a price to pay in computation time, dramatically magnified if a large number of consistency checks are requested. Notice that the sequential process of calculating the trajectory approximation prevents any attempt to derive conclusions about the trajectory value of the final point without first considering all the intermediate points. The above static sequential approach may spend too much effort in looking for precision when a less accurate result could be sufficient to detect inconsistency. Much of the consistency checks required by the algorithm of the previous section do not require such high precision and can be done with a variable number of intermediate points.

Therefore we propose an alternative dynamic approach where the interval approximation of the whole trajectory is done incrementally. The main idea is that, being impossible to decide, a priori, how many intermediate points are necessary to check consistency, the least number possible is attempted first, to avoid unnecessary time costs.

Therefore it is essential to link as soon as possible the initial point of the trajectory with the final point to obtain a first, possibly very imprecise, interval approximation of the true trajectory. Without this connection, consistency cannot be checked. If such first trajectory does not allow a decision regarding the existence (or not) of a consistent solution then new intermediate points are added together with the associated constraints (corresponding to the approximation step). These will eventually narrow, through propagation, the bounds of the trajectory. The uncertainty about the trajectory is thus incrementally reduced, allowing the algorithm to stop whenever a consistency check is solved (or, as we will see below, there is a reason to believe that it can never be solved due to the uncertainty of the parameters).

The incremental approach we propose consists of two parts: firstly, the initial point is linked with the final point of the trajectory; secondly, the trajectory uncertainty is incrementally reduced.

5.1. Linking the Initial with the Final Point of the Trajectory

The underlying goal of the algorithm is to consider the less possible number of intermediate points. The ideal would be to link directly the initial with the final point without considering any intermediate point. So, this direct connection is tried first.

As seen in Section 3.2, the process of linking any two points (by a validated method) involves a first phase to bind the error term that is used in a second approximation phase. The strategy used in the first phase is normally based on the Picard operator (Φ) in the following way:

Let $\mathbf{u}'=f(t,\mathbf{u})$ be an ODE system
Let t_i be the initial point, t_f the final point and $h = t_f - t_i$;
Let box D_i be the uncertainty bounding box of the trajectory at point t_i;
Let F be the interval extension of f;
If it is possible to find some box B with $D_i \subseteq B$ and such that:

$$\Phi(B) = D_i + [0,h]F([t_i,t_f],B) \text{ and } \Phi(B) \subseteq B \tag{5}$$

Then: there is a unique solution (between t_i and t_f) for the system with $\mathbf{u}(t_i) \in D_i$
        $\Phi(B)$ is a bounding box for the trajectory between the two points.
The traditional procedure now is to try different guesses for the box B and/or to consider smaller steps to reduce h. Notice that by reducing h, it is always possible to guarantee the condition (5) for any box B. In sequential approaches the step considered is usually small, and B is easily obtained. However, in our approach we are trying large steps and so, convergence is not guaranteed.

Our algorithm thus starts with an initial guess B_0 for box B and checks condition (5). If it is verified, it tries to narrow further the first guess by making $B_1 = \Phi(B_0)$ as a new guess for B, and this process is repeated till a fix point is reached. Otherwise, if condition (5) is false, the algorithm concludes that the step is too big and tries a smaller one (half the size of the current step).

Using the above algorithm the trajectory is bounded between t_0 and an intermediate point t_1 that is as close as possible (using this method) to the final point. Subsequently, in the second phase, a better enclosing for the value of the trajectory at the intermediate point ($\mathbf{u}(t_1) \in D_1$) is obtained by the constraint below corresponding to the Taylor approximation step of order k (we used k=4) between the two points with the bounded error term:

$$D_1 = D_0 + \sum_{i=1}^{i=k-1} \frac{(t_1 - t_0)^i}{i!} F^{(i-1)}(t_0, D_0) + \frac{(t_1 - t_0)^k}{k!} F^{(k-1)}([t_0,t_1]\Phi(B)) \tag{6}$$

The complete connection with the final point is obtained applying recursively the above algorithm each time trying to link directly the last intermediate point to the final point.

5.2. Reducing the Trajectory Uncertainty

Once connected the initial with the final point of the trajectory a first consistency check is performed. Let C_f be a box representing the values of the trajectory at the final point whose consistency is being checked and let D_f be the calculated box for the approximated trajectory value at that same point. The consistency check problem is already decidable if $D_f \cap C_f = \emptyset$ (it is inconsistent) or if $D_f \subseteq C_f$ (it is consistent). Otherwise no conclusions can be made, and a further refinement of the trajectory approximation is needed.

At this stage, the algorithm will incrementally consider new intermediate points until a stopping criteria is satisfied. This is a three step process: firstly it chooses where to introduce the new point; secondly it adds the constraints relating the new point with the already considered points; and finally it verifies the stopping criteria.

Choosing a new intermediate point. The new intermediate point should be placed so as to obtain the best refinement of the trajectory approximation. Unfortunately, it is not possible to predict the overall refinement, and a heuristic choice must be used. We have tried several heuristics (only mid points of existing intervals are considered) and the two most promising were:

a)   the mid point between the two most distant adjacent points already considered;
b)   the mid point between the two adjacent points already considered where the trajectory uncertainty between them had increased the most.

Adding the constraints for the new point. Once chosen a new intermediate point t_k to consider between t_i and t_j, the constraints associated with the approximation step (X) must be added, first to link t_i and t_k, and then to link t_k and t_j. In order to do it, new a priori enclosure bounds should be calculated (as described in 5.1) between t_i and t_k and between t_k and t_j. After adding these constraints the refinement of the trajectory approximation is achieved through constraint propagation.

The stopping criteria. The stopping criteria must evaluate whether it is worth considering further intermediate points for the refinement of the trajectory approximation when the consistency checking problem has not been already decided. Two cases must be considered:

a)   either the uncertainty on the parameters will not allow any conclusion, even with a "perfect" approximation method; or
b)   A conclusion on consistency can be obtained if the approximation is refined.

Clearly, in the first case the algorithm should stop and in the second case it should go on. However, it is difficult to decide which is the present case.

The heuristic we have used is based on the improvement (Δ) achieved in the last iteration, i.e. the difference between the uncertainty before and after considering the last intermediate point. It also measures how far way is the current approximation from a possible decision. This distance (δ) is obtained by examining the decision requirements ($D_f \cap C_f = \emptyset$ and $D_f \subseteq C_f$) and calculating the minimum uncertainty improvement that is necessary for D_f to verify one of them.

Assuming that in the next iterations the uncertainty improvement will be the same as the last one (Δ) then it will be necessary at least δ/Δ new points to reach a conclusion. So, the stopping criteria is as follows: if the number of already considered points plus the δ/Δ new points exceeds a predefined threshold then it stops; otherwise it continues.

6. Results

To illustrate the possibilities of our proposed approach we have used it to support the diagnosis of diabetes based on the model presented in Section 2. On arrival to the hospital two patients ingested a large amount of glucose. Table 1 shows the results of the blood exams performed during the first 3 hours of the test.

Table 1. The glucose and insulin blood concentrations measurements during a glucose tolerance test performed in two patients (deviations from the fasting level concentrations).

	after ingestion	1 hour later	2 hours later	3 hours later
Patient 1	g=80;h=0	g=-29.7;h=46.4	g=-24.8;h=-34.5	g=35.8;h=-1.5
Patient 2	g=80;h=0	g=18.1;h=41.4	g=-38.8;h=18.6	g=-28.0;h=-15.9

Based on this data and assuming that the acceptable bounds for the parameter values are +/- 50% of the typical normal values (p_1=0.0044, p_2=0.04, p_3=0.0044, p_4=0.03) it is necessary to decide if the patients are diabetic or normal. We would like to narrow the range of the parameters accordingly to the available data (through constraint propagation) so as to be able to make a safe decision.

The values of Table 1 were artificially chosen to coincide with the simulations of Section 2. Hence, perfect constraint propagation from this data on equations (1) and (2) would establish period T_1=179.9 for patient 1 and T_2=289.9 for patient 2, thus implying that the first patient is normal and the second diabetic. Nonetheless, such conclusion could not be reached with the usual consistency enforcing techniques, namely, local hull-consistency and box-consistency.

To enforce local hull-consistency a constraint network was defined for the validated approximation of the trajectory of the parametric ODE (as explained in sections 3 and 5) and the information about the observed trajectory values was added. This was implemented in the constraint programming language DECLIC [14] which allows the definition of interval constraints and offers a propagation mechanism for enforcing hull-consistency (and box-consistency as well) on each individual constraint. After propagation, the ranges of the parameter domains remained exactly the same, and so this consistency requirement was not sufficient to solve the problem.

To check whether a global box-consistency strategy would achieve any pruning, we have checked whether the original domains for the parameters were already box-consistent. To do this, we checked the consistency of both bounds of the domain of each variable with the domains of the other variables (see Section 4.1). As there are 4 parameters, 8 bounds were checked for box-consistency, i.e. we computed the trajectory obtained by instantiating each bound (one at a time) and verified whether it included the observed trajectory values. Table 2 shows the trajectory values, at minute 60, obtained for each bound instantiation, which clearly include the observed values for both patients (similar results were obtained for the other observed trajectory values). Therefore, no pruning on the model parameters could be achieved by box-consistency alone (nor could the patients be diagnosed).

Table 2. Trajectory values at minute 60 calculated for each parameter bound instantiation. The observed values were: g=-29.7 and h=46.4 for patient 1; g=18.1 and h=41.4 for patient 2.

$p_1=p_1^- \rightarrow$ (g=[-238..93];h=[-112..176])	$p_1=p_1^+ \rightarrow$ (g=[-209..79];h=[-104..156])
$p_2=p_2^- \rightarrow$ (g=[-142..47];h=[14..165])	$p_2=p_2^+ \rightarrow$ (g=[-204..75];h=[-104..141])
$p_3=p_3^- \rightarrow$ (g=[-243..97];h=[-117..176])	$p_3=p_3^+ \rightarrow$ (g=[-220..93];h=[-104..156])
$p_4=p_4^- \rightarrow$ (g=[-42..48];h=[21.. 55])	$p_4=p_4^+ \rightarrow$ (g=[-203..48];h=[-75..133])

Enforcing global hull-consistency with our proposed approach we obtained the bounds for the model parameters shown in Tables 3 and 4. Table 3 shows the narrowing of the parameter domains obtained by considering the trajectory values observed at minute 60 for both patients.

Table 3. First pruning of the parameter domains.

	Initial domains	Patient 1 g=-29.7 and h=46.4	Patient 2 g=18.1 and h=41.4
p_1	[0.0022..0.0066]	[0.0022..0.0066]	[0.0022..0.0066]
p_2	[0.0200..0.0600]	[0.0324..0.0490]	[0.0238..0.0360]
p_3	[0.0022..0.0066]	[0.0022..0.0066]	[0.0028..0.0066]
p_4	[0.0150..0.0450]	[0.0222..0.0415]	[0.0150..0.0179]
T	[119.9..359.9]	[137.8..233.5]	[239.5..329.8]
Decision	Unknown	Normal ??	Diabetic ??

The pruning obtained after this propagation is already enough to decide the right diagnosis for each patient (i.e. whether T < 240). However, if a more precise evaluation of the decision variable is needed (as some of the bounds of T were "borderline") then, another observed trajectory value may also be considered. Table 4 shows the results obtained adding the observations at minute 120.

Table 4. Final pruning obtained after the first two observations.

	Patient 1 g=-24.8 and h=-34.5	Patient 2 g=-38.8 and h=18.6
p_1	[0.0022..0.0066]	[0.0022..0.0066]
p_2	[0.0324..0.0490]	[0.0273..0.0319]
p_3	[0.0022..0.0066]	[0.0028..0.0052]
p_4	[0.0258..0.0341]	[0.0150..0.0179]
T	[151.7..216.7]	[255.3..308.2]
Decision	Normal !!	Diabetic!!

After considering the observations made on the two first hours of the glucose tolerance test, and using our global hull-consistent enforcing algorithm on the patient model, we could reach clear diagnoses: patient 1 is normal and patient 2 diabetic.

7. Conclusions

This paper addressed the application of interval constraints to problems involving parametric ODEs. The motivation is the need in many decision support applications (e.g. in medicine, economics or engineering) to narrow the possible ranges of the parameters of such ODEs in order to make safe decisions. Current analytical approaches are not adequate because many systems have no analytical solutions; traditional numerical simulation methods are also inadequate since they cannot provide safe ranges for the model parameters.

Although ODEs have been recently handled by constraint techniques, research done so far has focussed on the precision of the trajectories of such systems. Of course, such precision cannot be very accurate if the parameters of an ODE are also known with little precision. Because of their different focus, these approaches are not able to narrow the ODEs parameters, given some experimental or observed data, in order to fit a parametric ODE model with these observations.

Therefore we proposed in this paper a new consistency type, global hull-consistency and described an algorithm to enforce it. This algorithm is improved by the use of an incremental approach to compute trajectories.

We illustrate our method with a problem from the medical domain (diagnosis of diabetes). We show that the usual consistency techniques are not useful for the problem in hand and show how our approach correctly diagnoses the patients.

Being a preliminary implementation, we anticipate that further work is required to improve our current implementation. Moreover, we intend to study a more formal characterisation of the new consistency type that we have proposed. The results obtained so far demonstrate the need to improve on the current types of consistency checking in the interval domain, and we are confident that our proposal may be considered as a promising starting point.

References

1. Deville, Y., Janssen, M. and Van Hentenryck, P.: Consistency Techniques in Ordinary Differential Equations. In: Proceedings of the 4th International Conference on Principles and Practice of Constraint Programming – CP98. Springer, Pisa, Italy (1998) 162-176.
2. Cruz, J., Barahona, P. and Benhamou, F.: Integrating Deep Biomedical Models into Medical Decision Support Systems: An Interval Constraints Approach. In: Proceedings of the Joint European Conference on Artificial Intelligence in Medicine and Medical Decision Making, AIMDM'99, Springer, Aalborg, Denmark (1999).
3. Collavizza, H., Delobel. F. and Rueher, M.: A Note on Partial Consistencies over Continuous Domains. In: Proceedings of the 4th International Conference on Principles and Practice of Constraint Programming - CP98. Springer, Pisa, Italy (1998) 147-161.
4. Ackerman, E., Gatewood, L., Rosevar, J., and Molnar, G.: Blood Glucose Regulation and Diabetes. In: Concepts and Models of Biomathematics, Chapter 4, F. Heinmets, ed., Marcel Dekker, (1969) 131-156.
5. Hartman, P.: Ordinary Differential Equations. Wiley, New York (1964).
6. Moore R.E.: Interval Analysis. Prentice-Hall, Englewood Cliffs, NJ (1966).

7. Lohner, R. J.: Enclosing the solutions of ordinary initial and boundary value problems. In: Computer Arithmetic: Scientific Computation and Programming Languages. Wiley-Teubner Series in Computer Science, Stuttgart, (1987) 255-286.
8. Nedialkov, N. and Jackson, K.: Software Issues in Validated ODE Solving. Technical Report, Department of Computer Science, University of Toronto, Canada (1998).
9. Cleary, J.G.: Logical Arithmetic. In Future Generation Computing Systems, 2(2) (1987) 125-149.
10. Older, W.: Application of Relational Interval Arithmetic to Ordinary Differential Equations. In: Workshop on Constraint Languages and their use in Problem Modelling, Int'l Logic Programming Symposium, Ithaca, New York (1994).
11. Hickey, T.J.: CLP(F) and Constrained ODEs. In: Proceedings of the Workshop on Constraint Languages and their use in Problem Modelling, Jourdan, Lim, Yap ed., (1994) 69-79.
12. Lhomme, O.: Consistency Techniques for numeric CSPs. In: Procedures of IJCAI93, Chambery, France, (1993) 232-238.
13. Benhamou, F., McAllester, D. and Van Hentenryck, P.: CLP(intervals) revisited. In: Proceedings of the International Logic Programming Symposium. MIT Press, (1994).
14. Benhamou, F., Goualard, F., Granvilliers, L.: An Extension of the WAM for Cooperative Interval Solvers. Technical Report, Department of Computer Science, University of Nantes, France (1998).

Non-trivial Symbolic Computations in Proof Planning

Volker Sorge

Universität des Saarlandes Fachbereich Informatik
D-66041 Saarbrücken Germany
sorge@ags.uni-sb.de

Abstract. We discuss a pragmatic approach to integrate computer algebra into proof planning. It is based on the idea to separate computation and verification and can thereby exploit the fact that many elaborate symbolic computations are trivially checked. In proof planning the separation is realized by using a powerful computer algebra system during the planning process to do non-trivial symbolic computations. Results of these computations are checked during the refinement of a proof plan to a calculus level proof using a small, self-implemented system that gives us protocol information on its calculation. This protocol can be easily expanded into a checkable low-level calculus proof ensuring the correctness of the computation. We demonstrate our approach with the concrete implementation in the ΩMEGA system.

1 Introduction

In recent years there have been many attempts combining computer algebra systems (CAS) and deduction systems (DS), either for the purpose of enhancing the computational power of the DS [17, 18, 3] or in order to strengthen the reasoning capabilities of a CAS [1, 4]. For the former integration there exist basically three approaches: (1) to fully trust the CAS, (2) to use the CAS as an oracle and to try to reconstruct the proof in the DS with purely logical inferences, and (3) to generate protocol output during a CAS calculation and to use this protocol to verify the computation. Following approach (1) one cannot guarantee the correctness of the proof in the DS any longer. While the correctness is no issue in approach (2) it foregoes the efficiency of a CAS and replay the computation with purely logical reasoning might still impose a hard task to the DS. (3) is a compromise, where one can employ the computational strength of a CAS and additionally gain important hints easing the reconstruction and checking of the computation.

We have, indeed, successfully experimented with idea (3) by implementing a prototypical CAS ($\mu\mathcal{CAS}$) that is a small library of simple polynomial algorithms which give us protocol information on their computations [18, 23]. This protocol information is used to derive abstract proof plans that can be transformed into proofs of the ΩMEGA system [16]. Exploiting ΩMEGA's ability of step-by-step expansions of proof plans into natural deduction (ND) calculus proofs [13], the

H. Kirchner and C. Ringeissen (Eds.): FroCoS 2000, LNAI 1794, pp. 121–135, 2000.

computations can be machine-checked on a fine-grained calculus level. While this way of integrating a computer algebra system into ΩMEGA solves the correctness issue, it has the drawback that apart from μCAS there does not exist a full grown CAS that provides us with the necessary protocol output on its calculations.

In this paper we present a pragmatic approach working around this problem in proof planning. It is based on the idea presented in [17] that many hard symbolic computations are easy to check. As an example for this thesis one may consider symbolic integration, a surely non-trivial computation, whose verification is to simply differentiate the result of the integration algorithm. We exploit this fact within the proof planning component of ΩMEGA: Results of non-trivial symbolic computations are used during the proof planning process. The verification of these calculations is postponed until a complete proof plan is refined to a low level calculus proof and it is arranged in a way, that we can use the trivial direction of the verification. This is achieved by using MAPLE [22] for computations during the planning process and μCAS to aid the verification. The technique we present here is already successfully applied in the ΩMEGA system in the domains of limit theorems and optimization problems. For detailed report concerning proof planning limit theorems see [20] and for some earlier work on optimization problems see [18]. In this paper we will only be concerned with the details of these problems that involve the application of computer algebra.

The paper is organized as follows: We first give a general overview on proof planning and its special features in ΩMEGA in the next section. In Sec. 3 we present the general architecture of the integration of CAS into ΩMEGA and elaborate how the systems MAPLE and μCAS are used within the proof planning component of ΩMEGA in order to achieve our goal of separating non-trivial computation and easy verification. In Sec. 4 we tackle the problem of how to deal with different normal forms of different systems that can arise in our context. In Sec. 5 we illustrate our ideas with two examples from the domains of limit theorems and optimization problems. We finally discuss advantages and drawbacks of the presented approach in Sec. 6 before concluding in Sec. 7.

2   Proof Planning in ΩMEGA

The purpose of the ΩMEGA system is to provide assistance in the process of deriving, presenting and checking proofs in mathematics. It provides several means to prove theorems by applying tactics, by using external reasoners — such as automated theorem provers or CAS — or by automatically planning proofs with the help of a proof planner. ΩMEGA's basic logic is a variant of the natural deduction (ND) calculus [13] based on a higher order λ-calculus [8]. As ΩMEGA neither makes correctness assumptions about the incorporated external reasoners nor about tactics or planning methods a user specifies, it requires that proofs are finally checked in the basic calculus to ensure correctness. Thus, although proofs can be both constructed (either interactively or automatically) and represented on more abstract levels, ΩMEGA accepts a proof to be valid, only, if it can be successfully expanded into a proof object whose derivations are in the basic ND calculus that can be machine checked.

ΩMEGA's proof planner is designed to automatically plan proofs on an abstract level by constructing a sequence of methods that represents an abstract derivation for a given theorem from a set of assumptions. Methods are (partial) specifications of tactics [5] known from tactical theorem proving [15]. In ΩMEGA planning methods are declaratively represented in frame-like structures consisting of four major slots (two instances of methods are given in Sec. 5):

Premises and Conclusions are sets of sequents where the method serves to logically derive the conclusions from the premises. Single sequents can be annotated in STRIPS-style with $\oplus$ or $\ominus$ (cf. [10]) specifying whether a sequent is added to or removed from the planning state by the method application. In particular, a conclusion annotated with $\ominus$ is a goal that is closed by the method and a premise annotated with $\oplus$ represents a new open subgoal that will have to be proved after the application of the method.

Proof Schema is a schematic specification of the sub-proof the method abbreviates. The schemas of premise and conclusion sequents are used to be matched with actual formulas in a proof in order to determine the method's applicability. Moreover, the proof schema can contain lines that are introduced into the proof only when the application of the method is refined by expansion.

Application Conditions are constraints on the particular instantiations of the parameters of a method that have been introduced by matching of the proof schema. A method is applicable with an instantiation $\mathcal{I}$ of parameters if and only if for $\mathcal{I}$ none of the application conditions evaluates to *false*. Thus, application condition can either fail or return some values which can be bound to additional method parameters. This offers a platform to execute arbitrary functions, e.g., calls to external reasoners.

Once a proof plan has been found in ΩMEGA, it has to be refined by expanding it to a ND calculus-level proof to gain a proof object which can be checked in order to ensure its validity. This expansion is a recursive process, i.e., the expansion of a planning method or an abstract tactic yields a subproof which can again contain abstract proof steps that can be expanded further. The expansion is generally not carried out immediately but postponed until a full proof plan has been found. Yet, as long as a proof step has still an *abstract justification*, i.e., a justification that does not belong to the set of basic ND calculus rules, it is considered *planned*. ΩMEGA's main data structure for storing proofs and proof plans — the proof plan data structure, $\mathcal{PDS}$ [7] — offers facilities to execute this expansion as well as to contract expanded subproofs to shorter, abstract proofs again. The expansion of abstract proof steps is triggered either manually by the user or automatically by the proof checker and executed automatically by ΩMEGA.

3   Integration of Computer Algebra

In this section we first present the general architecture for the integration of computer algebra into ΩMEGA. For a more detailed introduction see also [18, 23]. Then we elaborate our new approach for integrating symbolic computations and their verification into proof planning in ΩMEGA.

3.1   Architecture

The integration of computer algebra into ΩMEGA is accomplished by the SAPPER system [23] which can be seen as a generic interface for connecting one or several computer algebra systems (see Figure 1). An incorporated CAS, like MAPLE [22] or μCAS [23], is treated as a slave to ΩMEGA which means that only the latter can call the former but not vice versa. From the technical point of view, ΩMEGA and the CASs are independent processes while the interface is a process providing a link for communication. Its role is to automate the broadcasting of messages by transforming output of one system into data that can be processed by the other.[1] The maintenance of processes and passing of messages is managed by the MATHWEB [11] environment into which ΩMEGA is embedded.

The role of SAPPER in the integration has two distinct aspects: Firstly, arbitrary CASs can be easily used as black box systems for term rewriting (similar to the approaches of [4, 3]) and SAPPER works as a simple bridge between the planner and the CASs. Secondly, SAPPER also offers means to use a CAS as a proof planner. That is, if the CAS can provide additional information on its computations, this information is recorded by SAPPER and translated into a sequence of tactics that can eventually verify the computation. Since there does not exist a state-of-the-art system that provides this information, we use our own μCAS system, that is a collection of simple algorithms for arithmetic simplification and polynomial manipulations including a plan generating mode (cf. [18]).

The two tasks of a CAS, rewriting and plan generation, are mirrored in the interface (cf. Fig. 1) that basically can be divided into two major parts; the *translator* and the *plan generator*. The former performs syntax translations between ΩMEGA and a CAS in both directions while the latter only transforms protocol output of μCAS to ΩMEGA proof plans. Figure 1 also depicts the different uses of the two CAS involved: while MAPLE is connected as a black box system, only, μCAS can be used both as black box and as plan generator.

While the translation part is very common, the plan generator is the actual specialty of SAPPER. It provides the machinery for the proof plan extraction from the specialized algorithms in μCAS. These are equipped with a *proof plan generating mode* that returns information on single steps of the computation within the algorithms. The output produced by the execution of a particular algorithm is recorded by the plan generator which converts it, according to additional information on the proof, into a proof plan. In order to produce meaningful information μCAS needs to have a certain knowledge about the proof methods and tactics available to ΩMEGA in its knowledge base. Thus references to logical objects (methods, tactics, theorems, or definitions) of the knowledge base are compiled a priori into the algebraic algorithms for documenting their calculations. SAPPER's plan generator uses produced protocol output to lookup tactics and theorems of an ΩMEGA theory (cf. Figure 1) in order to assemble a valid proof plan. How algorithms in μCAS have to be expanded and how the plan generation is performed is illustrated in Sec. 3.3.

---

[1] This is an adaptation of the general approach on combining systems in [9].

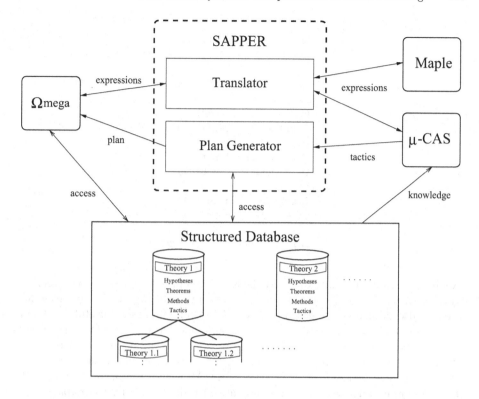

Fig. 1. Interface between ΩMEGA and computer algebra systems

3.2 Integration into Proof Planning

When integrating computer algebra into proof planning we have to keep in mind that all plans have to be expandable to ND calculus proofs. However, using a system whose computations are checkable, like μCAS, restrict us to the use of its rather simple algorithms which might not always be sufficient for the task at hand. What we really would like, is to combine the computational power of a CAS like MAPLE to perform non-trivial computations with the verification strength of μCAS, i.e., simple arithmetic.

Therefore, we try to exploit as much as possible the fact that many difficult symbolic computations are easy to verify. This is folklore in mathematics and has already been elaborated in [17]; the most prominent example for this is certainly symbolic integration which is still a hard task for many CASs. Results of symbolic integration algorithms are, however, easily checked, since it involves to differentiate the result and compare it with the original function, only. Other examples are computation of roots of functions or factorization of polynomials, which involve non-trivial algorithms, but the verification of the results only involves straight-forward arithmetic.

The separation of computation and verification can be easily achieved within proof planning: During the planning process the applicability of a method is

```
(addition (a+c b+d)
          (cond ((not (monomial a))
                 (addition (addition (simplify a) c) b+d))
                ((not (monomial b))
                 (addition a+c (addition (simplify b) d)))
                ((a =_lex b)
                 (tactic "mono-add")
                 ((add a b) + (addition c d)))
                ((a >_lex b)
                 (tactic "pop-first")
                 (a + (addition c b+d)))
                ((a <_lex b)
                 (tactic "pop-second")
                 (b + (addition a+c d)))))
```

Fig. 2. Polynomial addition in $\mu\mathcal{CAS}$.

solely determined by matching and checking the application conditions. As mentioned earlier, the latter can be used to execute arbitrary functions, therefore we can also implement conditions that call MAPLE and in case useful results are returned, bind these to some method parameters. During the planning process we are not concerned with the verification of the computation, and postpone it until the method is actually expanded. This is done by stating a rewriting step that is justified by the application of a CAS within the proof schema of the method, preferably in those lines that are introduced during the expansion of the method.

Thus, we design our planning methods in a way that MAPLE is called in one of the application conditions to perform the difficult computations during the planning process. The proof schema then contains the appropriate proof steps that enable the application of $\mu\mathcal{CAS}$ to verify MAPLE's computation during the refinement of a proof plan, in the easier direction.

3.3 Verification of Computations

To implement a plan generating mode is a simple task for simple CAS algorithms. Figure 2 shows the simplified version of the recursive addition algorithm in $\mu\mathcal{CAS}$'s arithmetic simplification module in a Lisp-like pseudo-code. $\mu\mathcal{CAS}$ works with polynomial-like representation of given terms together with a lexicographic monomial ordering. Thus the simplification algorithm of $\mu\mathcal{CAS}$ always transforms terms into lexicographically ordered polynomials and in case of rational functions into a fraction of two polynomials.

The algorithm in Fig. 2 is basically a case split where the first two cases ensure that the arguments are in some normalized form. The last three cases then take care of correctly adding the two argument polynomials. The predicates $=_{lex}$, $<_{lex}$, and $>_{lex}$ correspond to comparison of monomials according to the lexicographical term-ordering in $\mu\mathcal{CAS}$. The protocol output of the `tactic` function corresponds to names of tactics that are known to ΩMEGA and from which a proof plan can be assembled. This proof plan can be inserted into the given proof and further expanded using ΩMEGA's tactic expansion mechanism. Note, that such a proof plan only serves to verify the correctness of the result of one run of the algorithm and cannot be used to verify the correctness of the entire algorithm.

Still, the verification of a single computation is not totally trivial, since it involves to have the appropriate algorithm available in $\mu\mathcal{CAS}$, to extend this algorithm with a proof plan generating mode, and maybe to write some of the tactics for ΩMEGA that correspond to those in the algorithm. However, as we are mainly concerned with verifications involving arithmetic, the latter task should diminish over time, since many of the tactics involved can be reused.

$\mu\mathcal{CAS}$ could also be viewed as a collection of elaborate tactics for ΩMEGA. However, it is implemented as an independent system (in Common Lisp) and can be used like a regular CAS, even though with very limited power, only.

4 Dealing with Normal Forms

When using MAPLE for a computation within some method and $\mu\mathcal{CAS}$ to verify MAPLE's result we might have problems identifying the term resulting from $\mu\mathcal{CAS}$'s computation with the original term MAPLE was applied to. Thus, we have to take care of the problem of distinct normal forms of the systems involved during the expansion of a computation.[2]

Let Φ_0 be the original term in the proof, while Φ_{MAPLE} denotes the term which results from applying MAPLE to Φ_0, and let $\Phi_{\mu\mathcal{CAS}}$ be the term returned by $\mu\mathcal{CAS}$ applied to Φ_{MAPLE}. Furthermore, let $(\mathcal{T}_1, \ldots, \mathcal{T}_n)$ be the sequence of tactics computed by $\mu\mathcal{CAS}$ whose application to Φ_{MAPLE} yields the proof plan (1).[3]

$$\Phi_{\text{MAPLE}} \xrightarrow{\mathcal{T}_1} \Phi' \xrightarrow{\mathcal{T}_2} \ldots \xrightarrow{\mathcal{T}_n} \Phi_{\mu\mathcal{CAS}}. \tag{1}$$

We then have three cases to consider:

(a) Φ_0 and $\Phi_{\mu\mathcal{CAS}}$ coincide,
(b) Φ_0 and $\Phi_{\mu\mathcal{CAS}}$ are distinct, however Φ_0 occurs at some point during the expansion, and
(c) Φ_0 and $\Phi_{\mu\mathcal{CAS}}$ are distinct, and Φ_0 does not occur during the expansion.

Case (a) is trivial. Case (b) means that we have some $1 \leq i \leq n$, such that

$$\Phi_{\text{MAPLE}} \xrightarrow{\mathcal{T}_1} \Phi' \xrightarrow{\mathcal{T}_2} \ldots \xrightarrow{\mathcal{T}_i} \Phi_0 \xrightarrow{\mathcal{T}_{i+1}} \ldots \xrightarrow{\mathcal{T}_n} \Phi_{\mu\mathcal{CAS}}. \tag{2}$$

This problem can be easily solved by successively applying the single tactics and checking after each application whether the resulting term is already equivalent

---

[2] The form of MAPLE's result may vary even for equivalent arithmetic expressions (in two different runs of MAPLE), depending on the form of the input. For instance, MAPLE's simplification of $x + 2z + y - z$ yields $x + z + y$, while the same computation with input $x + y + 2z - z$ would yield $x + y + z$ in a different run.

[3] For the sake of clarity, we omit any context the terms Φ_j might be embedded in, i.e., we view the proof plan as rewriting steps of a sub-term of some arbitrary formula.

to Φ_0. In this case the proof can be concluded directly. The remainder of the tactic sequence, i.e., $(\mathcal{T}_{i+1}, \ldots, \mathcal{T}_n)$ in (2), is discarded.

Case (c) is less trivial since the produced tactics are not sufficient to fully justify the computation and thus we are left with a new proof problem, namely to derive the equality of $\Phi_{\mu CAS}$ and Φ_0. However, at this point we can make use of the lexicographic term ordering of μCAS: if $\Phi_{\mu CAS}$ and Φ_0 really constitute the same arithmetic expression, applying μCAS simplification algorithm to Φ_0 will yield $\Phi_{\mu CAS}$. Note, that this step might not only include trivial reordering of a sum but can contain more sophisticated arithmetic. The execution of the simplification algorithm will then return a sequence of tactics $(\mathcal{S}_1, \ldots, \mathcal{S}_m)$ that results in:

$$\Phi_{\text{MAPLE}} \xrightarrow{\mathcal{T}_1} \Phi' \xrightarrow{\mathcal{T}_2} \ldots \xrightarrow{\mathcal{T}_n} \Phi_{\mu CAS} \xleftarrow{\mathcal{S}_m} \ldots \xleftarrow{\mathcal{S}_1} \Phi_0 \qquad (3)$$

In praxis, we deal with this problem slightly different, since in ΩMEGA's tactic expansion mechanism calls to μCAS have to be carried out explicitly by expanding the according justification, and not implicitly during an expansion itself. Thus, we introduce a new subproof for the equality of $\Phi_{\mu CAS}$ and Φ_0:

$$\Phi_{\mu CAS} = \Phi_{\mu CAS} \qquad (\text{=Ref})$$
$$\Phi_{\mu CAS} = \Phi_0 \qquad (\text{CAS})$$

The first line is an instance of reflexivity of equality, an axiom of ΩMEGA's basic calculus. The equation of the second line serves then to apply a rule of equality substitution (=Subst) to finish the original expansion, resulting in proof plan (4).

$$\Phi_{\text{MAPLE}} \xrightarrow{\mathcal{T}_1} \Phi' \xrightarrow{\mathcal{T}_2} \ldots \xrightarrow{\mathcal{T}_n} \Phi_{\mu CAS} \xrightarrow{\text{=Subst}} \Phi_0. \qquad (4)$$

In order to completely verify the computation the justification (CAS) above must be expanded as well. This results in the second call to μCAS, yielding a proof plan equivalent to the right hand side of (3).

5   Examples

We illustrate our approach with two examples of proof planning methods that are used in different domains. The first is the Complex-Estimate method that is needed for the proof planning of limit theorems; the second is the method Polynomial-Root which is employed for solving optimization problems.

The use of proof planning to solve optimization problems is advancement of work already reported on in [18]. However, in [18] we were restricted to the use of μCAS and therefore could only tackle problems involving polynomials of degree at most two, due to lack of sophisticated algorithms for computing roots in μCAS. We are currently experimenting with a set of 30 different problems where the proofs of most of them can be planned automatically.

Proof planning of limit theorems is extensively described in [20]. Besides the application of computer algebra it also involves the use of a constraint solver in order to fully automate the planning process. However, we are not concerned with these details here since we will elaborate for neither example how the

Method: Complex-Estimate	
Premises	$L_1, \oplus L_2, \oplus L_3, \oplus L_4$
Appl. Cond.	$\sigma \longleftarrow GetSubst(a,b)$ $k, l \longleftarrow CAS\_split(a_\sigma, b)$
Conclusions	$\ominus L_7$
Proof Schema	$(L_1)\ \Delta \vdash \|a\| < e_1$ $(L_2)\ \Delta \vdash \|k\| < \mathbf{M}$ $\hspace{2cm}$ (OPEN) $(L_3)\ \Delta \vdash \|a_\sigma\| < \epsilon/2 * \mathbf{M}$ $\hspace{0.8cm}$ (OPEN) $(L_4)\ \Delta \vdash \|l\| < \epsilon/2$ $\hspace{1.8cm}$ (OPEN) $(L_5)\ \ \vdash b = b$ $\hspace{2.2cm}$ (=Ref) $(L_6)\ \ \vdash b = k * a_\sigma + l$ $\hspace{1.2cm}$ (CAS L_5) $(L_7)\ \Delta \vdash \|b\| < \epsilon$ $\hspace{0.7cm}$ (fix $L_6 L_1 L_2 L_3 L_4$)

Fig. 3. The Complex-Estimate Method

complete proof plans are found or look like. Instead, we rather concentrate on the application of those methods involving computer algebra. In ΩMEGA we can currently plan around 20 theorems from the limit domain [19] automatically.

5.1 The Complex-Estimate Method

Figure 3 depicts the planning method Complex-Estimate whose purpose is to estimate the magnitude of the absolute value of a complex term by estimating its simpler factors. The method formalizes the derivation of the conclusion L_7 from the premises L_1, L_2, L_3, and L_4. The annotations $\oplus$ indicate that the lines L_1 and L_7 have to be present in the current planning state for the method to be applicable, whereas L_2, L_3, and L_4 will be introduced as new open goals.

Thus, Complex-Estimate is handled by the planner as follows: If an open line in the planning state matches L_7 and if an assumption matching L_1 is available, Complex-Estimate's parameters a, b, e_1, ϵ are instantiated. Then the two application conditions are evaluated. The first, $GetSubst(a,b)$, computes a substitution in order to match sub-terms of the expressions a and b. If such a substitution exists, it is returned and bound to the parameter σ and the next application condition, $CAS\_split(a_\sigma, b)$, is evaluated. Here a_σ denotes the term resulting from the application of the substitution σ computed by $GetSubst$ to a. CAS_split calls MAPLE to factorize b. This factorization is done using MAPLE's functions quo and rem which both employ the Euclidean algorithm where quo returns the quotient of a polynomial division (corresponding to k in our method) and rem returns the remainder of that division (i.e., l in the method). The two functions are called in the following way: quo(b, a_σ, a') and rem(b, a_σ, a'). a' is a sub-term of a_σ that functions as the reference variable for the polynomial division in quo and rem respectively. Instead of using both MAPLE functions we could have supplied the call of quo with an additional formal parameter to store the remainder of the division automatically. However, using rem frees us from the requirement to choose the formal parameter distinct from any term occurring in both b and a_σ.

If the two application conditions are successfully evaluated, the instantiated method is introduced into the partial plan. The goal L_7 is removed as planning goal, and the new goals L_2, L_3, and L_4 are introduced instead.

When an application of the method is expanded later on, the complete sub-proof given in the proof schema is introduced, i.e., the sequents L_5 and L_6 are newly added to the proof and the justification of L_7 is updated. It is then justified by (`fix` $L_6 L_1 L_2 L_3 L_4$), where `fix` is an abstract justification itself that will be expanded during further refinement of the proof plan as well. Lines L_5 and L_6 serve to certify the correctness of the values computed by MAPLE by making this computation explicit. Here the equality of b and $k * a_\sigma + l$ is derived from the reflexivity of equality which is an axiom of the underlying calculus. The justification of line L_6 indicates both that the step has been introduced by the application of a CAS and that its expansion can be realized by using μCAS in plan generating mode.

L_5 and L_6 indicate that for the verification of MAPLE's computation it is necessary to certify that $k * a_\sigma + l$ can successfully be transformed into b. To perform this goal, we must use basic arithmetic, only, instead of the considerably harder polynomial division performed by MAPLE. Thus, we can use μCAS's simplification component which employs, among others, the algorithm of Fig. 2. For a concrete instance μCAS would return a sequence of tactics indicating single computational steps that have been performed inside the computer algebra algorithm. This proof plan is then inserted into the proof and further expanded to show the correctness of the computation.

As a concrete example we consider the application of **Complex-Estimate** in the proof of LIM+. LIM+ states that the limit of the sum of two functions is the sum of their limits and is formalized by:

$$\forall f. \forall g. \forall l_1. \forall l_2. \lim_{x \to a} f(x) = l_1 \wedge \lim_{x \to a} g(x) = l_2 \to \lim_{x \to a} (f(x) + g(x)) = l_1 + l_2.$$

During the proof planning process for LIM+, the assumption $\Delta \vdash |f(X_1) - l_1| < e_1$ is available at some point. Next the goal $\Delta \vdash |f(x) + (g(x) - (l_1 + l_2))| < \epsilon$ can be removed by applying the method **Complex-Estimate**. During matching the premise and conclusion lines the method variables a and b are instantiated by $(g(X_1) - l_2)$ and $f(x) + (g(x) - (l_1 + l_2))$, respectively. When the application conditions of **Complex-Estimate** are evaluated, *GetSubst* returns $[x/X_1]$ and *CAS_split* calls MAPLE to compute the instantiations k and l. The actual function calls passed to MAPLE are

`quo`$(f(x)+g(x)-(l_1+l_2),f(x)-l_1,f(x))$; `rem`$(f(x)+g(x)-(l_1+l_2),f(x)-l_1,f(x))$;

returning 1 and $(g(x) - l_2)$ as results for k and l, respectively. Because `quo` and `rem` are algorithms dealing with polynomials, only, functional notation is abolished in SAPPER's translator when MAPLE is called; e.g., $f(x)$ would be translated into f_x.

The goal $\Delta \vdash |f(x) + g(x) - (l_1 + l_2)| < \epsilon$ is then replaced by the new goals[4]

1. $\Delta \vdash |1| < \mathbf{M}$,   2. $\Delta \vdash |f(x) - l_1| < \epsilon/2 * \mathbf{M}$,   3. $\Delta \vdash |g(x) - l_2| < \epsilon/2$.

---

[4] Note, that $\mathbf{M}$ is a meta-variable, i.e., a place holder for instantiations of an existentially quantified variable, that will be instantiated later in the course of planning LIM+. It's actual value is, however, irrelevant to our example.

After a complete proof plan has been found for the LIM+ theorem, the plan can be refined to a calculus level proof. We elaborate this expansion for the `Complex-Estimate` method, focusing on the steps of the expansion dealing with the verification of MAPLE's computation:

$$
\begin{array}{lll}
f(x) + (g(x) - (l_1 + l_2)) & = & f(x) + (g(x) - (l_1 + l_2)) & (\text{=Ref}) \\
f(x) + (g(x) - (l_1 + l_2)) & = & ((1 * (f(x) - l_1)) + (g(x) - l_2)) & (\text{CAS})
\end{array}
$$

Here the second line is justified by the application of a CAS. In order to obtain a pure ND-level proof this line needs to be further expanded. However, since during the application of `Complex-Estimate` the values for l and k were computed by MAPLE, we do not have any additional information for an expansion. To justify the computation in more detail we use an algorithm within our μCAS system in plan generation mode that produces a trace output that gives more detailed information on single computational steps. Instead of simulating the complex algorithm for polynomial division within μCAS, we simply use an algorithm that simplifies the term on the right-hand side of the equation. Thus, μCAS verifies the result of MAPLE's computation with the help of a much simpler algorithm. The yielded proof plan consists of a sequence of tactics indicating single computational steps of the algorithm. Within the $\mathcal{PDS}$, the single step can be expanded to a plan with higher granularity. The newly introduced proof steps are:

$$
\begin{array}{lll}
f(x) + ((g(x) - l_1) - l_2) & = & f(x) + ((g(x) - l_1) - l_2) & (\text{=Ref}) \\
f(x) + ((g(x) - l_1) - l_2) & = & f(x) + (g(x) - (l_1 + l_2)) & (\text{CAS}) \\
\end{array}
$$

$$
\begin{array}{lll}
f(x) + (g(x) - (l_1 + l_2)) & = & f(x) + (g(x) - (l_1 + l_2)) & (\text{=Ref}) \\
f(x) + (g(x) - (l_1 + l_2)) & = & f(x) + ((g(x) - l_1) - l_2) & (\text{=Subst}) \\
f(x) + (g(x) - (l_1 + l_2)) & = & ((f(x) - l_1) + (g(x) - l_2)) & (\text{Pop-Second}) \\
f(x) + (g(x) - (l_1 + l_2)) & = & ((1 * (f(x) - l_1)) + (g(x) - l_2)) & (\text{Mult-1-Left})
\end{array}
$$

The lower four lines correspond to a step-by-step version of the computation as one might do it by hand. We can also observe a conflict of normalforms here, as described in the preceding section. μCAS's simplification algorithm only yields $f(x) + ((g(x) - l_1) - l_2)$ as a result and thus the upper two lines have to be introduced in the proof in order to justifiy the equality substitution (=Subst). The new (CAS) justification is expanded with another call to μCAS. However, we want to focus on the expansion of the original proof plan, i.e., of the lower four lines. So far the expansion of the original CAS justification has been exclusively done by μCAS proof plan generation mode. At this stage μCAS cannot provide any more details about the computation and the subsequent expansion of the next hierarchic level can be achieved without further use of a CAS. Let us for instance take a look at the expansion of the (Pop-Second) tactic which basically describes the reordering within a sum:

$$
\begin{array}{lll}
\dots & = & f(x) + ((g(x) - l_1) - l_2) & (\text{=Subst}) \\
\dots & = & (f(x) + (g(x) - l_1)) - l_2 & (\text{Associativity}_{+-}) \\
\dots & = & (f(x) + (-l_1 + g(x))) - l_2 & (\text{Commutativity}_{+-}) \\
\dots & = & ((f(x) - l_1) + g(x)) - l_2 & (\text{Associativity}_{+-}) \\
\dots & = & ((f(x) - l_1) + (g(x) - l_2)) & (\text{Associativity}_{+-})
\end{array}
$$

Here the tactics named Associativity$_{+-}$ and Commutativity$_{+-}$ correspond to the application of the obvious theorems as a rewrite rule. Now the little subproof introduced when expanding (Pop-Second) is already on the level of applications of basic laws of arithmetic. These tactics can, however, be expanded even further. Expanding, for example, the (Commutativity$_{+-}$) justification yields:

$$\forall a. \forall b. a - b = -b + a \qquad \text{(Theorem)}$$
$$\forall b. g(x) - b = -b + g(x) \qquad (\forall E \quad g(x))$$
$$g(x) - l_1 = -l_1 + g(x) \qquad (\forall E \quad l_1)$$

$$\ldots = (f(x) + (g(x) - l_1)) - l_2 \quad \text{(Associativity}_{+-})$$
$$\ldots = (f(x) + (-l_1 + g(x))) - l_2 \qquad (=\text{Subst})$$

This last expansion step details the application of commutativity of addition as rewrite step by deriving the right instance from the theorem of commutativity. We can proof check for correctness if the expansion is carried out up to this detailed level for the whole proof plan $\mu\mathcal{CAS}$ computed to justify the computation.[5] Provided the proof checker approves and we have correct proofs for all the applied theorems in our database, we have successfully verified the particular computation which guarantees the correctness of the overall proof.

5.2   The Polynomial-Root Method

In general, optimization problems are of the form $Opt_<(f, I)$ or $Opt_>(f, I)$, where f is a cost function consisting of a rational polynomial and some denomination, that either has to be minimized or maximized within the rational interval I. Thus, the goal is to show that the polynomial part of f has a total maximum or minimum within I. Generally f is not directly given but has to be computed from a set of given cost functions of economic processes and prices for resources. These manipulations are done by methods using $\mu\mathcal{CAS}$ computations. The calculation of extrema and subsequently of roots to justify total minimum or maximum properties are performed by methods using MAPLE and are verified by $\mu\mathcal{CAS}$.

One of these latter methods is Polynomial-Root shown in Fig. 4 whose purpose is to confirm the existence of a root of a polynomial. It therefore has one conclusion only and no premises, that is when the method is applied, it closes an open line, but does not introduce any new goals into the proof plan. The application conditions are handled as follows: if p is actually a polynomial the CAS_root function calls MAPLE's function roots to compute the rational roots of p. In case MAPLE's computation is successful the method is applied and the first of the returned values is bound to the method parameter a (whereas possible additional roots are kept for backtracking purposes). The value of a

---

[5] Actually, we have not yet expanded the subproof to ND calculus level. Since equality is a concept defined by Leibniz-equality in ΩMEGA, =Subst is a tactic which can be further expanded. However, we omitted the rather tedious details of this expansion here.

Method: Polynomial-Root	
Premises	
Appl. Cond.	$IsPolynomial(p)$ $a \longleftarrow CAS\_root(p,x)$ $\sigma \longleftarrow [x/a]$
Conclusions	$\ominus L_3$
Proof **Schema**	$(L_1) \quad \vdash 0 = 0 \hspace{2.5cm} (=\text{Ref})$ $(L_2) \; \Delta \vdash p_\sigma = 0 \hspace{2cm} (\text{CAS } L_1)$ $(L_3) \; \Delta \vdash \exists x. p = 0 \hspace{1.5cm} (\exists I\ L_2)$

Fig. 4. The `Polynomial-Root` Method

is then used to construct an appropriate substitution for p which is bound to the parameter σ. Because `Polynomial-Root` does not introduce any new lines during the planning process all the instantiations are needed for expansion of the method.

We observe the behavior of `Polynomial-Root` with a concrete example but focus only on the application and expansion of the method and do not elaborate on the actual optimization problem and how it is solved. Suppose we apply `Polynomial-Root` to an open line of the form: $\exists x. (x^3 + x^2 - 5x - 2) = 0$ the method variable p is then bound to $x^3 + x^2 - 5x - 2$ and MAPLE is called with `roots`$(x^3 + x^2 - 5x - 2, x)$. MAPLE returns 2 as the only rational root of p (together with its multiplicity which we can ignore) which is bound to x and finally the substitution σ is constructed as $[x/2]$.

When expanding the method's application, lines L_1 and L_2 are properly substituted and introduced into the proof, as shown on the right. For the expansion of the CAS justification in the second line μCAS is used to perform the step-by-step low-level verification of the result.

$$0 = 0 \hspace{1cm} (=\text{Ref})$$
$$((2^3 + (2^2 - 5 * 2)) - 2) = 0 \hspace{1cm} (\text{CAS})$$
$$\exists x. ((x^3 + (x^2 - 5x)) - 2) = 0 \hspace{1cm} (\exists I)$$

The expansion employs tactics that represent single computational steps on rational numbers. We omit, however, the details of this expansion since it works analogously to the one presented in Sec. 5.1.

6 Discussion

From our current experience, the presented approach is well suited for symbolic computations whose verification is relatively trivial, e.g., where only simple arithmetic needs to be employed. However, the method is not feasible for computations where the verification is as expensive or even more complicated than the computation itself. At least in the latter case it might be more practicable to immediatly specify the computation as a μCAS algorithm. Computations where the verification will be definitely non-trivial are those involving certain uniqueness properties of the result. For instance, when employing MAPLE to compute all roots of a function, it will be a hard task to verify that there exist no more roots than those actually computed. For further discussion of this point refer to [17].

Although we presented our ideas in this paper in the context of proof planning, we strongly believe that the approach could also work in tactical (interactive) theorem proving. One necessary prerequisite will be the existence of an explicit proof object for storing proof steps that contain calculations. These steps can then be verified with the help of the simple $\mu\mathcal{C}\mathcal{A}\mathcal{S}$ algorithm. Even if the proof object does not have the advanced facilities for step-wise expansion of proof steps the verification could be done by transforming $\mu\mathcal{C}\mathcal{A}\mathcal{S}$ output into tactics, and thereby primitive inferences, of the respective system. Those primitive inferences would not necessarily have to be incorporated into the proof object. For instance, one can imagine a tactic such as `Complex-Estimate` in TPS [2]. After successful proof construction it could be verified with the help of $\mu\mathcal{C}\mathcal{A}\mathcal{S}$ and remain as a single step in the proof object. This treatment would also ease automation.

For systems not maintaining explicit proof objects, such as HOL [14] or PVS [21], the approach of [17] would suit best. Here the symbolic computations are verified immediately by tactics build on primitive inferences of HOL. However, this approach directly implements the verification algorithms as tactics in the HOL system as correspondences to MAPLE's computation. Using a separate system for the verification, i.e., the generation of more or less detailed inference chains justifying computations, keeps the DS clean from special algorithms, methods and tactics employed for this task, only. Moreover, both SAPPER and $\mu\mathcal{C}\mathcal{A}\mathcal{S}$ are generic enough to be easily adapted for connection with other systems, whereas tactics can generally not that easily be exported.

7   Conclusion

We have presented an approach for integrating symbolic computations into logical derivations without foregoing the formal correctness requirements for proofs in deduction systems. The main idea underlying the approach is to employ the computational power of full grown CASs, such as MAPLE, during proof planning and thereby ease the derivation of certain witness terms. After a proof has been successfully planned we use a little self-tailored CAS, $\mu\mathcal{C}\mathcal{A}\mathcal{S}$, during the refinement phase of the plan to an actual natural deduction proof in order to verify the correctness of the original computation. This is based on the idea that for some computations the verification of a result is much simpler than the computation itself, and, indeed, for our examples it suffices to use arithmetic.

Although we hinted at how this method could also be successfully employed outside the proof planning context, it is certainly not the *ultima ratio* as a paradigm for integrating deduction and computer algebra. However, since there are currently no CASs available that are either provably correct or give us enough information on their computations we have chosen this pragmatic approach in the ΩMEGA system in order to derive proof plans which in turn could be used to justify the correctness of symbolic computations.

We demonstrated the presented technique with two examples from the domain of limit theorems and optimization problems. We are currently investigating how CAS such as GAP [12] and MAGMA [6] can be employed to proof plan theorems in group theory.

References

[1] A. Adams, H. Gottliebsen, S. Linton, and U. Martin. VSDITLU: a verifiable symbolic definite integral table look-up. In *Proc. of CADE–16, LNAI* 1632, p. 112–126. Springer, 1999.

[2] P. Andrews, M. Bishop, S. Issar, D. Nesmith, F. Pfenning, and H. Xi. TPS: A Theorem Proving System for Classical Type Theory. *J. of Autom. Reasoning*, 16(3):321–353, 1996.

[3] C. Ballarin, K. Homann, and J. Calmet. Theorems and Algorithms: An Interface between Isabelle and Maple. In *Proc. of ISSAC'95*, p. 150–157. ACM Press, 1995.

[4] A. Bauer, E. Clarke, and X. Zhao. Analytica: an Experiment in Combining Theorem Proving and Symbolic Computation. *J. of Autom. Reasoning*, 21(3):295–325, 1998.

[5] A. Bundy. The Use of Explicit Plans to Guide Inductive Proofs. In *Proc. of CADE–9, LNCS* 310. Springer, 1988.

[6] J. Cannon and C. Playout. *Algebraic Programming with Magma.* Springer, 1998.

[7] L. Cheikhrouhou and V. Sorge. $\mathcal{PDS}$ — A Three-Dimensional Data Structure for Proof Plans. In *Proc. of ACIDCA'2000*, 2000.

[8] A. Church. A Formulation of the Simple Theory of Types. *J. of Symbolic Logic*, 5:56–68, 1940.

[9] D. Clément, F. Montagnac, and V. Prunet. Integrated Software Components: a Paradigm for Control Integration. In *Proc. of the Europ. Symp. on Software Development Environments, LNCS* 509. Springer, 1991.

[10] R. Fikes and N. Nilsson. STRIPS: A new approach to the application of theorem proving to problem solving. *Artificial Intelligence*, 2:189–208, 1971.

[11] A. Franke, S. Hess, Ch. Jung, M. Kohlhase, and V. Sorge. Agent-Oriented Integration of Distributed Mathematical Services. *J. of Universal Computer Science*, 5(3):156–187, 1999. Special issue on Integration of Deduction System.

[12] The GAP Group, Aachen, St Andrews. *GAP – Groups, Algorithms, and Programming, Version 4*, 1998. http://www-gap.dcs.st-and.ac.uk/~{}gap.

[13] G. Gentzen. Untersuchungen über das Logische Schließen I und II. *Mathematische Zeitschrift*, 39:176–210, 405–431, 1935.

[14] M. Gordon and T. Melham. *Introduction to HOL.* Cambridge Univ. Press, 1993.

[15] M. Gordon, R. Milner, and Ch. Wadsworth. *Edinburgh LCF: A Mechanized Logic of Computation, LNCS* 78. Springer, 1979.

[16] The ΩMEGA Group. ΩMEGA: Towards a Mathematical Assistant. In *Proc. of CADE–14, LNAI* 1249, p. 252–255. Springer, 1997.

[17] J. Harrison and L. Théry. A Skeptic's Approach to Combining HOL and Maple. *J. of Autom. Reasoning*, 21(3):279–294, 1998.

[18] M. Kerber, M. Kohlhase, and V. Sorge. Integrating Computer Algebra Into Proof Planning. *J. of Autom. Reasoning*, 21(3):327–355, 1998.

[19] E. Melis. The "Limit" Domain. In *Proc. of the Fourth International Conference on Artificial Intelligence in Planning Systems*, p. 199–206, 1998.

[20] E. Melis and V. Sorge. Specialized External Reasoners in Proof Planning. Seki Report SR-00-01, Computer Science Department, Universität des Saarlandes, 2000.

[21] S. Owre, S. Rajan, J. Rushby, N. Shankar, and M. Srivas. PVS: Combining Specification, Proof Checking, and Model Checking. In *Computer-Aided Verification, CAV '96, LNCS* 1102, p. 411–414. Springer, 1996.

[22] D. Redfern. *The Maple Handbook: Maple V Release 5.* Springer, 1998.

[23] V. Sorge. Integration eines Computeralgebrasystems in eine logische Beweisumgebung. Master's thesis, Universität des Saarlandes, November 1996.

Integrating Computer Algebra and Reasoning through the Type System of Aldor

Erik Poll[1] and Simon Thompson[2]

[1] Computing Science Department
University of Nijmegen, The Netherlands
erikpoll@cs.kun.nl
[2] Computing Laboratory,
University of Kent at Canterbury, UK
S.J.Thompson@ukc.ac.uk

Abstract. A number of combinations of reasoning and computer algebra systems have been proposed; in this paper we describe another, namely a way to incorporate a logic in the computer algebra system Axiom. We examine the type system of Aldor – the Axiom Library Compiler – and show that with some modifications we can use the dependent types of the system to model a logic, under the Curry-Howard isomorphism. We give a number of example applications of the logic we construct and explain a prototype implementation of a modified type-checking system written in Haskell.

1 Introduction

Symbolic mathematical – or computer algebra – systems, such as Axiom [13], Maple and Mathematica, are in everyday use by scientists, engineers and indeed mathematicians, because they provide a user with techniques of, say, integration which far exceed those of the person themselves, and make routine many calculations which would have been impossible some years ago. These systems are, moreover, taught as standard tools within many university undergraduate programmes and are used in support of both academic and commercial research.

There are, however, drawbacks to the widespread use of automated support of complex mathematical tasks, which has been widely noted: Fateman [10] gives the graphic example of systems which will assume that $a \neq 0$ on the basis that $a = 0$ has not been established. This can have potentially disastrous consequences for the naive user of the system or indeed, if it occurs within a sufficiently complicated context, *any* user.

Symbolic mathematics systems are also limited by their reliance on algebraic techniques. As Martin [14] remarks, in performing operations of analysis it might be a precondition that a function be continuous; such a property cannot be guaranteed by a computer algebra system alone.

All this makes the combination of computer algebra with theorem proving a topic of considerable interest. Reasoning capabilities can allow a user to track assumptions, and thus to ensure that symbolic computations are *sound*, in contrast to the current situation in many CA systems.

H. Kirchner and C. Ringeissen (Eds.): FroCoS 2000, LNAI 1794, pp. 136–150, 2000.

Reasoning can also *extend* the capability of a CA system. A scenario might involve working with a particular monoid: if during the course of computation it can be shown, for instance, that the monoid is commutative then it is possible to use different, more efficient, simplification algorithms for expressions. The addition of reasoning here has made computation more efficient; in other situations - such as Martin's analysis example mentioned earlier – reasoning can allow computations to proceed where in general this would not be possible.

The literature contains a number of different strategies proposed for combining computer algebra and theorem proving; see, for instance, [4,6,3]. This paper describes another approach: we use the type system of the Axiom computer algebra system [13] to represent a logic, and thus to use the constructions of Axiom to handle the logic and represent proofs and propositions, in the same way as is done in theorem provers based on type theory such as Nuprl [7] or Coq [8].

This paper particularly explores the recent Axiom Library Compiler, Aldor [30], which is unusual among computer algebra systems in being strongly typed, and moreover in having a very powerful type system, including dependent types which are central to our work.

The implementation of dependent types in Aldor is somewhat nonstandard: there is no evaluation within type expressions, so that, for example, 'vectors of length 2+3' are distinct from 'vectors of length 5'; we show how this limits the expressivity of the dependent types. We describe a modification of the Aldor system which allow the types to represent the propositions of a constructive logic, under the Curry-Howard correspondence. We argue that this integrates a logic into the Aldor system, and thus permits a variety of logical extensions to Aldor, including adding pre- and post-conditions to function specifications, axiomatisations to categories of mathematical objects as well as the ability to reason about the objects in Aldor.

The structure of the paper is as follows. Section 2 introduces Aldor and in particular examines its system of types. In Section 3 we examine the issue of type equality in Aldor since it is central to our approach to embedding a logic in Aldor. The section also contains a number of strategies for modifying the Aldor compiler. We show how a logic can be defined in a modified variant of the Aldor system in Section 4 and Section 5 gives some example applications. We conclude with a discussion of related and future work.

2   An Introduction to Aldor

The Axiom Library Compiler, Aldor [30] (known in the past as AXIOM-XL and A$^\sharp$), provides the user with a powerful, general-purpose programming language in which to model the structures of mathematics. Aldor is compiled, in contrast to most computer algebra languages, and so it can provide much more efficient implementations of algorithms than interpreted languages.

The core of Aldor is a functional programming language which provides higher-order functions, generators (which bear a strong relationship to list comprehensions) and other features of modern functional languages like Standard

ML [17] and Haskell [21]. It is also strongly typed, in common with these languages and indeed the majority of modern programming languages. Under this type
discipline any type error – such as adding a character to a boolean operator –
can be caught at compile time rather than at run time. This has two consequent
advantages. First, a whole class of programming errors can be detected prior
to program execution, thus increasing the dependability of the compiled code.
Secondly, it means that it is possible to produce more efficient compiled code
since no run-time type tags on program data need to be maintained to support
type checking at run-time.

Since Aldor is designed with mathematics in mind, its type system is more
complex than those of most programming languages. Mathematicians take a
flexible approach to terminology, with the consequence that often the meaning
of a symbol or phrase is only determined by its context. This requires of a
programming language that symbols can be *overloaded*, and that sometimes
values need to be *coerced* from one type to another: from the integers to floating-
point numbers, for example.

More importantly this flexibility necessitates an entity like the collection of
integers to be seen in various different ways, depending on the context. In the
case of the integers this might be a set of values, a group, an integral domain,
a subset of the real numbers and so forth. To do this, the language allows types
and functions to be collected into *domains*, and the type of a domain, which is
described by a signature, is called a *category*.

Categories can be built on top of other categories, giving a version of inheritance between domains. Categories can also be parametrised by values including
domains; rather than implement a theory of parametric categories, Aldor takes
types to be values just like more traditional values like 23 and the Boolean value
'false'. This has far-reaching consequences for the language.

Current descriptions of Aldor, [30,29], give informal definitions of the type
system. We have given a formal description of the essence of the Aldor type
system in [22]. In the remainder of this section we summarise our approach in
that paper and the conclusions that are drawn there.

2.1 An Overview of the Type System of Aldor

Unusually among languages for computer algebra, but in keeping with the functional school, Aldor is strongly typed. Each declaration of a binding can be
accompanied by a declaration of the type of the value bound, as in the definition

```
a : Integer == 23;
```

The type of an expression can be declared explicitly to resolve any uses of overloaded identifiers. This cannot simply be done by the typing rules, since arbitrary
overloading is allowed, so that, for instance, a single identifier fun may be overloaded to have types Int -> Int, Int -> Bool and Bool -> Int so that neither
the type of the argument nor the type of result expected can disambiguate an
application of fun.

Some 'courtesy' coercions are provided by the system automatically: these convert between multiple values (*à la* LISP), cross products and tuples. It is also possible to make explicit conversions – by means of the `coerce` function – from integers to floating point numbers and so forth.

As mentioned earlier, Aldor treats types as values. In particular, a type such as `Integer` has itself a type. The type of types is called `Type`. Having this type of all types means that the system supports functions over types, such as the identity function over (the type of) types:

```
idType (ty : Type) : Type == ty;
```

and explicit polymorphism, as in the polymorphic identity function which takes two arguments. The first is a type `ty` and the second is a value of that type which is returned as the result.

```
id (ty : Type, x : ty) : ty == x;                              (id)
```

Aldor permits functions to have dependent types, in which the type of a function result depends upon the value of a parameter. An example is the function which sums the values of vectors of integers. This has the type

```
vectorSum : (n:Integer) -> Vector(n) -> Integer
```

in which the result of a function application, say

```
vectorSum(34)
```

has the type `Vector(34) -> Integer` because its argument has the value `34`. In a similar way, when the `id` function of definition (id) is applied, its result type is determined by the type which is passed as its first argument. We discuss this aspect of the language in more detail in Section 2.3.

The system is not fully functional, containing as it does variables which denote storage locations. The presence of updatable variables inside expressions can cause side-effects which make the elucidation of types considerably more difficult. There is a separate question about the role of 'mathematical' variables in equations and the like, and the role that they play in the type system of Aldor.

Categories and domains provide a form of data abstraction and are addressed in more detail in Section 2.5.

The Aldor type system can thus be seen to be highly complex and we shall indeed see that other features such as macros (see Section 2.5) complicate the picture further.

2.2   Formalising the Type System of Aldor

This section outlines the approach we have taken in formalising the type system of Aldor. Our work is described in full in [22]; for reasons of space we can only give a summary here.

The typing relation is formally described by typing judgements of the form

$$\Gamma \vdash t : T$$

which is read 't has the type T in the context Γ'. A context here consists of a list of variable declarations, type definitions and so on. Contexts represent the collection of bindings which are in scope at a point in a program text. Note that t might have more than one type in a given context because of overloading of identifiers in Aldor, and so it would be perfectly legitimate for a well-formed context Γ to imply that $t : T$ and $t : T'$ where T and T' are different types.

Complex typing judgements are derived using deduction rules that codify conditions for a typing judgement to hold. For example,

$$\frac{\Gamma \vdash f : S \text{->} T \qquad \Gamma \vdash s : S}{\Gamma \vdash f(s) : T} \text{ (function elim)}$$

describes the type-correct application of a function. This deductive approach is standard; we have adapted it to handle particular features of Aldor such as overloading, first-class types and categories.

Our discussion in [22] examines the essential features of the full type system of Aldor; in this paper we concentrate on those aspects of the language relevant to our project. These are dependent function and product types; equality between types; and categories and domains, and we look at these in turn now.

2.3   Dependent Types

As we have already seen with the examples of id and vectorSum, the Aldor language contains dependent types. To recap, the function vectorSum defines a sum function for vectors of arbitrary length and has the type

```
vectorSum : (n:Integer) -> Vector(n) -> Integer
```

Similarly one can define a function **append** to join two vectors together

```
append : (n:Integer,m:Integer,Vector(n),Vector(m)) -> Vector(n+m)
```

The typing rule for dependent function elimination modifies the rule (function elim) so that the values of the arguments are *substituted* in the result type, thus

$$\frac{\Gamma \vdash f : (x : S) \text{->} T \qquad \Gamma \vdash s : S}{\Gamma \vdash f(s) : T[x := s]} \text{ (dependent function elim)}$$

Given vectors of length two and three, vec2 and vec3, we can join them thus

```
append(2,3,vec2,vec3) : Vector(2+3)
```

where 2 and 3 have been substituted for n and m respectively.

We would expect to be able to find the sum of this vector by applying vectorSum 5, thus

```
(vectorSum 5) append(2,3,vec2,vec3)
```

but this will fail to type check, since the argument is of type `Vector(2+3)`, which is not equal to the expected type, namely `Vector(5)`. This is because no evaluation takes place in type expressions in Aldor (nor indeed in the earlier version of Axiom). We examine this question in the next section, and in Section 3 we discuss how the Aldor type mechanism can be modified to accommodate a more liberal evaluation strategy within the type checker. Similar remarks apply to dependent product types in which the type of a field can depend on the value of another field.

2.4 Equality of Types in Aldor

When are two types in Aldor equal? The definition of type equality in any programming language is non-trivial, but in the presence of dependent types and types as values it becomes a subtle matter.

Type equality is fundamental to type checking, as can be seen in the rule (function elim): the effect of the rule in a type-checker is to say that the application $f(s)$ is only legitimate if f has type S->T, s has type S', *and the types S and S' are equal.* Non-identical type expressions can denote identical types for a number of reasons.

- A name can be given to a type, as in

  `myInt : Type == Int;`

  and in many situations `myInt` and `Int` will be treated as identical types. [This is often called δ-equality.]
- The bound variables in a type should be irrelevant and Aldor treats them as so. This means that the types

  `vectorSum : (n:Integer) -> Vector(n) -> Integer`
  `vectorSum : (int:Integer) -> Vector(int) -> Integer`

  should be seen as identical. [α-equality]
- Types are values like any other in Aldor, and so can be evaluated. In particular a function over types like `idType` will be used in expressions such as `idType Int`. It would be expected that this would evaluate to `Int` and thus be seen as equivalent. [β-equality]
- In the presence of dependent types, expressions of any type whatsoever can be subexpressions of type expressions, as in `Vector(2+3)`. Equality between these subexpressions can be lifted to types, making `Vector(2+3)` equal to `Vector(5)`. [Value-equality]

Our report on the type system examines the practice of equality in the Aldor system and shows it to be complex. The Aldor system implements α-equality in nearly all situations, but δ-equality is not implemented in a uniform way. Over types neither β-equality nor value-equality is implemented, so that type equality in Aldor is a strong relation, in that it imposes finer distinctions than notions like β- or value-equality.

A rationale for the current definition in Aldor is that it is a simple notion of type equality which is strong enough to implement a weak form of type dependency in which arguments to types are themselves (literal) types which are not used in a computational way. This form of dependency is useful in the module system of Aldor where it can be used to formulate mathematical notions like 'the ring of polynomials in one variable over a field F' where the field F is a parameter of the type.

Our approach to integrating reasoning into Aldor requires a weaker notion of type equality, which we explore in Section 3.

2.5   Categories and Domains

Aldor is designed to be a system in which to represent and manipulate mathematical objects of various kinds, and support for this is given by the Aldor type system. One can specify what it is to be a monoid, say, by defining the Category[1] called Monoid, thus

```
Monoid : Category == BasicType with {                    (Mon)
    * : (%,%) -> %;
    1 : %; }
```

This states that for a structure over a type '%' to be a monoid it has to supply two bindings; in other words a Category describes a signature. The first name in the signature is '*' and is a binary operation over the type '%'; the second is an element of '%'.

In fact we have stated slightly more than this, as Monoid *extends* the category BasicType which requires that the underlying type carries an equality operation.

```
BasicType : Category == with {
    = : (%,%) -> Boolean; }
```

We should observe that this Monoid category does not impose any constraints on bindings to '*' and '1': we shall revisit this example in Section 5.2 below.

Implementations of a category are abstract data types which are known in Aldor as domains, and are defined as was the value a at the start of Section 2.1, e.g.

```
IntegerAdditiveMonoid : Monoid == add {
    Rep == Integer;
    (x:%) * (y:%) : % == per((rep x) + (rep y));
    1 : %          == per 0; }
```

The category of the object being defined – Monoid – is the type of the domain which we are defining, IntegerAdditiveMonoid. The definition identifies a representation type, Rep, and also uses the conversion functions rep and per which have the types

---

[1] There is little relation between Aldor's notion of category and the notion from category theory!

```
rep : % -> Rep                    per : Rep -> %
```

The constructs `Rep`, `rep` and `per` are implemented using the macro mechanism of Aldor, and so are eliminated before type checking. In our report [22] we show how definitions of domains can be type checked without macro expansion, which allows, for instance, more accurate error diagnosis.

Categories can also be parametric, and depend upon value or type parameters; an example is the ring of polynomials over a given field mentioned earlier.

2.6   Conclusion

This section has given a brief overview of Aldor and its type system. It has shown that the notion of type equality in Aldor is a strong one, which makes distinctions between types which could naturally be considered equivalent. This is especially relevant when looking at the effect of type equality on the system of dependent types. In the sections to come we show how a logic can be incorporated into Aldor by modifying the notion of type equality in Aldor.

3   Modifying Type Equality in Aldor

Section 2.4 describes type equality in Aldor and argues that it is a strong notion which distinguishes between type terms which can naturally be identified. In this section we examine various ways of modifying type equality including the way we have chosen to do this in our prototype implementation.

3.1   Using the Existing System

It is possible to use the existing Aldor system to mimic a different – weaker – type equality by explicitly casting values to new types, using the `pretend` function of Aldor.[2] This effectively sidesteps the type checker by asserting the type of an expression which is accepted by the type checker without verification.

For instance, the vector example of Section 2.3 can be made to type check in Aldor by annotating it thus

```
(vectorSum 5) (append(2,3,vec2,vec3) pretend Vector(5))
```

or thus

```
((vectorSum 5) pretend (Vector(2+3) -> Integer)) append(2,3,vec2,vec3)
```

This achieves a result, but at some cost. Wherever we expect to need some degree of evaluation, that has to be shadowed by a type cast; these casts are also potentially completely unsafe.

---

[2] The `pretend` function is used in the definition of `rep` and `per` in the current version of Aldor; a more secure mechanism would be preferable.

3.2   Coercion Functions

Another possibility is to suggest that the current mechanism for coercions in Aldor is modified to include coercion functions which would provide conversion between type pairs such as Vector(2+3) and Vector(5), extending the coercion mechanism already present in Aldor. This suggestion could be implemented but we envisage two difficulties with it.

- In all but the simplest of situations we will need to supply uniformly-defined *families* of coercions rather than single coercions. This will substantially complicate an already complex mechanism.
- Coercions are currently not applied transitively: the effect of this is to allow us to model single steps of evaluation but not to take their transitive closure.

Putting these two facts together force us to conclude that effectively mimicking the evaluation process as coercions is not a reasonable solution to the problem of modifying type checking.

3.3   Adding Full Evaluation

To deal with the problem of unevaluated subexpressions in types, we have implemented a prototype version of Aldor using Haskell [23]. In this implementation all type expressions are fully evaluated to their *normal form* as a part of the process of type checking. To give an example, the rule (function elim) will be interpreted thus:

$f(s)$ is well-formed if and only if f has type S->T, s has type S', *and the normal forms of S and S' are equal modulo α-equality.*

The effect of this modification is to force the type checker to perform evaluation of expressions at compile time. Clearly this can cause the type checker to diverge in general, since in, for instance, an application of the form vectorSum(e) an arbitrary expression e:Nat will have to be evaluated.

More details of the prototype implementation of Aldor in Haskell are given in the technical report [23].

3.4   Controlling Full Evaluation

A number of existing type systems, Haskell among them, have undecidable type systems [12] which can diverge at compile time. In practice this is not usually a problem as the pathologies lie outside the 'useful' part of the type system. This may well be the case with Aldor also, but it is also possible to design a subset of the language, Aldor--, whose type system is better behaved.

There is considerable current interest in defining *terminating* systems of recursion [27,16]. A system like this is sufficient to guarantee the termination of expressions chosen for evaluation as part of the type checking process. The main effect of the restricted system is to force recursion to be structural (in a general sense); in practice this is acceptable, particularly in the subset of the language used within type expressions.

4   Logic within Aldor

In this section we discuss the Curry-Howard isomorphism between propositions and types, and show that it allows us to embed a logic within the Aldor type system, if dependent types are implemented to allow evaluation within type contexts.

4.1   The Curry-Howard Correspondence

Under the Curry-Howard correspondence, logical propositions can be seen as types, and proofs can be seen as members of these types. Accounts of constructive type theories can be found in notes by Martin-Löf [15] amongst others [19,26]. Central to this correspondence are dependent types, which allow the representation of predicates and quantification.

Central to the correspondence is the idea that a constructive proof of a proposition gives enough evidence to witness the fact that the proposition stands.

- A proof of a conjunction $A \wedge B$ has to prove each half of the proposition, so has to provide witnessing information for each conjunct; this corresponds precisely to a product type, in Aldor notation written as (A, B), members of which consist of pairs of elements, one from each of the constituent types.
- A proof of an implication $A \Rightarrow B$ is a proof transformer: it transforms proofs of A into proofs of B; in other words it is a function from type A to type B, i.e. a function of type A->B.
- In a similar way a proof of a universal statement $(\forall x : A)B(x)$ is a function taking an element a of A into a proof of $B(a)$; in other words it is an element of the *dependent* function type (x:A) -> B.
- Similar interpretations can be given to the other propositional operators and the existential quantifier.

We can summarise the correspondence in a table

Programming		**Logic**
Type		Formula
Program		Proof
Product/record type	(...,...)	Conjunction
Sum/union type	\/	Disjunction
Function type	->	Implication
Dependent function type	(x:A) -> B(x)	Universal quantifier
Dependent product type	(x:A,B(x))	Existential quantifier
Empty type	Exit	Contradictory proposition
One element type	Triv	True proposition
...		...

Predicates (that is dependent types) can be constructed using the constructs of a programming language. A direct approach is to give an explicit (primitive recursive) definition of the type, which in Aldor might take the form

```
lessThan(n:Nat,m:Nat) : Type ==                    (lessThan)
    if m=0 then        Exit
    else (if n=0 then Triv
                else lessThan(n-1,m-1));
```

The equality predicate can be implemented by means of a primitive operation which compares the normal forms of the two expressions in question.

4.2 A Logic within Aldor

We need to examine whether the outline given in Section 4.1 amounts to a proper embedding of a logic within Aldor. We shall see that it places certain requirements on the definition and the system.

Most importantly, for a definition of the form (lessThan) to work properly as a definition of a predicate we need an application like lessThan(9,3) to be reduced to Exit, hence we need to have evaluation of type expressions. This is a modification of Aldor which we are currently investigating, as outlined in Section 2.3. In the case of (lessThan) the evaluation can be limited, since the scheme used is recognisable as terminating by, for instance, the algorithm of [16].

The restriction to terminating (well-founded) recursions is also necessary for consistency of the logic. For the logic to be consistent, we need to require that not all types are inhabited, which is clearly related to the power of the recursion schemes allowed in Aldor. One approach is to expect users to check this for themselves: this has a long history, beginning with Hoare's axiomatisation of the function in Pascal, but we would expect this to be supported with some automated checking of termination, which ensures that partially or totally undefined proofs are not permitted.

Consistency also depends on the strength of the type system itself; a sufficiently powerful type system will be inconsistent as shown by Girard's paradox [11].

5   Applications of an Integrated Logic

Having identified a logic within Aldor, how can it be used? There are various applications possible; we outline some here and for others one can refer to the number of implementations of type theories which already exist, including Nuprl [7] and Coq [8].

5.1 Pre- and Post-Conditions

A more expressive type system allows programmers to give more accurate types to common functions, such as the function which indexes the elements of a list.

```
index : (l:List(t))(n:Nat)((n < length l) -> t)
```

An application of index has *three* arguments: a list 1 and a natural number n
– as for the usual index function – and a third argument of type (n < length
1), that is a *proof* that n is a legitimate index for the list in question. This
extra argument becomes a *proof obligation* which must be discharged when the
function is applied to elements 1 and n.

In a similar vein, it is possible to incorporate post-conditions into types, so
that a sorting algorithm over lists might have the type

```
sort : ((1:List(t))(List(t),Sorted(1))
```

and so return a sorted list together with a proof that the list is Sorted.

5.2   Adding Axioms to the Categories of Aldor

In definition (Mon), Section 2.5, we gave the category of monoids, Monoid, which
introduces two operation symbols, * and 1. A monoid consists not only of two
operations, but of operations with properties. We can ensure these properties
hold by extending the definition of the category to include three extra components which are proofs that 1 is a left and right unit for * and that * is
associative, where we assume that '≡' is the equality predicate:

```
Monoid : Category == BasicType with {                              (MonL)
    * : (%,%) -> %;
    1 : %;

    leftUnit   : (g:%) -> (1*g ≡ g);
    rightUnit  : (g:%) -> (g*1 ≡ g);
    assoc      : (g:%,h:%,j:%) -> ( g*(h*j) ≡ (g*h)*j );
}
```

For example, the declaration of leftUnit has the logical interpretation that
leftUnit is a proof of the statement 'for all g in the monoid (%), 1*g is equal
to g'.

The equality predicate is implemented as follows: the type a ≡ b contains
a value if and only if a and b have the same normal form. The extension operation (i.e. the with in the definition above) over categories will lift to become
operations of extension over the extended 'logical' categories such as (MonL).

5.3   Commutative Monoids

In the current library for Axiom it is not possible to distinguish between general
monoids and commutative monoids: both have the same signature. With logical
properties it is possible to distinguish the two:

```
CommutativeMonoid : Category == Monoid with {
comm : (g:%,h:%) -> ( g*h ≡ h*g );
}
```

To be a member of this category, a domain needs to supply an extra piece of evidence, namely that the multiplication is commutative; with this evidence the structure can be treated in a different way than if it were only known to be a monoid. This process of discovery of properties of an mathematical structure corresponds exactly to a mathematician's experience. Initially a structure might be seen as a general monoid, and only after considerable work is it shown to be commutative; this proof gives entry to the new domain, and thus allows it to be handled using new approaches and algorithms.

5.4   Different Degrees of Rigour

One can interpret the obligations given in Sections 5.1 and 5.2 with differing degrees of rigour. Using the `pretend` function we can conjure up proofs of the logical requirements of (MonL); even in this case they appear as important documentation of requirements, and they are related to the lightweight formal methods of [9].

Alternatively we can build fully-fledged proofs as in the numerous implementations of constructive type theories mentioned above, or we can indeed adopt an intermediate position of proving properties seen as 'crucial' while asserting the validity of others.

6   Conclusion

We have described a new way to combine – or rather, to integrate – computer algebra and theorem proving. Our approach is similar to [3] and [4] in that theorem proving capabilities are incorporated in a computer algebra system. (In the classification of possible combinations of computer algebra and theorem proving of [6], all these are instance of the "subpackage" approach.) But the way in which we do this is completely different: we exploit the expressiveness of the type system of Aldor, using the Curry-Howard isomorphism that also provides the basis of theorem provers based on type theory such as Nuprl [7] or Coq [8]. This provides a logic as part of the computer algebra system. Also, having the same basis as existing theorem provers such as the ones mentioned above makes it easier to interface with them.

So far we have worked on a formal description of the core of the Aldor type system [22], and on a pilot implementation of a typechecker for Aldor which does evaluation in types which can be used as a logic [23]. This pilot forms the model for modifications to the Aldor system itself, as well as giving a mechanism for interfacing Aldor with other systems like the theorem prover Coq, complementary to recent work on formalising the Aldor system within Coq [1]. The logic is being used in a mathematical case study of symbolic asymptotics [25].

It is interesting to see a convergence of interests in type systems from a number of points of view, namely

- computer algebra,
- type theory and theorem provers based on type theory,
- functional programming.

For instance, there seem to be many similarities between structuring mechanisms used in these different fields: [5] argues for functors in the sense of the programming language ML as the right tool for structuring mathematical theories in Mathematica, and [24] notes similarities between the type system of Aldor, existential types [18], and Haskell classes [28]. More closely related to our approach here, it is interesting to note that constructive type theorists have added inductive types [20], giving their systems a more functional flavour, while functional programmers are showing an interest in dependent types [2] and languages without non-termination [27]. We see our work as part of that convergence, bringing type-theoretic ideas together with computer algebra systems, and thus providing a bridge between symbolic mathematics and theorem proving.

Acknowledgements. We are grateful to Stephen Watt of the University of Western Ontario and to Ursula Martin and her research group at the University of St Andrews for feedback on these ideas. We would also like to thank NAG for granting us access to the Aldor compiler, and in particular to Mike Dewar for his help in facilitating this. We are indebted to Dominique Duval who first introduced us to the type system of Aldor, and to EPSRC for supporting her visit to UKC under the MathFIT programme. Finally we are grateful to Thérèse Hardin of LIP6, Paris, for her comments on this work.

References

1. Guillaume Alexandre. *De* ALDOR *à Zermelo*. PhD thesis, Université Paris VI, 1998.
2. Lennart Augustsson. Cayenne – a language with dependent types. ACM Press, 1998.
3. Andrej Bauer, Edmund Clarke, and Xudong Zhao. Analytica - an experiment in combining theorem proving and symbolic computation. In *AISMC-3*, volume 1138 of *LNCS*. Springer, 1996.
4. Bruno Buchberger. Symbolic Computation: Computer Algebra and Logic. In F. Baader and K.U. Schulz, editors, *Frontiers of Combining Systems*. Kluwer, 1996.
5. Bruno Buchberger, Tudor Jebelean, Franz Kriftner, Mircea Marin, Elena Tomuta, and Daniela Vasaru. A survey of the Theorema project. In *Proceedings of ISSAC'97 (International Symposium on Symbolic and Algebraic Computation)*, pages 384–391. ACM, 1997.
6. Jaques Calmet and Karsten Homann. Classification of communication and cooperation mechanisms for logical and symbolic computation systems. In *FroCos'96*. Kluwer, 1996.
7. Robert L. Constable et al. *Implementing Mathematics with the Nuprl Proof Development System*. Prentice-Hall Inc., 1986.
8. C. Cornes et al. The Coq proof assistant reference manual, version 5.10. Rapport technique RT-0177, INRIA, 1995.

9. Martin Dunstan and Tom Kelsey. Lightweight Formal Methods for Computer Algebra Systems. ISSAC'98, 1998.
10. Richard Fateman. Why computer algebra systems can't solve simple equations. *ACM SIGSAM Bulletin*, 30, 1996.
11. Jean-Yves Girard. Intérpretation fonctionelle et élimination des coupures dans l'arithmétique d'ordre supérieure. Thèse d'Etat, Université Paris VII, 1972.
12. Fritz Henglein. Type Inference with Polymorphic Recursion. *ACM Transactions on Programming Languages and Systems*, 15, 1993.
13. Richard D. Jenks and Robert S. Sutor. *Axiom: The Scientific Computation System*. Springer, 1992.
14. Ursula Martin. Computers, reasoning and mathematical practice. In Helmut Schwichtenberg, editor, *Computational Logic, Marktoberdorf 1997*. Springer, 1998.
15. Per Martin-Löf. *Intuitionistic Type Theory*. Bibliopolis, Naples, 1984. Based on a set of notes taken by Giovanni Sambin of a series of lectures given in Padova, June 1980.
16. D. McAllester and K. Arkondas. Walther recursion. In M.A. Robbie and J.K. Slaney, editors, *CADE 13*. Springer, 1996.
17. Robin Milner, Mads Tofte, and Robert Harper. *The Definition of Standard ML*. MIT Press, 1990.
18. John C. Mitchell and Gordon D. Plotkin. Abstract types have existential type. *ACM Trans. on Prog. Lang. and Syst.*, 10(3):470–502, 1988.
19. Bengt Nordström, Kent Petersson, and Jan M. Smith. *Programming in Martin-Löf's Type Theory — An Introduction*. Oxford University Press, 1990.
20. Christine Paulin-Mohring. Inductive definitions in the system Coq. In *TLCA*, volume 664 of *LNCS*. Springer, 1993.
21. John Peterson and Kevin Hammond, editors. *Report on the Programming Language Haskell, Version 1.4*. htttp://www.haskell.org/report/, 1997.
22. Erik Poll and Simon Thompson. The Type System of Aldor. Technical Report 11-99, Computing Laboratory, University of Kent at Canterbury, 1999.
23. Chris Ryder and Simon Thompson. Aldor meets Haskell. Technical Report 15-99, Computing Laboratory, University of Kent at Canterbury, 1999.
24. Philip S. Santas. A type system for computer algebra. *Journal of Symbolic Computation*, 19, 1995.
25. J.R. Shackell. Symbolic asymptotics and the calculation of limits. *Journal of Analysis*, 3:189–204, 1995. Volume commemorating Maurice Blambert.
26. Simon Thompson. *Type Theory and Functional Programming*. Addison Wesley, 1991.
27. David Turner. Elementary strong functional programming. In Pieter Hartel and Rinus Plasmeijer, editors, *Functional programming languages in education (FPLE)*, *LNCS 1022*. Springer-Verlag, Heidelberg, 1995.
28. Philip Wadler and Stephen Blott. Making *ad hoc* polymorphism less *ad hoc*. In *Proceedings of the 16th ACM Symposium on Principles of Programming Languages*. ACM Press, 1989.
29. Stephen M. Watt et al. A First Report on the $A^\#$ Compiler. In *ISSAC 94*. ACM Press, 1994.
30. Stephen M. Watt et al. *AXIOM: Library Compiler User Guide*. NAG Ltd., 1995.

Combinations of Model Checking and Theorem Proving*

Tomás E. Uribe

Computer Science Department,
Stanford University, Stanford, CA. 94305-9045
uribe@cs.stanford.edu

Abstract. The two main approaches to the formal verification of reactive systems are based, respectively, on model checking (algorithmic
verification) and theorem proving (deductive verification). These two approaches have complementary strengths and weaknesses, and their combination promises to enhance the capabilities of each. This paper surveys
a number of methods for doing so. As is often the case, the combinations
can be classified according to how tightly the different components are
integrated, their range of application, and their degree of automation.

1 Introduction

Formal verification is the task of proving mathematical properties of mathematical models of systems. *Reactive systems* are a general model for systems
that have an ongoing interaction with their environment. Such systems do not
necessarily terminate, so their computations are modeled as infinite sequences
of states and their properties specified using *temporal logic* [48]. The verification problem is that of determining if a given reactive system satisfies a given
temporal property.

Reactive systems cover a wide range of hardware and software artifacts, including concurrent and distributed systems, which can be particularly difficult
to design and debug. (Sequential programs and programs that terminate are a
special case of this general model.) Reactive systems can be classified according
to their number of possible states. For *finite-state systems*, the states are given
by a finite number of finite-state variables. This includes, in particular, hardware systems with a fixed number of components. *Infinite-state systems* feature
variables with unbounded domains, typically found in software systems, such as
integers, lists, trees, and other datatypes. (Note that in both cases the computations are infinite sequences of states.)

The verification of temporal properties for finite-state systems is decidable:
model checking algorithms can automatically decide if a temporal property holds

* This research was supported in part by the National Science Foundation under grant
CCR-98-04100, by the Defense Advanced Research Projects Agency under contract
NAG2-892, by the Army under grants DAAH04-96-1-0122 and DAAG55-98-1-0471,
and by the Army under contract DABT63-96-C-0096 (DARPA).

H. Kirchner and C. Ringeissen (Eds.): FroCoS 2000, LNAI 1794, pp. 151–170, 2000.
© Springer-Verlag Berlin Heidelberg 2000

for a finite-state system. Furthermore, they can produce a *counterexample computation* when the property does not hold, which can be very valuable in determining the corresponding error in the system being verified or in its specification. However, model checking suffers from the *state explosion problem*, where the number of states to be explored grows exponentially in the size of the system description, particularly as the number of concurrent processes grows.

The verification problem for general infinite-state systems is undecidable, and finite-state model checking techniques are not directly applicable. (We will see, however, that particular decidable classes of infinite-state systems can be model checked using specialized tools.) First-order logic is a convenient language for expressing relationships over the unbounded data structures that make systems infinite-state. Therefore, it is natural to use theorem proving tools to reason formally about such data and relationships. *Deductive verification*, based on general-purpose theorem proving, applies to a wide class of finite- and infinite-state reactive systems. It provides *relatively complete* proof systems, which can prove any temporal property that indeed holds over the given system, provided the theorem proving tools used are expressive and powerful enough [42]. Unfortunately, if the property fails to hold, deductive methods normally do not give much useful feedback, and the user must try to determine whether the fault lies with the system and property being verified or with the failed proof.

Table 1. Deductive vs. algorithmic verification

	Algorithmic Methods	Deductive Methods	Combination
Automatic / decidable?	yes	no	sometimes
Generates counterexamples?	yes	no	sometimes
Handles general infinite-state systems?	no	yes	yes

The strengths and weaknesses of model checking and deductive verification, as discussed above, are summarized in Table 1. Given their complementary nature, we would like to combine the two methods in such a way that the desirable features of each are retained, while minimizing their shortcomings. Thus, combinations of model checking and theorem proving usually aim to achieve one or more of the following goals:

- More automatic (or, less interactive) verification of infinite-state systems for which model checking cannot be directly applied.
- Verifying finite-state systems that are larger than what stand-alone model checkers can handle.
- Conversely, verifying infinite-state systems with control structures that are too large or complex to check with purely deductive means.

- Generating counterexamples (that is, falsifying properties) for infinite-state systems for which classical deductive methods can only produce proofs.
- Formalizing, and sometimes automating, verification steps that were previously done in an informal or manual way.

Outline: Section 2 presents background material, including the basic model for reactive systems. We then describe the main components of verification systems based on model checking (Section 3) and theorem proving (Section 4). In Section 5 we describe abstraction and invariant generation, which are important links between the two. This defines the basic components whose combination we discuss in the remaining sections.

In Section 6, we describe combination methods that use the components as "black boxes," that is, which do not require the modification of their internal workings. In Section 7, we present combination approaches that require a tighter integration, resulting in new "hybrid" formalisms.

References to combination methods and related work appear throughout the paper, which tries to present a general overview of the subject. However, since much of the work on formal verification during the last decade is related to the subject of this paper, the bibliography can only describe a subset of the work in the field. Furthermore, this paper is based on [58], which presents the author's own view on the subject, and thus carries all the biases which that particular proposal may have. For this, we apologize in advance.

2   Preliminaries

2.1   Kripke Structures and Fair Transition Systems

A *reactive system* $\mathcal{S} : \langle \Sigma, \Theta, R \rangle$ is given by a set of *states* Σ, a set of *initial states* $\Theta \subseteq \Sigma$, and a *transition relation* $R \subseteq \Sigma \times \Sigma$. If $\langle s_1, s_2 \rangle \in R$, the system can move from s_1 to s_2. A system $\mathcal{S}$ can be identified with the corresponding *Kripke structure*, or *state-space*. This is the directed graph whose vertices are the elements of Σ and whose edges connect each state to its successor states.

If Σ is finite, $\mathcal{S}$ is said to be *finite-state*. Describing large or infinite state-spaces explicitly is not feasible. Therefore, the state-space is *implicitly* represented in some other form: a hardware description, a program, or an ω-automaton.

Fair transition systems are a convenient formalism for specifying both finite- and infinite-state reactive systems [44]. They are "low-level" in the sense that many other formalisms can be translated or compiled into them.

The state-space of the system is determined by a set of *system variables* $\mathcal{V}$, where each variable has a given domain (e.g., booleans, integers, recursive datatypes, or reals). The representation relies on an *assertion language*, usually based on first-order logic, to represent sets of states.

Definition 1 (Assertion). *A first-order formula whose free variables are a subset of $\mathcal{V}$ is an* assertion, *and represents the set of states that satisfy it. For an assertion ϕ, we say that $s \in \Sigma$ is a ϕ-state if $s \models \phi$, that is, ϕ holds given the values of $\mathcal{V}$ at the state s.*

In practice, an assertion language other than first-order logic can be used. The basic requirements are the ability to represent predicates and relations, and automated support for validity and satisfiability checking (which need not be complete). Examples of other suitable assertion languages include *ordered binary decision diagrams* (OBDD's) [12] and their variants, for finite-state systems (see Section 3), and the abstract domains used in invariant generation (see Section 5.1).

The initial condition is now expressed as an assertion, characterizing the set of possible system states at the start of a computation. The transition relation R is described as a set of *transitions* $\mathcal{T}$. Each transition $\tau \in \mathcal{T}$ is described by its *transition relation* $\tau(\mathcal{V}, \mathcal{V}')$, a first-order formula over the set of system variables $\mathcal{V}$ and a *primed set* $\mathcal{V}'$, indicating their values at the next state.

Definition 2 (Transition system). *A transition system* $\mathcal{S} : \langle \mathcal{V}, \Theta, \mathcal{T} \rangle$ *is given by a set of system variables* $\mathcal{V}$, *an initial condition* Θ, *expressed as an assertion over* $\mathcal{V}$, *and a set of transitions* $\mathcal{T}$, *each an assertion over* $(\mathcal{V}, \mathcal{V}')$.

In the associated Kripke structure, each state in the state-space Σ *is a possible valuation of* $\mathcal{V}$. *We write* $s \models \varphi$ *if assertion* φ *holds at state* s, *and say that* s *is a* φ-state. *A state* s *is initial if* $s \models \Theta$. *There is an edge from* s_1 *to* s_2 *if* $\langle s_1, s_2 \rangle$ *satisfy* τ *for some* $\tau \in \mathcal{T}$.

The Kripke structure is also called the *state transition graph* for $\mathcal{S}$. Note that if the domain of a system variable is infinite, the state-space is infinite as well, even though the *reachable state-space*, the set of states that can be reached from Θ, may be finite. We can thus distinguish between *syntactically* finite-state systems and *semantically* finite-state ones. Establishing whether a system is semantically finite-state may not be immediately obvious (and is in fact undecidable), so we prefer the syntactic characterization.

The global transition relation is the disjunction of the individual transition relations: $R(s_1, s_2)$ iff $\tau(s_1, s_2)$ holds for some $\tau \in \mathcal{T}$. For assertions ϕ and ψ and transition τ, we write

$$\{\phi\}\, \tau\, \{\psi\} \stackrel{\text{def}}{=} (\phi(V) \wedge \tau(\mathcal{V}, \mathcal{V}')) \to \psi(\mathcal{V}') \ .$$

This is the *verification condition* that states that every τ-successor of a ϕ-state must be a ψ-state.

A *run* of $\mathcal{S}$ is an infinite path through the Kripke structure that starts at an initial state, i.e., a sequence of states $(s_0, s_1, \ldots)$ where $s_0 \in \Theta$ and $R(s_i, s_{i+1})$ for all $i \geq 0$. If $\tau(s_i, s_{i+1})$ holds, then we say that transition τ is *taken* at s_i. A transition is *enabled* if it can be taken at a given state.

Fairness and Computations: Our computational model represents concurrency by *interleaving*: at each step of a computation, a single action or transition is executed [44]. The transitions from different processes are combined in all possible ways to form the set of computations of the system. *Fairness* expresses the constraint that certain actions cannot be forever prevented from occurring— that is, that they do have a fair chance of being taken. Describing R as a set of transition relations is convenient for modeling fairness:

Definition 3 (Fair transition system). *A* fair transition system (FTS) *is one where each transition is marked as* just *or* compassionate. *A* just *(or* weakly fair*) transition cannot be continually enabled without ever being taken; a* compassionate *(or* strongly fair*) transition cannot be enabled infinitely often but taken only finitely many times. A* computation *is a run that satisfies these fairness requirements (if any exist).*

To ensure that R is total on Σ, so that sequences of states can always be extended to infinite sequences, we assume an *idling transition*, with transition relation $\mathcal{V} = \mathcal{V}'$. The set of all computations of a system $\mathcal{S}$ is written $\mathcal{L}(\mathcal{S})$, a language of infinite strings whose alphabet is the set of states of $\mathcal{S}$.

2.2   Temporal Logic

We use temporal logic to specify properties of reactive systems [48,44]. *Linear-time temporal logic* (LTL) describes sets of sequences of states, and can thus capture *universal properties* of systems, which are meant to hold for all computations. However, LTL ignores the branching structure of the system's state-space, and thus cannot express *existential properties*, which assert the existence of particular kinds of computations. The logic CTL* includes both the branching-time *computation tree logic* (CTL) and LTL, and is strictly more expressive than both. Due to space limitations, we refer the reader to [44,16] for the corresponding definitions.

A temporal formula is $\mathcal{S}$-*valid* if it holds at all the initial states of the kripke structures of $\mathcal{S}$, considering only the fair runs of the system. For LTL, a convenient definition of $\mathcal{S}$-validity can be formulated in terms of the set $\mathcal{L}(\varphi)$ of *models* of φ, which is the set of all infinite sequences that satisfy φ:

Proposition 1 (LTL system validity). $\mathcal{S} \models \varphi$ *for an LTL formula φ if and only if all the computations of $\mathcal{S}$ are models of φ, that is, $\mathcal{L}(\mathcal{S}) \subseteq \mathcal{L}(\varphi)$.*

3   Finite-State Model Checking

Given a reactive system $\mathcal{S}$ and a temporal property φ, the verification problem is to establish whether $\mathcal{S} \models \varphi$. For finite-state systems, *model checking* [14,50] answers this question by a systematic exploration of the state-space of $\mathcal{S}$, based on the observation that checking that a formula is true in a particular model is generally easier than checking that it is true in all models; the Kripke structure of $\mathcal{S}$ is the particular model in question, and φ is the formula being checked.

Model checkers, such as SPIN [32], SMV [45], Murφ [24], and those in STeP [7], take as input what is essentially a finite-state fair transition system and a temporal formula in some subset of CTL*, and automatically check that the system satisfies the property. (See [16] for a recent and comprehensive introduction to model checking.)

The complexity of model checking depends on the size of the formula being checked (linear for CTL and exponential for LTL and CTL*) and the size of the system state-space (linear for all three logics). While the temporal formulas of interest are usually small, the size of the state-space can grow exponentially in the size of its description, e.g., as a circuit, program, or fair transition system. This is known as the *state explosion problem*, which limits the practical application of model checking tools.

Symbolic Model Checking: Model checking techniques that construct and explore states of the system, one at a time, are called *explicit-state*. In contrast, *symbolic model checking* combats the state-explosion problem by using specialized formalisms to represent sets of states. Ordered Binary Decision Diagrams (OBDD's) [12] are an efficient data structure for representing boolean functions and relations. They can be used to represent the transition relation of finite-state systems, as well as subsets of the systems' state-space. The efficient algorithms for manipulating OBDD's can be used to compute predicate transformations, such as pre- and post-condition operations, over the transition relation and large, symbolically represented sets of states [45].

Symbolic model checking extends the size of finite-state systems that can be analyzed, and is particularly successful for hardware systems. However, it is still restricted to finite-state systems of fixed size. The size of the OBDD representation can grow exponentially in the number of boolean variables, leading to what can be called the *OBDD explosion problem*, where the model checker runs out of memory before the user runs out of time. In these cases, efficient explicit-state model checkers such as Murφ [24] are preferred.

We arrive now at our first general combination scheme, which we can informally call "SMC(X);" here, symbolic model checking is parameterized by the constraint language used to describe and manipulate sets of states. For instance, extensions of OBDD's are used to move from bit-level to word-level representations [15]; more expressive assertions are used in [36]. Similarly, finite-state *bounded model checking* [5] abandons BDD's and relies instead on the "black-box" use of a propositional validity checker. This method can find counterexamples in cases for which BDD-based symbolic model checking fails.

4   Deductive Verification

Figure 1 presents the *general invariance rule*, G-INV, which proves the $\mathcal{S}$-validity of formulas of the form $\Box p$ for an assertion p [44]. The premises of the rule are first-order verification conditions. If they are valid, the temporal conclusion must hold for the system $\mathcal{S}$.

An assertion is *inductive* if it is preserved by all the system transitions and holds at all initial states. The invariance rule relies on finding an inductive auxiliary assertion φ that *strengthens* p, that is, φ implies p.

The soundness of the invariance rule is clear: if ϕ holds initially and is preserved by all transitions, it will hold for every reachable state of $\mathcal{S}$. If p is implied by φ, then p will also hold for all reachable states. Rule G-INV is also *relatively complete*: if p is an invariant of $\mathcal{S}$, then the strengthened assertion φ always

$$
\begin{array}{l}
\text{For assertions } \varphi \text{ and } p, \\
\text{I1. } \Theta \;\rightarrow\; \varphi \\
\text{I2. } \{\varphi\}\; \tau\; \{\varphi\} \text{ for each } \tau \in \mathcal{T} \\
\text{I3. } \varphi \;\rightarrow\; p \\
\hline
\quad\quad \mathcal{S} \models \Box p
\end{array}
$$

Fig. 1. General invariance rule G-INV

exists [44]. Assuming that we have a complete system for proving valid assertions, then we can prove the $\mathcal{S}$-validity of any $\mathcal{S}$-valid temporal property. Note, however, that proving invariants is undecidable for general infinite-state systems, and finding a suitable φ can be non-trivial.

Other verification rules can be used to verify different classes of temporal formulas, ranging from safety to progress properties. Together, these rules are also relatively complete and yield a direct proof of any $\mathcal{S}$-valid temporal property [42]. However, they may require substantial user guidance to succeed, and do not produce counterexample computations when the property fails.

When we require that a premise of a rule or a verification condition be valid, we mean for it to be $\mathcal{S}$-valid; therefore, verification conditions can be established with respect to invariants of the system, which can be previously proved or generated automatically. In general, axioms and lemmas about the system or the domain of computation can also be used. As we will see, this simple observation is an important basis for combined verification techniques.

4.1 Decision Procedures (and Their Combination)

Most verification conditions refer to particular theories that describe the domain of computation, such as linear arithmetic, lists, arrays and other data types. *Decision procedures* provide specialized and efficient validity checking for particular theories. For instance, equality need not be axiomatized, but its consequences can be efficiently derived by congruence closure. Similarly, specialized methods can efficiently reason about integers, lists, bit vectors and other datatypes frequently used in system descriptions.

Thus, using appropriate decision procedures can be understood as specializing the assertion language to the particular domains of computation used by the system being verified. However, this domain of computation is often a *mixed* one, resulting in verification conditions that do not fall in the decidable range of any single decision procedure. Thus, combination problems in automated deduction are highly relevant to the verification task. The challenge is to integrate existing decision procedures for the different decidable fragments and their combination into theorem-proving methods that can be effectively used in verification. See Bjørner [6] for examples of such combinations.

4.2 Validity Checking and First-Order Reasoning

Decision procedures usually operate only at the *ground* level, where no quantification is allowed. This is sufficient in many cases; for instance, the Stanford Validity Checker (SVC) [2] is an efficient checker specialized to handle large ground formulas, including uninterpreted function symbols, that occur in hardware verification.

However, program features such as parameterization and the *tick* transition in real-time systems introduce quantifiers in verification conditions. Fortunately, the required quantifier instantiations are often "obvious," and use instances that can be provided by the decision procedures themselves. Both the Extended Static Checking system (ESC) [23] and the Stanford Temporal Prover (STeP) [6,7,9] feature integrations of first-order reasoning and decision procedures that can automatically prove many verification conditions that would otherwise require the use of an interactive prover.

Finally, we note that automatic proof methods for first-order logic and certain specialized theories are necessarily incomplete, not guaranteed to terminate, or both. In practice, automatic verification tools abandon completeness and focus instead on quickly deciding particular classes of problems that are both tractable and likely to appear when verifying realistic systems.

To achieve completeness, interactive theorem proving must be used, and the above techniques must be integrated into interactive theorem proving frameworks, as done, for instance, in STeP and PVS [47]. Combining decision procedures, validity checkers and theorem provers, and extending them to ever-more expressive assertion languages, are ongoing challenges.

5   Abstraction and Invariant Generation

Abstraction is a fundamental and widely-used verification technique. Together with modularity, it is the basis of most combinations of model checking and deductive verification [37], as we will see in Sections 6 and 7.

Abstraction reduces the verification of a property φ over a *concrete system* $\mathcal{S}$, to checking a related property $\varphi^{\mathcal{A}}$ over a simpler *abstract system* $\mathcal{A}$. It allows the verification of infinite-state systems by constructing abstract systems that can be model checked. It can also mitigate the state explosion problem in the finite-state case, by constructing abstract systems with a more manageable state-space.

Abstract interpretation [18] provides a general framework and a methodology for automatically producing abstract systems given a choice of the abstract domain $\Sigma_{\mathcal{A}}$. The goal is to construct abstractions whose state-space can be represented, manipulated and approximated in ways that could not be directly applied to the original system. Originally designed for deriving safety properties in static program analysis, this framework has recently been extended to include reactive systems and general temporal logic, e.g., [39,20].

One simple but useful instance of this framework is based on *Galois connections*. Two functions, $\alpha : 2^{\Sigma_c} \mapsto \Sigma_{\mathcal{A}}$ and $\gamma : \Sigma_{\mathcal{A}} \mapsto 2^{\Sigma_c}$, connect the lattice of sets of concrete states and an abstract domain $\Sigma_\mathcal{A}$, which we assume to be a complete boolean lattice. The *abstraction function* α maps each set of concrete states to an abstract state that represents it. The *concretization function* $\gamma : \Sigma_\mathcal{A} \to 2^{\Sigma_c}$ maps each abstract state to the set of concrete states that it represents.

In general, the abstract and concrete systems are described using different assertion languages, specialized to the respective domains of computation (see Section 4.1). We say that an assertion or temporal formula is *abstract* or *concrete* depending on which language it belongs to.

For an abstract sequence of states $\pi^\mathcal{A} : a_0, a_1, \ldots,$ its concretization $\gamma(\pi^\mathcal{A})$ is the set of sequences $\{s_0, s_1, \ldots \mid s_i \in \gamma(a_i) \text{ for all } i \geq 0\}$. The *abstraction* $\alpha(S)$ of a set of concrete states S is

$$\alpha(S) \stackrel{\text{def}}{=} \bigwedge^{\mathcal{A}} \{a \in \Sigma_\mathcal{A} \mid S \subseteq \gamma(a)\} \ .$$

This is the smallest point in the abstract domain that represents all the elements of S. In practice, it is enough to soundly over-approximate such sets [17].

Definition 4 (Abstraction and concretization of CTL* properties). *For a concrete CTL* temporal property φ, its* abstraction $\alpha^t(\varphi)$ *is obtained by replacing each assertion f in φ by an abstract assertion $\alpha^-(f)$ that characterizes the set of abstract states $\alpha^-(f) : \bigvee^{\mathcal{A}} \{a \in \Sigma_\mathcal{A} \mid \gamma(a) \subseteq f\}$. Conversely, given an abstract temporal property $\varphi^\mathcal{A}$, its* concretization $\gamma(\varphi^\mathcal{A})$ *is obtained by replacing each atom a in $\varphi^\mathcal{A}$ by an assertion that characterizes $\gamma(a)$.*

We can now formally define *weak property preservation*, where properties of the abstract system can be transferred over to the concrete one:

Definition 5 (Weak preservation). $\mathcal{A}$ *is a* weakly preserving abstraction *of $\mathcal{S}$ relative to a class of concrete temporal properties $\mathcal{P}$ if for any property $\varphi \in \mathcal{P}$,*

1. *If $\mathcal{A} \models \alpha^t(\varphi)$ then $\mathcal{S} \models \varphi$. Or, equivalently:*
2. *For any abstract temporal property $\varphi^\mathcal{A}$ where $\gamma(\varphi^\mathcal{A}) \in \mathcal{P}$, if $\mathcal{A} \models \varphi^\mathcal{A}$ then $\mathcal{S} \models \gamma(\varphi^\mathcal{A})$.*

Note that the failure of $\varphi^\mathcal{A}$ for $\mathcal{A}$ does *not* imply the failure of ϕ for $\mathcal{S}$. *Strong* property preservation ensures the transfer of properties from $\mathcal{S}$ to $\mathcal{A}$ as well; however, it severely limits the degree of abstraction that can be performed, so weak preservation is more often used.

There are two general applications of this framework: In the first, we can transfer any property of a correct abstraction over to the concrete system, independently of any particular concrete property to be proved. Thus, this is called the *bottom-up* approach. In the second, given a concrete property to be proved, we try to find an abstract system that satisfies the corresponding abstract property. The abstraction is now tailored to a specific property, so this is called the *top-down* approach. We return to this classification in Section 7.

The question now is how to determine that a given abstract system is indeed a sound abstraction of $\mathcal{S}$. The following theorem expresses sufficient conditions for this, establishing a *simulation relation* between the two systems:

Proposition 2 (Weakly preserving ∀CTL* abstraction). *Consider systems* $S : \langle \Sigma_C, \Theta_C, R_C \rangle$, *and* $A : \langle \Sigma_A, \Theta_A, R_A \rangle$ *such that for a concretization function* $\gamma : \Sigma_A \to 2^{\Sigma_C}$ *the following hold:*

1. INITIALITY: $\Theta_C \subseteq \gamma(\Theta_A)$.
2. CONSECUTION: *If* $R_C(s_1, s_2)$ *for some* $s_1 \in \gamma(a_1)$ *and* $s_2 \in \gamma(a_2)$, *then* $R_A(a_1, a_2)$.

Then A *is a weakly preserving abstraction of* S *for* $\forall CTL^*$.

Informally, the conditions ensure that A can do everything that S does, and perhaps some more. Note that this proposition is limited to universal properties and does not consider fairness. The framework can, however, be extended to include existential properties and take fairness into account—see [20,58].

5.1 Invariant Generation

Once established, invariants can be very useful in all forms of deductive and algorithmic verification, as we will see in Section 6. Given a sound ∀CTL*-preserving abstraction A of S, if $\Box \varphi^A$ is an invariant of A, then $\Box \gamma(\varphi^A)$ is an invariant of S. Thus, in particular, the concretization of the reachable state-space of the abstract system is an invariant of the concrete one; furthermore, any *over-approximation* of this state-space is also an invariant. This is the basis for most automatic invariant generation methods based on abstract interpretation, which perform the following steps:

1. Construct an abstract system A over some suitable domain of computation;
2. Compute an over-approximation of the state-space of A, expressed as an abstract assertion or a set of constraints;
3. Concretize this abstract assertion, to produce an invariant over the concrete domain.

Widening, a classic abstract interpretation technique, can be used to speed up or ensure convergence of the abstract fixpoint operations, by performing safe over-approximations. This is the approach taken in [8] to automatically generate invariants for general infinite-state systems. The abstract domains used include set constraints, linear arithmetic, and polyhedra. These methods are implemented as part of STeP [7].

6   Loosely Coupled Combinations

The preceding sections have presented model checking, deductive verification, and abstraction, including the special case of invariant generation. Numerous stand-alone tools that perform these tasks are available. This section surveys combination methods that can use these separate components as "black boxes," while Section 7 describes methods that require a closer integration.

6.1   Modularity and Abstraction

Given the complementary nature of model checking and theorem proving, as discussed in Section 1, a natural combination is to decompose the verification problem into sub-problems that fall within the range of application of each of the methods. Since theorem proving is more expressive and model checking is more automatic, the most reasonable approach is to use deductive tools to reduce the main verification problem to subgoals that can be model checked.

Abstraction is one of the two main methods for doing this: abstractions are deductively justified and algorithmically model checked. The other is *modular verification*. Here, the system being verified is split into its constituent components, which are analyzed independently, provided assumptions on their environment. The properties of the individual components are then combined, usually following some form of *assumption-guarantee reasoning*, to derive properties of the complete system. This avoids the state-space explosion and allows the re-use of properties and components. Different specialized tools can be applied to different modules; finite-state modules can be model checked, and infinite-state modules can be verified deductively.

Abstraction and modularity are orthogonal to each other: modules can be abstracted, and abstractions can be decomposed. Together, they form a powerful basis for scaling up formal verification [37,52].

6.2   General Deductive Environments

A general-purpose theorem prover can formalize modular decomposition and assume-guarantee reasoning, formalize and verify the correctness of abstractions, apply verification rules, and put the results together. It can also handle *parameterization*, where an arbitrary number of similar modules are composed, and provide decision procedures and validity checkers for specialized domains (see Section 4.1). This often includes OBDD's for finite-state constraint solving and model checking.

Provers such as HOL, the Boyer-Moore prover, and PVS, have been equipped with BDD's and used in this way. From the theorem-proving point of view, this is a tight integration: model checking becomes a proof rule or a tactic. However, it is usually only used at the leaves of the proof tree, so the general verification method remains loosely coupled.

The above theorem provers do not commit to any particular system representation, so the semantics of reactive systems and temporal properties must be formalized within their logic. The STeP system [7] builds in the notion of fair transition systems and LTL, and provides general model checking and theorem proving as well (and more specialized tools, which we will see below). The Symbolic Model Prover (SyMP) [4] is an experimental deductive environment oriented to model checking, featuring a modular system specification language and interfaces to SMV and decision procedures.

Verification environments such as the above, which include model checking and theorem proving under a common deductive environment, support the following tasks:

Debugging Using Model Checking: A simple but important application of model checking in a deductive environment is to check (small) finite instances of an infinite-state or parameterized system specification. In this way, many errors can quickly be found before the full deductive verification effort proceeds.

Incremental Verification: As mentioned in Section 4, verification conditions need not be valid in general, but only valid with respect to the invariants of the system. Therefore, deductive verification usually proceeds by proving a series of invariants of increasing strength, where each is used as a lemma to prove subsequent ones [44]. The database of system properties can also help justify new reductions or abstractions, which in turn can be used to prove new properties.

Automatically generated invariants can also be used to establish verification conditions. Different abstract domains generate different classes of invariants; in some cases, expressing the invariants so that they can be effectively used by the theorem proving machinery is a non-trivial task, and presents another combination challenge. So is devising invariant generation methods that take advantage of system properties already proven.

Invariants can also be used to constrain the set of states explored in symbolic model checking [49], for extra efficiency. Here, too, it may be necessary to translate the invariant into a form useful to the model checker, e.g. if an individual component or an abstracted finite-state version is being checked.

Formal Decomposition: System abstraction and modular verification are often performed manually and then proved correct. If done within a general theorem proving environment, these steps can be formally checked. For instance, Müller and Nipkow [46] use the Isabelle theorem prover to prove the soundness of an I/O-automata abstraction, which is then model checked. Hungar [33] constructs model-checkable abstractions based on data independence and modularity. Abstraction is also used by Rajan *et. al.* [51] to obtain subgoals that can be model checked, where the correctness of the abstraction is proved deductively. Kurshan and Lamport [38] use deductive modular decomposition to reduce the correctness of a large hardware system to that of smaller components that can be model checked.

6.3   Abstraction Generation Using Theorem Proving

Many of the works cited above use theorem proving to prove that an abstraction given *a priori* is correct. This abstraction is usually constructed manually, which can be a time-consuming and error-prone task. An attractive alternative is to use theorem proving to *construct* the abstraction itself.

The domain most often used for this purpose is that of *assertion-based abstractions*, also known as *predicate* or *boolean abstractions*. Here, the abstract state-space is the complete boolean algebra over a finite set of assertions $B : \{b_1, \ldots, b_n\}$, which we call the *basis* of the abstraction. For a point p in the boolean algebra, its *concretization* $\gamma(p)$ is the set of concrete states that satisfy p (see Section 5). For an abstract assertion $f^{\mathcal{A}}$, a concrete assertion that characterizes $\gamma(f^{\mathcal{A}})$ is obtained simply by replacing each boolean variable in $f^{\mathcal{A}}$ by the corresponding basis element. Similarly, for an abstract temporal formula φ^A, its concretization $\gamma(\varphi^A)$ is obtained by replacing each assertion in φ^A by its concretization. Since the abstract system is described in terms of logical formulas, off-the-shelf theorem proving can be used to construct and manipulate it.

Graf and Saidi [28] presented the first automatic procedure for generating such abstractions, using theorem proving to explicitly generate the abstract state-space. Given an abstract state s, an approximation of its successors is computed by deciding which assertions are implied by the postcondition of $\gamma(s)$. This is done using a tactic of the PVS theorem prover [47]. If the proof fails, the next-state does not include any information about the corresponding assertions, thus safely *coarsening* the abstraction: the abstract transition relation over-approximates the concrete one.

An alternative algorithm is presented by Colón and this author in [17], using the decision procedures in STeP. Rather than performing an exhaustive search of the reachable abstract states while constructing $\mathcal{A}$, this algorithm transforms $\mathcal{S}$ to $\mathcal{A}$ directly, leaving the exploration of the abstract state-space to an off-the-shelf model checker. Thus, this procedure is applicable to systems whose abstract state-space is too large to enumerate explicitly, but can still be handled by a symbolic model checker. The price paid by this approach, compared to [28], is that a coarser abstraction may be obtained.

Bensalem *et. al.* [3] present a similar framework for generating abstractions. Here, the invariant to be proved is assumed when generating the abstraction, yielding a better abstract system.

From the combination point of view, all of these approaches have the advantage of using a validity checker purely as a black box, operating only under the assumption of soundness; more powerful checkers will yield better abstractions. For instance, SVC [2] (see Section 4.1) has been used to generate predicate abstractions as well [21]. As with invariant generation, these methods can be parameterized by the abstract assertion language and validity checker used. Note that some user interaction is still necessary, in the choice of assertions for the abstraction basis; but the correctness of the resulting abstraction is guaranteed. A related method for generating abstract systems is presented in [40], intended for the practical analysis and debugging of complex software systems.

These finite-state abstractions can be used to generate invariants; as noted in Section 5.1, the concretization of the reachable state-space of $\mathcal{A}$ is an invariant of the original system (but may be an unwieldy formula); on the other hand, previously proven invariants of $\mathcal{S}$ can be used as lemmas in the abstraction generation process, yielding finer abstract systems.

7   Tight Combinations

We now describe and classify more tightly-coupled combination methods that are based on abstraction; its counterpart, modularity, is only briefly mentioned in the interest of satisfying space constraints.

Abstraction Refinement: Recall that weak property preservation (Section 5) only guarantees that $S \models \phi$ whenever $A \models \varphi^A$; if the abstract property fails, it is possible that φ does hold for S, but the abstraction was not *fine enough* to prove it. Thus, in general, abstractions must be *refined* in order to prove the desired property. In predicate abstraction, refinement occurs by performing new validity checks over the existing basis, or adding new assertions to the basis.

A generic abstraction-based verification procedure for proving $S \models \phi$ proceeds as follows: First, an initial weakly-preserving abstraction A of S is given by the user or constructed automatically. If model checking $A \models \varphi^A$ succeeds, the proof is complete. Otherwise, A is used as the starting point for producing a finer abstraction A', which is still weakly-preserving for S but satisfies more properties. This process is repeated until φ is proved (if it indeed holds for S).

At each step, the model checker produces an *abstract* counterexample, which does not necessarily correspond to any concrete computation. However, it can help choose the next refinement step, or serve as the basis for finding a concrete counterexample that indeed falsifies φ. Finding effective procedures to distinguish between these two cases and perform the refinement remains an intriguing research problem, which is addressed by some of the methods described below. (In the general case, we are still faced with an undecidable problem, so no method can guarantee that the process will terminate.)

Methods that combine deductive and algorithmic verification through abstraction can be classified according to the following criteria:

- whether the abstraction is constructed *a priori* or or dynamically refined (*static* vs. *dynamic*);
- whether the abstraction generation is tailored specifically for the property of interest (*bottom-up* vs. *top-down*);
- whether the process is automatic or interactive.

Focusing on a particular formula allows for coarser abstract systems, which satisfy fewer properties but can be easier to construct and model check.

The automatic abstraction generation algorithms of Section 6.3 are generally bottom-up, but can be focused towards a particular temporal property by including its atomic subformulas as part of the assertion basis. The tighter combinations described below are mostly top-down, using a mixture of model checking and theorem proving that is specialized to the particular temporal property being proved.

7.1   Diagram-Based Formalisms

We begin by describing a number of *diagram-based* verification formalisms that have been developed as part of the STeP project [7]. They offer deductive-algorithmic proof methods that are applicable to arbitrary temporal properties.

Static Abstractions: GVD's: The *Generalized Verification Diagrams* (GVD's) of Browne *et. al.* [11] provide a graphical representation of the verification conditions needed to establish an arbitrary temporal formula, extending the specialized diagrams of Manna and Pnueli [43].

A GVD serves as a proof object, but is also a weakly-preserving assertion-based abstraction of the system, where the basis is the set of formulas used in the diagram [41]. Each node in the diagram is labeled with an assertion f, and corresponds to an abstract state representing the set of states that satisfy f. The diagram Φ is identified with a set of computations $\mathcal{L}(\Phi)$, and a set of verification conditions associated with the diagram show that $\mathcal{L}(\mathcal{S}) \subseteq \mathcal{L}(\Phi)$. This proves, deductively, that Φ is a correct abstraction of $\mathcal{S}$. The proof is completed by showing that $\mathcal{L}(\Phi) \subseteq \mathcal{L}(\varphi)$. This corresponds to model checking φ over Φ, when Φ is seen as an abstract system, and is done algorithmically, viewing the diagram as an ω-automata.

Diagrams can be seen as a flexible generalization of the classic deductive rules. (As noted in, e.g., [3,37], the success of a deductive rule such as G-INV of Section 4 implies the existence of an abstraction for which the proven property can be model checked.) Since the diagram is tailored to prove a particular property φ, this can be classified as a *top-down* approach. Since the formalism does not include refinement operations in the case that the proof attempt fails, it is *static*.

Dynamic Abstractions: Deductive Model Checking: *Deductive Model Checking* (DMC), presented by Sipma *et. al.* [57], is a method for the interactive model checking of possibly infinite-state systems. To prove a property φ, DMC searches for a counterexample computation by refining an abstraction of the $(\mathcal{S}, \neg\varphi)$ *product graph*.

The DMC procedure interleaves the theorem proving and model checking steps, refining the abstraction as the model checking proceeds. This focuses the theorem proving effort to those aspects of the system that are relevant to the property being proved. On the other hand, the expanded state-space is restricted by the theorem proving, so that space savings are possible even in the case of finite-state systems.

Sipma [56] describes DMC, GVD's, and their application to the verification of real-time and hybrid systems. The *fairness diagrams* of de Alfaro and Manna [22] present an alternate dynamic refinement method that combines the top-down and bottom-up approaches.

7.2   Model Checking for Infinite-State Systems

Abstraction is also the basis of model checking algorithms for decidable classes of infinite-state systems, such as certain classes of real-time and hybrid systems. In some cases, the exploration of a finite quotient of the state-space is sufficient. For others, the convergence of fixpoint operations is ensured by the right choice of abstract assertion language, such as polyhedra [30] or Presburger arithmetic [13]. The underlying principles are explored in [26,31].

A number of "local model checking" procedures for general infinite-state systems, such as that of Bradfield and Stirling [10], are also hybrid combinations of deductive verification rules and model checking. Another top-down approach is presented by Damm *et. al.* [19] as a "truly symbolic" model checking procedure, analyzing the separation between data and control. A tableau-based procedure for ∀CTL generates the required verification conditions in a top-down, local manner, similarly to DMC.

Model Checking and Static Analysis: The relationship between model checking and abstract interpretation continues to be the subject of much research. Static program analysis methods based on abstract interpretation have been recently re-formulated in terms of model checking. For instance, Schmidt and Steffen [55] show how many program analysis techniques can be understood as the model checking of particular kinds of abstractions.

ESC [23] and Nitpick [34] are tools that automatically detect errors in software systems by combining static analysis and automatic theorem proving methods. Another challenge is to further combine program analysis techniques with formal verification and theorem proving (see [52]).

Finally, we note that there is also much work on applying abstraction to *finite-state* systems, particularly large hardware systems. Abstracting from the finite-state to an *infinite-state* abstract domain has proved useful here, namely, using uninterpreted function symbols and symbolically executing the system using a decidable logic, as shown, e.g., by Jones *et. al.* [35].

7.3   Integrated Approaches

A number of verification environments and methodologies have been proposed that combine the above ingredients in a systematic way.

As mentioned above, STeP [7] includes deductive verification, generalized verification diagrams, symbolic and explicit-state model checking, abstraction generation, and automatic invariant generation, which share a common system and property specification language. To this, modularity and compositional verification are being added as well—see Finkbeiner *et. al.* [27].

Dingel and Filkorn [25] apply abstraction and error trace analysis to infinite-state systems. The abstract system is generated automatically given a *data abstraction* that maps concrete variables and functions to abstract ones. If an abstract counterexample is found that does not correspond to a concrete one, an *assumption* that excludes this counterexample is generated. This is a temporal formula that should hold for the concrete system S. The model checker, which takes such assumptions into account, is then used again. The process is iterated until a concrete counterexample is found, or model checking succeeds under a given set of assumptions. In the latter case, the assumptions are deductively verified over the concrete system. If they hold, the proof is complete.

Rusu and Singerman [53] present a framework that combines abstraction, abstraction refinement and theorem proving specialized to the case of invariants, where the different components are treated as "black boxes." After an assertion-based abstraction is generated (using the method of Graf and Saidi [28]), abstract counterexamples are analyzed to refine the abstraction or produce a concrete counterexample. Conjectures generated during the refinement process are given to the theorem prover, and the process repeated.

A similar methodology is proposed by Saidi and Shankar [54]. The ongoing Symbolic Analysis Laboratory (SAL) project at SRI [52] proposes a collection of multiple different analysis tools, including theorem provers, model checkers, abstraction and invariant generators, that communicate through a common language in a blackboard architecture.

An Abstraction-Based Proposal: Table 2 summarizes the classification of the abstraction-based methods discussed in this section. For general infinite-state systems, the automatic methods are incomplete, and the complete methods are interactive.

Table 2. Classification of some combination methods based on abstraction

Method	refinement?	uses φ?	automatic?
Static Analysis and Invariant Generation (Abstract Interpretation)	static	bottom-up	automatic
(Generalized) Verification Diagrams (includes verification rules)	static	top-down	interactive
Deductive Model Checking	dynamic	top-down	interactive
Fairness Diagrams	dynamic	[both]	interactive
Infinite-state Model Checking	dynamic	top-down	[both]

In [58], this author proposes combining these approaches by exploiting their common roots in abstraction. Each different proof attempt (including failed ones) and static analysis operation provides additional information about the system being verified. This information can be captured, incrementally, as an *extended finite-state abstraction*, which includes information about fairness constraints and well-founded orders over the original system. Once generated, these abstractions can be combined and re-used, given a (correspondingly extended) model checker to reason about them.

Thus, abstractions can serve as the repository for all the information about the system that is shared by the different components. An important challenge is to manage and combine abstractions from different domains, finding a common language to express them.

The tight integration found in DMC's and GVD's (Section 7.1) is an obstacle for their implementation using off-the-shelf tools. However, it allows for more detailed user input and feedback. Given tools whose input is expressive and flexible enough, such as a model checker for the extended abstractions, it will be possible to implement such tightly coupled methods in a more modular way.

The abstraction framework has the advantage of leaving the combinatorial problems to the automatic tools, as well as getting the most out of the automatic theorem-proving tools; the user can then focus on defining the right abstractions and guiding their refinement, until a proof or counterexample are found. We believe that such approaches will lead to more automated, lightweight and useful verification tools.

Acknowledgements. The author thanks: the FROCOS organizers, for their kind invitation; the STeP team, and in particular Nikolaj Bjørner, Anca Browne, Michael Colón, Bernd Finkbeiner, Zohar Manna, and Henny Sipma, for their inspiration and feedback—but even though much of the work described above is theirs, they are not to be held responsible for the views expressed herein; and the Leyva-Uribe family, for its hospitality and patience while the final version of this paper was prepared.

References

1. R. Alur and T. A. Henzinger, editors. *Proc. 8th Intl. Conf. on Computer Aided Verification*, vol. 1102 of *LNCS*. Springer, July 1996.

2. C. Barrett, D. L. Dill, and J. Levitt. Validity checking for combinations of theories with equality. In *1st Intl. Conf. on Formal Methods in Computer-Aided Design*, vol. 1166 of *LNCS*, pp. 187–201, Nov. 1996.

3. S. Bensalem, Y. Lakhnech, and S. Owre. Computing abstractions of infinite state systems compositionally and automatically. In *Proc. 10th Intl. Conf. on Computer Aided Verification*, vol. 1427 of *LNCS*, pp. 319–331. Springer, July 1998.

4. S. Berezin and A. Groce. *SyMP: The User's Guide.* Comp. Sci. Department, Carnegie-Mellon Univ., Jan. 2000.

5. A. Biere, A. Cimatti, E. M. Clarke, M. Fujita, and Y. Zhu. Symbolic model checking using SAT procedures instead of BDDs. In *Design Autom. Conf. (DAC'99)*, 1999.

6. N. S. Bjørner. *Integrating Decision Procedures for Temporal Verification.* PhD thesis, Comp. Sci. Department, Stanford Univ., Nov. 1998.

7. N. S. Bjørner, A. Browne, E. S. Chang, M. Colón, A. Kapur, Z. Manna, H. B. Sipma, and T. E. Uribe. STeP: Deductive-algorithmic verification of reactive and real-time systems. In [1], pp. 415–418.

8. N. S. Bjørner, A. Browne, and Z. Manna. Automatic generation of invariants and intermediate assertions. *Theoretical Comp. Sci.*, 173(1):49–87, Feb. 1997.

9. N. S. Bjørner, M. E. Stickel, and T. E. Uribe. A practical integration of first-order reasoning and decision procedures. In *Proc. of the 14th Intl. Conf. on Automated Deduction*, vol. 1249 of *LNCS*, pp. 101–115. Springer, July 1997.

10. J. C. Bradfield and C. Stirling. Local model checking for infinite state spaces. *Theoretical Comp. Sci.*, 96(1):157–174, Apr. 1992.

11. A. Browne, Z. Manna, and H. B. Sipma. Generalized temporal verification diagrams. In *15th Conf. on the Foundations of Software Technology and Theoretical Comp. Sci.*, vol. 1026 of *LNCS*, pp. 484–498. Springer, 1995.

12. R. E. Bryant. Graph-based algorithms for Boolean function manipulation. *IEEE Transactions on Computers*, C-35(8):677–691, Aug. 1986.

13. T. Bultan, R. Gerber, and W. Pugh. Symbolic model checking of infinite state systems using Presburger arithmetic. In Grumberg [29], pp. 400–411.

14. E. M. Clarke and E. A. Emerson. Design and synthesis of synchronization skeletons using branching time temporal logic. In *Proc. IBM Workshop on Logics of Programs*, vol. 131 of *LNCS*, pp. 52–71. Springer, 1981.

15. E. M. Clarke, M. Fujita, and X. Zhao. Hybrid decision diagrams. Overcoming the limitations of MTBDDs and BMDs. In *IEEE/ACM Intl. Conf. on Computer-Aided Design*, pp. 159–163, Nov. 1995.

16. E. M. Clarke, O. Grumberg, and D. Peled. *Model Checking.* MIT Press, Dec. 1999.

17. M. A. Colón and T. E. Uribe. Generating finite-state abstractions of reactive systems using decision procedures. In *Proc. 10th Intl. Conf. on Computer Aided Verification*, vol. 1427 of *LNCS*, pp. 293–304. Springer, July 1998.

18. P. Cousot and R. Cousot. Abstract interpretation: A unified lattice model for static analysis of programs by construction or approximation of fixpoints. In *4th ACM Symp. Princ. of Prog. Lang.*, pp. 238–252. ACM Press, 1977.

19. W. Damm, O. Grumberg, and H. Hungar. What if model checking must be truly symbolic. In *TACAS'95*, vol. 1019 of *LNCS*, pp. 230–244. Springer, May 1995.

20. D. R. Dams. *Abstract Interpretation and Partition Refinement for Model Checking.* PhD thesis, Eindhoven Univ. of Technology, July 1996.

21. S. Das, D. L. Dill, and S. Park. Experience with predicate abstraction. In *Proc. 11th Intl. Conf. on Computer Aided Verification*, vol. 1633 of *LNCS*. Springer, 1999.

22. L. de Alfaro and Z. Manna. Temporal verification by diagram transformations. In [1], pp. 287–299.

23. D. L. Detlefs, K. R. M. Leino, G. Nelson, and J. B. Saxe. Extended static checking. Tech. Report 159, Compaq SRC, Dec. 1998.

24. D. L. Dill. The Murφ verification system. In [1], pp. 390–393.

25. J. Dingel and T. Filkorn. Model checking of infinite-state systems using data abstraction, assumption-commitment style reasoning and theorem proving. In *Proc. 7th Intl. Conf. on Computer Aided Verif.*, vol. 939 of *LNCS*, pp. 54–69, July 1995.

26. E. A. Emerson and K. S. Namjoshi. On model checking for non-deterministic infinite-state systems. In *Proc. 13th IEEE Symp. Logic in Comp. Sci.*, pp. 70–80. IEEE Press, 1998.

27. B. Finkbeiner, Z. Manna, and H. B. Sipma. Deductive verification of modular systems. In *COMPOS'97*, vol. 1536 of *LNCS*, pp. 239–275. Springer, Dec. 1998.

28. S. Graf and H. Saidi. Construction of abstract state graphs with PVS. In Grumberg [29], pp. 72–83.

29. O. Grumberg, editor. *Proc. 9th Intl. Conf. on Computer Aided Verification*, vol. 1254 of *LNCS*. Springer, June 1997.

30. T. A. Henzinger and P. Ho. HyTech: The Cornell hybrid technology tool. In *Hybrid Systems II*, vol. 999 of *LNCS*, pp. 265–293. Springer, 1995.

31. T. A. Henzinger and R. Majumdar. A classification of symbolic transition systems. In *Proc. of the 17th Intl. Conf. on Theoretical Aspects of Comp. Sci. (STACS 2000)*, LNCS. Springer, 2000.

32. G. J. Holzmann. *Design and Validation of Computer Protocols.* Prentice Hall, Engelwood Cliffs, NJ, 1991.

33. H. Hungar. Combining model checking and theorem proving to verify parallel processes. In *Proc. 5th Intl. Conf. on Computer Aided Verification*, vol. 697 of *LNCS*, pp. 154–165. Springer, 1993.

34. D. Jackson and C. A. Damon. Nitpick reference manual. Tech. report, Carnegie-Mellon Univ., 1996.

35. R. B. Jones, J. U. Skakkebæk, and D. L. Dill. Reducing manual abstraction in formal verification of out-of-order execution. In G. Gopalakrishnan and P. Windley, editors, *2nd Intl. Conf. on Formal Methods in Computer-Aided Design*, vol. 1522 of *LNCS*, pp. 2–17, Nov. 1998.

36. Y. Kesten, O. Maler, M. Marcus, A. Pnueli, and E. Shahar. Symbolic model checking with rich assertional languages. In Grumberg [29], pp. 424–435.

37. Y. Kesten and A. Pnueli. Modularization and abstraction: The keys to practical formal verification. In *Mathematical Foundations of Comp. Sci.*, vol. 1450 of *LNCS*, pp. 54–71, Aug. 1998.

38. R. P. Kurshan and L. Lamport. Verification of a multiplier: 64 bits and beyond. In *Proc. 5th Intl. Conf. on Computer Aided Verification*, vol. 697 of *LNCS*, pp. 166–179. Springer, 1993.

39. C. Loiseaux, S. Graf, J. Sifakis, A. Bouajjani, and S. Bensalem. Property preserving abstractions for the verification of concurrent systems. *Formal Methods in System Design*, 6:1–35, 1995.

40. M. Lowry and M. Subramaniam. Abstraction for analytic verification of concurrent software systems. In *Symp. on Abstraction, Reformulation, and Approx.*, May 1998.

41. Z. Manna, A. Browne, H. B. Sipma, and T. E. Uribe. Visual abstractions for temporal verification. In A. Haeberer, editor, *Algebraic Methodology and Software Technology (AMAST'98)*, vol. 1548 of *LNCS*, pp. 28–41. Springer, Dec. 1998.

42. Z. Manna and A. Pnueli. Completing the temporal picture. *Theoretical Comp. Sci.*, 83(1):97–130, 1991.

43. Z. Manna and A. Pnueli. Temporal verification diagrams. In M. Hagiya and J. C. Mitchell, editors, *Proc. Intl. Symp. on Theoretical Aspects of Computer Software*, vol. 789 of *LNCS*, pp. 726–765. Springer, 1994.

44. Z. Manna and A. Pnueli. *Temporal Verification of Reactive Systems: Safety*. Springer, New York, 1995.

45. K. L. McMillan. *Symbolic Model Checking*. Kluwer Academic Pub., 1993.

46. O. Müller and T. Nipkow. Combining model checking and deduction for I/O-automata. In *TACAS'95*, vol. 1019 of *LNCS*, pp. 1–12. Springer, May 1995.

47. S. Owre, S. Rajan, J. M. Rushby, N. Shankar, and M. K. Srivas. PVS: Combining specification, proof checking, and model checking. In [1], pp. 411–414.

48. A. Pnueli. The temporal logic of programs. In *Proc. 18th IEEE Symp. Found. of Comp. Sci.*, pp. 46–57. IEEE Computer Society Press, 1977.

49. A. Pnueli and E. Shahar. A platform for combining deductive with algorithmic verification. In [1], pp. 184–195.

50. J. Queille and J. Sifakis. Specification and verification of concurrent systems in CESAR. In M. Dezani-Ciancaglini and U. Montanari, editors, *Intl. Symp. on Programming*, vol. 137 of *LNCS*, pp. 337–351. Springer, 1982.

51. S. Rajan, N. Shankar, and M. K. Srivas. An integration of model checking with automated proof checking. In *Proc. 7th Intl. Conf. on Computer Aided Verification*, vol. 939 of *LNCS*, pp. 84–97, July 1995.

52. J. Rushby. Integrated formal verification: Using model checking with automated abstraction, invariant generation, and theorem proving. In *Theoretical and Practical Aspects of SPIN Model Checking*, vol. 1680 of *LNCS*, pp. 1–11, July 1999.

53. V. Rusu and E. Singerman. On proving safety properties by integrating static analysis, theorem proving and abstraction. In *TACAS'99*, vol. 1579 of *LNCS*. Springer, Mar. 1999.

54. H. Saidi and N. Shankar. Abstract and model check while you prove. In *Proc. 11th Intl. Conf. on Computer Aided Verification*, vol. 1633 of *LNCS*. Springer, 1999.

55. D. A. Schmidt and B. Steffen. Program analysis as model checking of abstract interpretations. In *Proc. 5th Static Analysis Symp.*, LNCS. Springer, Sept. 1998.

56. H. B. Sipma. *Diagram-based Verification of Discrete, Real-time and Hybrid Systems*. PhD thesis, Comp. Sci. Department, Stanford Univ., Feb. 1999.

57. H. B. Sipma, T. E. Uribe, and Z. Manna. Deductive model checking. *Formal Methods in System Design*, 15(1):49–74, July 1999.

58. T. E. Uribe. *Abstraction-based Deductive-Algorithmic Verification of Reactive Systems*. PhD thesis, Comp. Sci. Department, Stanford Univ., Dec. 1998. Tech. Report STAN-CS-TR-99-1618.

Compiling Multi-paradigm Declarative Programs into Prolog*

Sergio Antoy[1] and Michael Hanus[2]

[1] Department of Computer Science, Portland State University,
P.O. Box 751, Portland, OR 97207, U.S.A., antoy@cs.pdx.edu
[2] Institut für Informatik, Christian-Albrechts-Universität Kiel,
Olshausenstr. 40, D-24098 Kiel, Germany, mh@informatik.uni-kiel.de

Abstract. This paper describes a high-level implementation of the concurrent constraint functional logic language Curry. The implementation, directed by the lazy pattern matching strategy of Curry, is obtained by transforming Curry programs into Prolog programs. Contrary to previous transformations of functional logic programs into Prolog, our implementation includes new mechanisms for both efficiently performing concurrent evaluation steps and sharing common subterms. The practical results show that our implementation is superior to previously proposed similar implementations of functional logic languages in Prolog and is competitive w.r.t. lower-level implementations of Curry in other target languages.

An noteworthy advantage of our implementation is the ability to immediately employ in Curry existing constraint solvers for logic programming. In this way, we obtain with a relatively modest effort the implementation of a declarative language combining lazy evaluation, concurrency and constraint solving for a variety of constraint systems.

1 Introduction

The multi-paradigm language Curry [12,18] seamlessly combines features from functional programming (nested expressions, lazy evaluation, higher-order functions), logic programming (logical variables, partial data structures, built-in search), and concurrent programming (concurrent evaluation of expressions with synchronization on logical variables). Moreover, the language provides both the most important operational principles developed in the area of integrated functional logic languages: "residuation" and "narrowing" (see [10] for a survey on functional logic programming).

Curry's operational semantics (first described in [12]) combines lazy reduction of expressions with a possibly non-deterministic binding of free variables occurring in expressions. To provide the full power of logic programming, (equational) constraints can be used in the conditions of function definitions. Basic

* This research has been partially supported by the DAAD/NSF under grant INT-9981317, the German Research Council (DFG) under grant Ha 2457/1-1, and by a grant from Portland State University.

H. Kirchner and C. Ringeissen (Eds.): FroCoS 2000, LNAI 1794, pp. 171–185, 2000.

constraints can be combined into complex constraint expressions by a concurrent conjunction operator that evaluates constraints concurrently. Thus, purely functional programming, purely logic programming, and concurrent (logic) programming are obtained as particular restrictions of this model [12].

In this paper, we propose a high-level implementation of this computation model in Prolog. This approach avoids the complex implementation of an abstract machine (e.g., [16]) and is able to reuse existing constraint solvers available in Prolog systems. In the next section, we review the basic computation model of Curry. The transformation scheme for compiling Curry programs into Prolog programs is presented in Section 3. Section 4 contains the results of our implementation. Section 5 discusses related work and contains our conclusions.

2   The Computation Model of Curry

This section outlines the computation model of Curry. A formal definition can be found in [12,18].

The basic computational domain of Curry is, similarly to functional or logic languages, a set of *data terms* constructed from constants and data constructors. These are introduced by data type declarations such as:[1]

```
data Bool   = True | False
data List a = []    | a : List a
```

True and False are the Boolean constants. [] (empty list) and : (non-empty list) are the constructors for polymorphic lists (a is a type variable ranging over all types and the type List a is usually written as [a] for conformity with Haskell). A *data term* is a well-formed expression containing variables, constants and data constructors, e.g., True:[] or [x,y] (the latter stands for x:(y:[])).

Functions are operations on data terms whose meaning is specified by (*conditional*) *rules* of the general form "$l \mid c = r$ where vs free". l has the form $f\,t_1 \ldots t_n$, where f is a function, $t_1, \ldots, t_n$ are data terms and each variable occurs only once. The *condition* c is a constraint. r is a well-formed *expression* that may also contain function calls. vs is the list of *free variables* that occur in c and r, but not in l. The condition and the where part can be omitted if c and vs are empty, respectively. A *constraint* is any expression of the built-in type Constraint. Primitive constraints are equations of the form $e_1 =:= e_2$. A conditional rule is applied only if its condition is satisfiable. A *Curry program* is a set of data type declarations and rules.

Example 1. Together with the above data type declarations, the following rules define operations to concatenate lists and to find the last element of a list:

```
conc []     ys = ys
conc (x:xs) ys = x : conc xs ys
```

---

[1] Curry has a Haskell-like syntax [25], i.e., (type) variables and function names start with lowercase letters and the names of type and data constructors start with an uppercase letter. Moreover, the application of f to e is denoted by juxtaposition ("$f\ e$").

```
last xs | conc ys [x] =:= xs  = x    where x,ys free
```

If "`conc ys [x] =:= xs`" is solvable, then x is the last element of list xs.   □

Functional programming: In functional languages, the interest is in computing *values* of expressions, where a value does not contain function symbols (i.e., it is a data term) and should be equivalent (w.r.t. the program rules) to the initial expression. The value can be computed by replacing instances of rules' left sides with corresponding instances of right sides. For instance, we compute the value of "`conc [1] [2]`" by repeatedly applying the rules for concatenation to this expression:

```
conc [1] [2]  →  1:(conc [] [2])  →  [1,2]
```

Curry is based on a lazy (outermost) strategy, i.e., the selected function call in each reduction step is outermost among all reducible function calls. This strategy supports computations with infinite data structures and a modular programming style with separation of control aspects. Moreover, it yields optimal computations [5] and a demand-driven search method [15] for the logic part of a program which will be discussed next.

Logic programming: In logic languages, an expression (or constraint) may contain free variables. A logic programming system should compute solutions, i.e., find values for these variables such that the expression (or constraint) is reducible to some value (or satisfiable). Fortunately, this requires only a minor extension of the lazy reduction strategy. The extension deals with non-ground expressions and variable instantiation: if the value of a free variable is demanded by the left-hand sides of some program rules in order to continue the computation (i.e., no program rule is applicable if the variable remains unbound), the variable is bound to all the demanded values. For each value, a separate computation is performed. For instance, if the function f is defined by the rules

```
f 0 = 2
f 1 = 3
```

(the integer numbers are considered as an infinite set of constants), then the expression "`f x`", with x a free variable, is evaluated to 2 by binding x to 0, or it is evaluated to 3 by binding x to 1. Thus, a single computation step may yield a single new expression (*deterministic step*) or a disjunction of new expressions together with the corresponding bindings (*non-deterministic step*). For inductively sequential programs [3] (these are, roughly speaking, function definitions with one demanded argument), this strategy is called *needed narrowing* [5]. Needed narrowing computes the shortest successful derivations (if common subterms are shared) and minimal sets of solutions. Moreover, it is fully deterministic for expressions that do not contain free variables.

Constraints: In functional logic programs, it is necessary to solve equations between expressions containing defined functions (see Example 1). In general, an *equation* or *equational constraint* $e_1 =:= e_2$ is satisfied if both sides e_1 and e_2 are reducible to the same value (data term). As a consequence, if both sides are undefined (non-terminating), then the equality does not hold.[2] Operationally, an equational constraint $e_1 =:= e_2$ is solved by evaluating e_1 and e_2 to unifiable data terms, where the lazy evaluation of the expressions is interleaved with the binding of variables to constructor terms [21]. Thus, an equational constraint $e_1 =:= e_2$ without occurrences of defined functions has the same meaning (unification) as in Prolog. Curry's basic kernel only provides equational constraints. Constraint solvers for other constraint structures can be conceptually integrated without difficulties. The practical realization of this integration is one of the goals of this work.

Concurrent computations: To support flexible computation rules and avoid an uncontrolled instantiation of free argument variables, Curry gives the option to *suspend a function call* if a demanded argument is not instantiated. Such functions are called *rigid* in contrast to *flexible* functions—those that instantiate their arguments when the instantiation is necessary to continue the evaluation of a call. As a default easy to change, Curry's constraints (i.e., functions with result type `Constraint`) are flexible whereas non-constraint functions are rigid. Thus, purely logic programs (where predicates correspond to constraints) behave as in Prolog, and purely functional programs are executed as in lazy functional languages, e.g., Haskell.

To continue a computation in the presence of suspended function calls, constraints are combined with the *concurrent conjunction* operator &. The constraint c_1 & c_2 is evaluated by solving c_1 and c_2 concurrently.

A design principle of Curry is the clear separation of sequential and concurrent activities. Sequential computations, which form the basic units of a program, are expressed as usual functional (logic) programs and are composed into concurrent computation units via concurrent conjunctions of constraints. This separation supports the use of efficient and optimal evaluation strategies for the sequential parts. Similar techniques for the concurrent parts are not available. This is in contrast to other more fine-grained concurrent computation models like AKL [19], CCP [27], or Oz [28].

Monadic I/O: To support real applications, the monadic I/O concept of Haskell [29] has been adapted to Curry to perform I/O in a declarative manner. In the monadic approach to I/O, an interactive program is considered as a function computing a sequence of actions which are applied to the outside world. An *action* has type "`IO` α", which means that it returns a result of type α whenever it is applied to a particular state of the world. For instance, `getChar`, of type "`IO Char`", is an action whose execution, i.e., application to a world, reads a character from the standard input. Actions can be composed only sequentially in a program and their composition is executed whenever the main program is executed. For instance, the action `getChar` can be composed with the action `putChar` (which has type `Char -> IO ()` and writes a character to the terminal) by the sequential composition operator `>>=` (which has type `IO` α `-> (`α `-> IO` β`) -> IO` β). Thus, "`getChar >>= putChar`" is a compo-

---

[2] This notion of equality, known as *strict equality* [9,22], is the only reasonable notion of equality in the presence of non-terminating functions.

sed action which prints the next character of the input stream on the screen. The second composition operator, >>, is like >>=, but ignores the result of the first action. Furthermore, done is the "empty" action which does nothing (see [29] for more details). For instance, a function which takes a string (list of characters) and produces an action that prints the string to the terminal followed by a new line is defined as follows:

```
putStrLn []     = putChar '\n'
putStrLn (c:cs) = putChar c >> putStrLn cs
```

In the next section, we will describe a transformation scheme to implement this computation model in Prolog.

3   A Transformation Scheme for Curry Programs

As mentioned above, the evaluation of nested expressions is based on a lazy strategy. The exact strategy is specified via definitional trees [3], a data structure for the efficient selection of the outermost reducible expressions. Direct transformations of definitional trees into Prolog (without an implementation of concurrency features) have been proposed in [2,4,11,21]. Definitional trees deal with arbitrarily large patterns and use the notion of "position" (i.e., a sequence of positive integers) to specify the subterm where the next evaluation step must be performed. We avoid this complication and obtain a simpler transformation by first compiling definitional trees into case expressions as described, e.g., in [14]. Thus, each function is defined by exactly one rule in which the right-hand side contains case expressions to specify the pattern matching of actual arguments. For instance, the function conc in Example 1 is transformed into:

```
conc xs ys = case xs of []     -> ys
                        (z:zs) -> z : conc zs ys
```

A case expression is evaluated by reducing its first argument to a *head normal form*, i.e., a term which has no defined function symbol at the top, and matching this reduced term with one of the patterns of the case expression. Case expressions are used for both rigid and flexible functions. Operationally, case expressions are used for rigid functions only, whereas flexcase expressions are used for flexible functions. The difference is that a case expression suspends if the head normal form is a free variable, whereas a flexcase expression (don't know non-deterministically) instantiates the variable to the different constructors in the subsequent patterns.

To implement functions with overlapping left-hand sides (where there is no single argument on which a case distinction can be made), there is also a *disjunctive expression* "e_1 or e_2" meaning that both alternatives are don't know non-deterministically evaluated.[3] For instance, the function

```
0 * x = 0
x * 0 = 0
```

---

[3] In the implementation described in this paper, don't know non-determinism is implemented via backtracking as in Prolog.

is transformed into the single rule

```
x * y = or (flexcase x of 0 -> 0)
            (flexcase y of 0 -> 0)
```

under the assumption that "*" is a flexible operation.

Transformation schemes for programs where all the functions are flexible have been proposed in [2,4,11,21]. These proposals are easily adaptable to our representation using `case` and `or` expressions. The challenge of the implementation of Curry is the development of a transformation scheme that provides both the suspension of function calls and the concurrent evaluation of constraints (which will be discussed later).

3.1   Implementing Concurrent Evaluations

Most of the current Prolog systems support coroutining and the delaying of literals [23] if some arguments are not sufficiently instantiated. One could use these features to provide the suspension, when required, of calls to rigid functions. However, in conditional rules it is not sufficient to delay the literals corresponding to suspended function calls. One has to wait until the condition has been completely proved to avoid introducing unnecessary computations or infinite loops. The following example helps in understanding this problem.

Example 2. Consider the function definitions

```
f x y | g x =:= y  = h y
g [] = []
h [] = []
h (z:zs) = h zs
```

where `g` is rigid and `h` is flexible. To evaluate the expression "`f x y`" (where x and y are free variables), the condition "`g x =:= y`" must be proved. Since g is rigid, this evaluation suspends and the right-hand side is not evaluated. However, if we only delay the evaluation of the condition and proceed with the right-hand side, we run into an infinite loop by applying the last rule forever. This loop is avoided if x is eventually instantiated by another thread of the entire computation.   □

To explain how we solve this problem we distinguish between sequential and concurrent computations. A sequential computation is a sequence of calls to predicates. When a call is activated, it may return for two reasons: either the call's computation has completed or the call's computation has been suspended or delayed. In a sequential computation, we want to execute a call only if the previous call has completed. Thus, we add an input argument and an output argument to each predicate. Each argument is a variable that is either uninstantiated or bound to a constant—by convention the symbol `eval` that stand for "fully evaluated". We use these arguments as follows. In a sequential computation, the call to a predicate is executed if and only if its input argument is instantiated to `eval`. Likewise, a computation has completed if and only if its output argument is instantiated to `eval`. As one would expect, we chain the output argument of a call to the input argument of the next call to ensure the sequentiality of a computation.

The activation or delay of a call is easily and efficiently controlled by `block` declarations.[4] For instance, the `block` declaration ":- block f(?,?,?,-,?)" specifies that a call to `f` is delayed if the fourth argument is a free variable. According to the scheme just described, we obtain the following clauses for the rules defining the functions `f` and `g` above:[5]

```
:- block f(?,?,?,-,?).
f(X,Y,Result,Ein,Eout) :- eq(g(X),Y,Ein,E1), h(Y,Result,E1,Eout).

:- block g(?,?,-,?).
g(X,Result,Ein,Eout) :- hnf(X,HX,Ein,E1), g_1(HX,Result,E1,Eout).
:- block g_1(-,?,?,?), g_1(?,?,-,?).
g_1([],[],E,E).
```

The predicate `hnf` computes the head normal form of its first argument. If argument `HX` of `g` is bound to the head normal form of `X`, we can match this head normal form against the empty list with the rule for `g_1`.

We use `block` declarations to control the rigidity or flexibility of functions, as well. Since `g` is a rigid function, we add the block declaration `g_1(-,?,?,?)` to avoid the instantiation of free variables. A computation is initiated by setting argument `Ein` to a constant, i.e., expression (`f x y`) is evaluated by goal `f(X,Y,Result,eval,Eout)`. If `Eout` is bound to `eval`, the computation has completed and `Result` contains the computed result (head normal form).

Based on this scheme, the concurrent conjunction operator `&` is straightforwardly implemented by the following clauses (the constant `success` denotes the result of a successful constraint evaluation):

```
&(A,B,success,Ein,Eout) :- hnf(A,HA,Ein,E1), hnf(B,HB,Ein,E2),
                           waitconj(HA,HB,E1,E2,Eout).

?- block waitconj(?,?,-,?,?), waitconj(?,?,?,-,?).
waitconj(success,success,_,E,E).
```

As one can see, predicate `waitconj` waits for the solution of both constraints.

The elements of our approach that most contribute to this simple transformation of Curry programs into Prolog programs are the implementation of concurrency and the use of both `case` and `or` expressions. Each function is transformed into a corresponding predicate to compute the head normal form of a call to this function. As shown above, this predicate contains additional arguments for storing the head normal form and controlling the suspension of function calls. Case expressions are implemented by evaluating the case argument to head

---

[4] An alternative to `block` is `freeze` which leads to a simpler transformation scheme. However, our experiments indicate that `freeze` is a more expensive operation (at least in Sicstus-Prolog Version 3#5). Using `freeze`, the resulting Prolog programs were approximately six times slower than using the scheme presented in this paper.

[5] As usual in the transformation of functions into predicates, we transform n-ary functions into $n+1$-ary predicates where the additional argument contains the result of the function call.

normal form. We use an auxiliary predicate to match the different cases. The difference between `flexcase` and `case` is only in the block declaration for the case argument. "`or`" expressions are implemented by alternative clauses and all other expressions are implemented by calls to predicate `hnf`, which computes the head normal form of its first argument. Thus, function `conc` in Example 1 is transformed into the following Prolog clauses:

```
:- block conc(?,?,?,-,?).
conc(A,B,R,Ein,Eout) :- hnf(A,HA,Ein,E1), conc_1(HA,B,R,E1,Eout).

:- block conc_1(-,?,?,?,?), conc_1(?,?,?,-,?).
conc_1([]     ,Ys,R,Ein,Eout) :- hnf(Ys              ,R,Ein,Eout).
conc_1([Z|Zs],Ys,R,Ein,Eout) :- hnf([Z|conc(Zs,Ys)],R,Ein,Eout).
```

Should `conc` be a flexible function, the block declaration `conc_1(-,?,?,?,?)` would be omitted, but the rest of the code would be unchanged. The definition of `hnf` is basically a case distinction on the different top-level symbols that can occur in an expression and a call to the corresponding function if there is a defined function at the top (compare [11,21]).

Although this code is quite efficient due to the first argument indexing of Prolog implementations, it can be optimized by partially evaluating the calls to `hnf`, as discussed in [4,11]. Further optimizations could be done if it is known at compile time that the evaluation of expressions will not cause any suspension (e.g., when all arguments are ground at run time). In this case, the additional two arguments in each predicate and the block declarations can be omitted and we obtain the same scheme as proposed in [11]. This requires static analysis techniques for Curry which is an interesting topic for further research.

3.2   Implementing Sharing

Every serious implementation of a lazy language must implement the sharing of common subterms. For instance, consider the rule

```
double x = x + x
```

and the expression "`double (1+2)`". If the two occurrences of the argument x in the rule's right-hand side are not shared, the expression 1+2 is evaluated twice. Thus, sharing the different occurrences of a same variable avoids unnecessary computations and is the prerequisite for optimal evaluation strategies [5]. In low level implementations, sharing is usually obtained by graph structures and destructive assignment of nodes [26]. Since a destructive assignment is not available in Prolog, we resort to Prolog's sharing of logic variables. This idea has been applied, e.g., in [8,11,20] where the predicates implementing functions are extended by a free variable that, after the evaluation of the function call, is instantiated to the computed head normal form. Although this avoids the multiple evaluation of expressions, it introduce a considerable overhead when no common subterms occur at run time—in some cases more than 50%, as reported in [11]. Therefore, we have developed a new technique that causes no overhead in all practical experiments we performed. As seen in the example above, sharing is

only necessary if terms are duplicated by a variable having multiple occurrences in a condition and/or right-hand side. Thus, we share these occurrences by
a special **share** structure containing the computed result of this variable. For
instance, the rule of **double** is translated into

```
double(X,R,E0,E1) :- hnf(share(X,EX,RX)+share(X,EX,RX),R,E0,E1).
```

In this way, each occurrence of a left-hand side variable X with multiple occurrences in the right-hand side is replaced by **share(X,EX,RX)**, where RX contains
the result computed by evaluating X. EX is bound to some constant if X has been
evaluated. EX is necessary because expressions can also evaluate to variables in
a functional logic language. Then, the definition of **hnf** is extended by the rule:

```
hnf(share(X,EX,RX),RX,E0,E1) :- !,
   (nonvar(EX) -> E1=E0
              ; hnf(X,HX,E0,E1), EX=eval, propagateShare(HX,RX)).
```

where **propagateShare(HX,RX)** puts **share** structures into the arguments of HX
(yielding RX) if HX is bound to a structure and the arguments are not already
shared.

This implementation scheme has the advantage that the Prolog code for rules without multiple variable occurrences remains unchanged and consequently
avoids the overhead for such rules (in contrast to [8,11,20]). The following table
shows the speedup (i.e., the ratio of runtime without sharing over runtime with
sharing), the number of reduction steps without (RS1) and with sharing (RS2),
and the number of shared variables (SV) in the right-hand side of rules of programs we benchmarked. It is worth to notice that the speedup for the first two
goals reported in [11], which uses a different technique, is 0.64 (i.e., a slowdown)
and 3.12. These values show the superiority of our technique.

Example:	Speedup	RS1	RS2	# SV
10000≤10000+10000 =:= True	1.0	20002	20002	0
double(double(one 100000)) =:= x	4.03	400015	100009	1
take 25 fibs	6650.0	196846	177	3
take 50 primes	15.8	298070	9867	2
quicksort (quicksort [...])	8.75	61834	3202	2
mergesort [...]	91.5	303679	1057	14

Program analysis techniques are more promising with our scheme than with
[8,11,20]. For instance, no **share** structures must be introduced for argument
variables that definitely do not contain function calls at run time, e.g., arguments
that are always uninstantiated or bound to constructor terms.

3.3   Constraints

Equational constraints, denoted e_1=:=e_2, are solved by lazily evaluating each
side to unifiable data terms. In our translation, we adopt the implementation of
this mechanism in Prolog presented in [21]. Basically, equational constraints are
solved by a predicate, **eq**, which computes the head normal form of its arguments
and performs a variable binding if one of the arguments is a variable.

```
:- block eq(?,?,-,?).
eq(A,B,Ein,Eout) :- hnf(A,HA,Ein,E1), hnf(B,HB,E1,E2),
                    eq_hnf(HA,HB,E2,Eout).

:- block eq_hnf(?,?,-,?).
eq_hnf(A,B,Ein,Eout) :- var(A), !, bind(A,B,Ein,Eout).
eq_hnf(A,B,Ein,Eout) :- var(B), !, bind(B,A,Ein,Eout).
eq_hnf(c(X_1,...,X_n),c(Y_1,...,Y_n),Ein,Eout) :- !,
    hnf((X_1=:=Y_1)&...&(X_n=:=Y_n),_,Ein,Eout).  % ∀n-ary constr. c

bind(X,Y,E,E) :- var(Y), !, X=Y.
bind(X,c(Y_1,...,Y_n),Ein,Eout) :- !,  % ∀n-ary constructors c
    occurs_not(X,Y_1),..., occurs_not(X,Y_n), X=c(X_1,...,X_n),
    hnf(Y_1,HY_1,Ein,E_1), bind(X_1,HY_1,E_1,E_2),
    ...
    hnf(Y_n,HY_n,E_2n-2,E_2n-1), bind(X_n,HY_n,E_2n-1,Eout).
```

Due to the lazy semantics of the language, the binding is performed incrementally. We use an auxiliary predicate, bind, which performs an occur check followed by an incremental binding of the goal variable and the binding of the arguments.

Similarly, the evaluation of an expression e to its normal form, which is the intended meaning of e, is implemented by a predicate, nf, that repeatedly evaluates all e's subexpressions to head normal form.

Apart from the additional arguments for controlling suspensions, this scheme is identical to the scheme proposed in [21]. Unfortunately, this scheme generally causes a significant overhead when one side of the equation is a variable and the other side evaluates to a large data term. In this case, the incremental instantiation of the variable is unnecessary and causes the overhead, since it creates a new data structure and performs an occur check. We avoid this overhead by evaluating to normal form, if possible, the term to which the variable must be bound. To this aim, we replace bind with bind_trynf in the clauses of eq_hnf together with the following new clause:

```
bind_trynf(X,T,Ein,Eout) :- nf(T,NT,Ein,E1),
        (nonvar(E1) -> occurs_not(X,NT), X=NT, Eout=E1
         ; bind(X,T,Ein,Eout)).
```

If the evaluation to normal form does not suspend, the variable X is bound to the normal form by X=NT, otherwise the usual predicate for incremental binding is called. Although this new scheme might cause an overhead due to potential re-evaluations, this situation did not occur in all our experiments. In some practical benchmarks, we have measured a speedup up to a factor of 2.

The compilation of Curry programs into Prolog greatly simplifies the integration of constraint solvers for other constraint structures, if the underlying Prolog system offers solvers for these structures. For instance, Sicstus-Prolog includes a solver for an arithmetic constraint over reals, which is denoted by enclosing the constraint between curly brackets. E.g., goal {3.5=1.7+X} binds X to 1.8. We make these constraints available in Curry by translating them into the corresponding constraints of Sicstus-Prolog. For instance, the inequational constraint

$e_1 < e_2$ is translated as follows. First, e_1 and e_2, which might contain user-defined functions or might be variables, are evaluated to their (head) normal forms, say e_1' and e_2'. Then, the goal $\{e_1' < e_2'\}$ is called. With this technique, all constraint solvers available in Sicstus-Prolog become available in Curry.

3.4 Further Features

Curry supports standard higher-order constructs such as lambda abstractions and partial applications. In Prolog, the higher-order features of Curry are implemented according to Warren's original proposal [30] to translate higher-order constructs into first-order logic programming. A lambda abstraction is eliminated by transforming it into a top-level definition of a new function. Consequently, the fundamental higher-order construct is a binary function, apply, which applies its first argument, a function, to its second argument, the function's intended argument. For each n-ary function or constructor f, we introduce $n - 1$ constructors with the same name. This enables us to implement the application function with the following Prolog clauses:

```
apply(f(X_1,...,X_k),X,f(X_1,...,X_k,X),E,E).     % 0 ≤ k < n-1
apply(f(X_1,...,X_{n-1}),X,H,E0,E) :- hnf(f(X_1,...,X_{n-1},X),H,E0,E).
```

Note that predicate apply should be called only for partial applications or applications where it is known at compile time that the first argument is not a defined function or a constructor. In other words, all first-order calls are directly translated without using apply as shown in the previous sections. This implementation of apply has the advantage that the unique matching clause is found in constant time due to the first argument indexing of Prolog systems. Although the number of apply clauses could be high for large applications, and there are alternative schemes that avoid this problem (e.g., [24]), we have found that this scheme causes no problems for programs with several hundred functions.

Monadic I/O is easily implemented by introducing a special constructor (denoted by "$io") to hold the result of an I/O action. For instance, getChar is implemented as a procedure which reads a character, c, from standard input and returns the term "$io c" whenever it is evaluated. With this approach, both sequential composition operators >>= and >> for actions are defined by:

```
($io x) >>= fa = fa x
($io _) >>  b  = b
```

Thus, the first action is evaluated to head normal form before the second action is applied. This simple implementation has, however, a pitfall. The result of an I/O action should not be shared, otherwise I/O actions will not be executed as intended. For instance, the expressions "putChar 'X' >> putChar 'X'" and "let a = putChar 'X' in a >> a" are equivalent but would produce different results with sharing. Luckily, the intended behavior can be obtained by a slight change of the definition of hnf so that terms headed by $io are not shared.

The primitives of Curry to encapsulate search and define new search strategies [17] cannot be directly implemented in Prolog due to its fixed backtracking strategy. However, one can implement some standard depth-first search strategies of Curry via Prolog's findall and bagof primitives.

4   Experimental Results

We have developed a compiler from Curry programs into Prolog programs (Sicstus-Prolog Version 3#5) based on the principles described in this paper. The practical results are quite encouraging. For instance, the execution of the classic "naive reverse" benchmark is executed at the speed of approximately 660,000 rule applications per second on a Linux-PC (Pentium II, 400 Mhz) with Sicstus-3 (without native code). Note that Curry's execution with a lazy strategy is costlier than Prolog's execution. Although the development of the compiler is relatively simple, due to the transformation schemes discussed in the paper, our implementation is competitive w.r.t. other high-level and low-level implementations of Curry and similar functional logic languages. We have compared our implementation to a few other implementations of declarative multi-paradigm languages available to us. The following table shows the results of benchmarks for various features of the language.

Program	Prolog	Toy	Java-1	Java-2	UPV-Curry
rev180	50	110	1550	450	43300
twice120	30	60	760	190	40100
qqsort20	20	20	230	45	72000
primes50	80	90	810	190	>2000000
lastrev120	70	160	2300	820	59700
horse	5	10	50	15	200
account	10	n.a.	450	670	2050
chords	220	n.a.	4670	1490	n.a.
Average speedup:		1.77	23.39	13.55	1150.1

All benchmarks are executed on a Sun Ultra-2. The execution times are measured in milliseconds. The column "Prolog" contains the results of the implementation presented in this paper. "Toy" [7] is an implementation of a narrowing-based functional logic language (without concurrency) which, like ours, compiles into Prolog. This implementation is based on the ideas described in [21]. "Java-1" is the compiler from Curry into Java described in [16]. It uses JDK 1.1.3 to execute the compiled programs. "Java-2" differs from the former by using JDK 1.2. This system contains a Just-in-Time compiler. Finally, UPV-Curry [1] is an implementation of Curry based on an interpreter written in Prolog that employs an incremental narrowing algorithm.

Most of the programs, which are small, test various features of Curry. "rev180" reverses a list of 180 elements with the naive reverse function. "twice120" executes the call "twice (rev l)", where twice is defined by "twice xs = conc xs xs" and l is a list of 120 elements. "qqsort20" calls quicksort (defined with higher-order functions) twice on a list of 20 elements. "primes50" computes the infinite list of prime numbers and extracts the first 50 elements. "lastrev120" computes the last element x of a list by solving the equation "conc xs [x] =:= rev [...]". "horse" is a simple puzzle that needs some search. "account" is a simulation of a bank account that uses the concurrency features of Curry. "chords", the largest of our benchmarks, is a musical application [15] that uses encapsulated search, laziness, and monadic I/O.

The comparison with Toy shows that our implementation of the concurrency features does not cause a significant overhead compared to a pure-narrowing-based language. Furthermore, the "account" example, which heavily uses concurrent threads, demonstrates that our implementation is competitive with an implementation based on Java threads. Although the table indicates that our implementation is superior to other available systems, implementations compiling to C or machine languages may be more efficient. However, the development effort of these lower level implementations is much higher.

5   Related Work and Conclusions

The idea of implementing functional logic programs by transforming them into logic programs is not new. An evaluation of different implementations is presented in [11], where it is demonstrated that functional logic programs based on needed narrowing are superior to other narrowing-based approaches. There are several proposals of compilation of needed narrowing into Prolog [4,11,21]. All these approaches lack concurrent evaluations. Moreover, the implementation of sharing, similar in all these approaches, is less efficient than in our proposal, as can be verified in the comparison table (see columns "Prolog" and "Toy").

Naish [24] has proposed NUE-Prolog, an integration of functions into Prolog programs obtained by transforming function definitions into Prolog clauses with additional "when" declarations. when declarations, which are similar in scope to the block declarations that we propose, suspend the function calls until the arguments are sufficiently instantiated. The effect of this suspension is that all functions are rigid—flexible functions are not supported. Functions intended to be flexible must be encoded as predicates by flattening. This approach has the drawback that optimal evaluation strategies [5] cannot be employed for the logic programming part of a program. Strict and lazy functions can be freely mixed, which makes the meaning of programs harder to understand (e.g., the meaning of equality in the presence of infinite data structures). NUE-Prolog uses a form of concurrency for suspending function calls, as we do. But it is more restrictive in that there is no possibility to wait for the complete evaluation of an expression. This leads to the undesired behavior discussed in Example 2.

Apart from the efficiency and simplicity of our transformation scheme of Curry into Prolog programs, the use of Prolog as a target language has further advantages. A high-level implementation more easily accomodates the inclusion of additional features. For instance, the implementation of a standard program tracer w.r.t. Byrd's box model [6] requires only the addition of four clauses to each program and two predicate calls for each implemented function. The most important advantage is the reuse of existing constraint solvers available in Prolog, as shown in Section 3.3. Thus, with a limited effort, we obtain a usable implementation of a declarative language that combines constraint solving over various constraint domains, concurrent evaluation and search facilities from logic programming with higher-order functions and laziness from functional programming. The combination of laziness and search is attractive because it offers a modular implementation of demand-driven search strategies, as shown in [15].

Since the compilation time of our implementation is reasonable,[6] this Prolog-based implementation supports our current main development system for Curry programs.[7] This system has been used to develop large distributed applications with sophisticated graphical user interfaces and Web-based information servers that run for weeks without interruption (see [13] for more details). By taking advantage of both the features of our system and already developed code, we can make available on the Internet constraint programming applications in minutes.

References

1. M. Alpuente, S. Escobar, and S. Lucas. UPV-Curry: an Incremental Curry Interpreter. In *Proc. of 26th Seminar on Current Trends in Theory and Practice of Informatics (SOFSEM'99)*, pp. 327–335. Springer LNCS 1725, 1999.
2. S. Antoy. Non-Determinism and Lazy Evaluation in Logic Programming. In *Proc. Int. Workshop on Logic Program Synthesis and Transformation (LOPSTR'91)*, pp. 318–331. Springer Workshops in Computing, 1991.
3. S. Antoy. Definitional Trees. In *Proc. of the 3rd International Conference on Algebraic and Logic Programming*, pp. 143–157. Springer LNCS 632, 1992.
4. S. Antoy. Needed Narrowing in Prolog. Technical Report 96-2, Portland State University, 1996.
5. S. Antoy, R. Echahed, and M. Hanus. A Needed Narrowing Strategy. Journal of the ACM (to appear). Previous version in *Proc. 21st ACM Symposium on Principles of Programming Languages*, pp. 268–279, 1994.
6. L. Byrd. Understanding the Control Flow of Prolog Programs. In *Proc. of the Workshop on Logic Programming*, Debrecen, 1980.
7. R. Caballero-Roldán, J. Sánchez-Hernández, and F.J. López-Fraguas. User's Manual for TOY. Technical Report SIP 97/57, Universidad Complutense de Madrid, 1997.
8. P.H. Cheong and L. Fribourg. Implementation of Narrowing: The Prolog-Based Approach. In K.R. Apt, J.W. de Bakker, and J.J.M.M. Rutten, editors, *Logic programming languages: constraints, functions, and objects*, pp. 1–20. MIT Press, 1993.
9. E. Giovannetti, G. Levi, C. Moiso, and C. Palamidessi. Kernel LEAF: A Logic plus Functional Language. *Journal of Computer and System Sciences*, Vol. 42, No. 2, pp. 139–185, 1991.
10. M. Hanus. The Integration of Functions into Logic Programming: From Theory to Practice. *Journal of Logic Programming*, Vol. 19&20, pp. 583–628, 1994.
11. M. Hanus. Efficient Translation of Lazy Functional Logic Programs into Prolog. In *Proc. Fifth International Workshop on Logic Program Synthesis and Transformation*, pp. 252–266. Springer LNCS 1048, 1995.

---

[6] Since the transformation of Curry programs into Prolog is fast, the overall compilation time mainly depends on the time it takes to compile the generated Prolog code. In our tests, it takes only a few seconds even for programs with approximately 2,000 lines of source code.

[7] http://www-i2.informatik.rwth-aachen.de/~hanus/pacs/

12. M. Hanus. A Unified Computation Model for Functional and Logic Programming. In *Proc. of the 24th ACM Symposium on Principles of Programming Languages*, pp. 80–93, 1997.
13. M. Hanus. Distributed Programming in a Multi-Paradigm Declarative Language. In *Proc. of the International Conference on Principles and Practice of Declarative Programming (PPDP'99)*, pp. 376–395. Springer LNCS 1702, 1999.
14. M. Hanus and C. Prehofer. Higher-Order Narrowing with Definitional Trees. *Journal of Functional Programming*, Vol. 9, No. 1, pp. 33–75, 1999.
15. M. Hanus and P. Réty. Demand-driven Search in Functional Logic Programs. Research Report RR-LIFO-98-08, Univ. Orléans, 1998.
16. M. Hanus and R. Sadre. An Abstract Machine for Curry and its Concurrent Implementation in Java. *Journal of Functional and Logic Programming*, Vol. 1999, No. 6, 1999.
17. M. Hanus and F. Steiner. Controlling Search in Declarative Programs. In *Principles of Declarative Programming (Proc. Joint International Symposium PLILP/ALP'98)*, pp. 374–390. Springer LNCS 1490, 1998.
18. M. Hanus (ed.). Curry: An Integrated Functional Logic Language (Vers. 0.6). Available at http://www-i2.informatik.rwth-aachen.de/~hanus/curry, 1999.
19. S. Janson and S. Haridi. Programming Paradigms of the Andorra Kernel Language. In *Proc. 1991 Int. Logic Programming Symposium*, pp. 167–183. MIT Press, 1991.
20. J.A. Jiménez-Martin, J. Marino-Carballo, and J.J. Moreno-Navarro. Efficient Compilation of Lazy Narrowing into Prolog. In *Proc. Int. Workshop on Logic Program Synthesis and Transformation (LOPSTR'92)*, pp. 253–270. Springer Workshops in Computing Series, 1992.
21. R. Loogen, F. Lopez Fraguas, and M. Rodríguez Artalejo. A Demand Driven Computation Strategy for Lazy Narrowing. In *Proc. of the 5th International Symposium on Programming Language Implementation and Logic Programming*, pp. 184–200. Springer LNCS 714, 1993.
22. J.J. Moreno-Navarro and M. Rodríguez-Artalejo. Logic Programming with Functions and Predicates: The Language BABEL. *Journal of Logic Programming*, Vol. 12, pp. 191–223, 1992.
23. L. Naish. *Negation and Control in Prolog*. Springer LNCS 238, 1987.
24. L. Naish. Adding equations to NU-Prolog. In *Proc. of the 3rd Int. Symposium on Programming Language Implementation and Logic Programming*, pp. 15–26. Springer LNCS 528, 1991.
25. J. Peterson et al. Haskell: A Non-strict, Purely Functional Language (Version 1.4). Technical Report, Yale University, 1997.
26. S.L. Peyton Jones. *The Implementation of Functional Programming Languages*. Prentice Hall, 1987.
27. V.A. Saraswat. *Concurrent Constraint Programming*. MIT Press, 1993.
28. G. Smolka. The Oz Programming Model. In *Computer Science Today: Recent Trends and Developments*, pp. 324–343. Springer LNCS 1000, 1995.
29. P. Wadler. How to Declare an Imperative. *ACM Computing Surveys*, Vol. 29, No. 3, pp. 240–263, 1997.
30. D.H.D. Warren. Higher-order extensions to PROLOG: are they needed? In *Machine Intelligence 10*, pp. 441–454, 1982.

Modular Redundancy for Theorem Proving

Miquel Bofill[1], Guillem Godoy[2], Robert Nieuwenhuis[2], and Albert Rubio[2]*

[1] Universitat de Girona, Dept. IMA,
Lluís Santaló s/n, 17071 Girona, Spain
mbofill@ima.udg.es
[2] Technical University of Catalonia, Dept. LSI,
Jordi Girona 1, 08034 Barcelona, Spain
{ggodoy,roberto,rubio}@lsi.upc.es

Abstract. We introduce a notion of modular redundancy for theorem proving. It can be used to exploit redundancy elimination techniques (like tautology elimination, subsumption, demodulation or other more refined methods) in combination with arbitrary existing theorem provers, in a refutation complete way, even if these provers are not (or not known to be) complete in combination with the redundancy techniques when applied in the usual sense.

1 Introduction

The concept of *saturation* in theorem proving is nowadays a well-known, widely recognized useful concept. The main idea of saturation is that a theorem proving procedure does not need to compute the *closure* of a set of formulae w.r.t. a given inference system, but only the closure *up to redundancy*. Examples of early notions of redundancy (in the context of resolution) are the elimination of tautologies and subsumption. Bachmair and Ganzinger gave more general abstract notions of redundancy for inferences and formulae (see, e.g., [BG94]), where, roughly, something is redundant if it is a logical consequence of smaller logical formulae (w.r.t. a given ordering on formulae). In provers based on ordered resolution, paramodulation or superposition w.r.t. a total reduction ordering on ground terms >, apart from standard methods like removing tautologies, subsumption, or demodulation w.r.t. >, these abstract redundancy notions cover a large number of other practical techniques [NN93,GNN98].

However, in other contexts refutation completeness is lost if such redundancy elimination methods are applied. For example, the elimination of certain classes of tautologies is incompatible with logic programming with equality and arbitrary selection rules [Lyn97] and it is also incompatible with *strict superposition* [BG98]. Another situation where even less redundancy can be used is in the context of paramodulation with non-monotonic orderings [BGNR99], which is needed when monotonic orderings are not available, like in Knuth-Bendix completion of arbitrary terminating term rewrite systems, or in deduction modulo

---

* All authors partially supported by the ESPRIT Basic Research Action CCL-II, ref. WG # 22457 and the CICYT project HEMOSS ref. TIC98-0949-C02-01.

H. Kirchner and C. Ringeissen (Eds.): FroCoS 2000, LNAI 1794, pp. 186–199, 2000.

certain equational theories E for which no monotonic *E-compatible* orderings exist. Apart from these situations where it is known that the standard redundancy techniques cannot be used, there is also a large number of other cases where the compatibility with redundancy is unknown, or where for the completeness proof techniques applied it is unclear how to extend them to include redundancy, or where the theorem proving method is simply not based on the closure under a set of inference rules.

For all these cases our aim is to provide here modular notions of redundancy. Assume we have an unsatisfiable initial set of formulae S_0, and we want to prove its unsatisfiability by some refutation complete prover $\mathcal{P}$ that is a "black box", i.e., apart from its refutation completeness nothing is known about it. In particular, nothing is known about its compatibility w.r.t. redundancy elimination methods. In principle, the prover $\mathcal{P}$ may be based on any technique, but in our context it will be useful that $\mathcal{P}$ soundly generates new formulae that are consequences from its input (this happens, for instance, if $\mathcal{P}$ is based on a sound inference system).

Now the problem we address is: how, and up to which degree, can we still exploit redundancy elimination methods in this general context?

One instance of our ideas roughly amounts to the following. Apart from the prover $\mathcal{P}$, there is a *redundancy module* $\mathcal{R}$ that works with a given arbitrary well-founded ordering $>$ on formulae (this ordering $>$ may be completely different from the one $\mathcal{P}$ uses, if $\mathcal{P}$ is based on some ordered inference system). The redundancy module $\mathcal{R}$ keeps track of a set of formulae U that is assumed to be unsatisfiable. Initially, U is S_0. Each time a new formula is generated by $\mathcal{P}$, this is communicated to $\mathcal{R}$, who uses this information to try to prove the redundancy of some formula F of U, i.e., it checks whether F follows from smaller (w.r.t. $>$) formulae $G_1, \ldots, G_n$ that are in U or have been generated by $\mathcal{P}$ (or are logical consequences of these). Each time this is the case, a new set U is obtained by replacing F by $G_1, \ldots, G_n$. This fact is then communicated to $\mathcal{P}$, whose aim becomes to refute the new U instead of the previous one. In a naive setting, $\mathcal{P}$ could simply restart with the new U, but of course one can do better. Instead, $\mathcal{P}$ can apply several techniques (that may already be implemented in $\mathcal{P}$) in order to exploit the new situation. For example, $\mathcal{P}$ can apply *orphan murder* techniques: the descendants of F that have already been computed by $\mathcal{P}$ can be eliminated (although some of them can be kept if we want to completely avoid the repeated computation of formulae; for this, some additional bookkeeping is required). In this procedure, U is a finite set of formulae that decreases w.r.t. the well-founded multiset extension $\gg$ of $>$ each time it is modified. Hence at some point of the procedure U will not be modified any more and its limit U_∞ is obtained. If $\mathcal{P}$ computes all inferences from U_∞, then the refutation completeness of the general procedure follows.

Our techniques have been conceived with first-order clausal logic in mind. However, note that we do not rely on any particular logic.

This paper is structured as follows. In Section 2 we give a formal description of prover modules, redundancy modules, their combination and some first results. In Section 3 we explain how a redundancy module with the right properties can be built. Section 4 explains for three cases how to build a prover module in terms of an existing prover. First, this is done for an (almost) arbitrary prover, in a rather naive way where not much power of the redundancy module is exploited. Second, it is done for a prover based on the closure with respect to a complete inference system. And third, it is done for saturation-based provers (which is the case in most applications, like the ones of [Lyn97], [BG98], and [BGNR99]). Finally in Section 5 we give some conclusions.

2   Modular Redundancy

We now give a formal description of prover modules, redundancy modules, their combination and some first results.

Definition 1. *A* prover module *is a procedure satisfying:*

- *it takes as input a set S_0 of formulae*
- *at any point of its execution, it may receive messages of the form "replace F by $G_1, \ldots, G_n$", where F and $G_1, \ldots, G_n$ are formulae*
- *at any point of its execution, it may produce messages of the form "F has been deduced", where F is a formula*
- *it may halt with output "unsatisfiable".*

Definition 2. *Let $\mathcal{P}$ be a prover module, and consider a given execution of $\mathcal{P}$. We denote by S_0 the input set of formulae, and for each $k > 0$ by S_k we denote the set $S_k = S_{k-1} \cup \{G_1, \ldots, G_n\}$, where "replace F by $G_1, \ldots, G_n$" is the k-th message it receives.*

Furthermore, we denote by U_0 the input set of formulae, and for each $k > 0$ by U_k we denote the set $U_k = (U_{k-1} \setminus \{F\}) \cup \{G_1, \ldots, G_n\}$, where "replace F by $G_1, \ldots, G_n$" is the k-th message it receives.

Definition 3. *A prover module $\mathcal{P}$ is* correct *w.r.t. the given logic if:*

- *$S_k \models F$ whenever $\mathcal{P}$ emits an output message "F has been deduced" after receiving its k-th message.*
- *if $\mathcal{P}$ halts with output "unsatisfiable" after receiving its k-th message then S_k is unsatisfiable.*

Definition 4. *A prover module $\mathcal{P}$ is* refutation complete *if, for every execution where it receives exactly k messages, it halts with output "unsatisfiable" whenever U_k is unsatisfiable.*

Definition 5. *A* redundancy module *is a procedure that satisfies:*

- *it takes as input a set of formulae*
- *at any point of its execution, it may produce messages of the form "replace F by $G_1, \ldots, G_n$", where F and $G_1, \ldots G_n$ are formulae*

− *at any point of its execution, it may receive messages of the form "F has been deduced", where F is a formula.*

Definition 6. *Let $\mathcal{R}$ be a redundancy module, and consider a given execution of $\mathcal{R}$. We denote by R_0 the input set of formulae, and for each $k \geq 0$ by R_k we denote the set $R_k = R_{k-1} \cup \{F\}$, where "F has been deduced" is the k-th received message.*

Definition 7. *A redundancy module is* correct *if $R_k \models \{G_1, \ldots, G_n\}$ and G_1, ..., $G_n \models F$ whenever it outputs a message "replace F by $G_1, \ldots, G_n$" after receiving its k-th message.*

Definition 8. *A redundancy module is* finitary *if for every execution the number of output messages produced is finite.*

A prover module $\mathcal{P}$ can be combined with a redundancy module $\mathcal{R}$ in the expected way to obtain what we will call a theorem prover: both modules receive the same input set of formulae, and the output messages of each one of them are the received messages of the other one. The theorem prover halts if and only if $\mathcal{P}$ halts. Let us remark that it is of course not our aim to discuss priority or synchronization issues or the like between both modules; one can simply assume that $\mathcal{R}$ uses a finite amount of cpu time initially and after each received message; if this time is finished $\mathcal{R}$ (possibly after sending a message) transfers control to $\mathcal{P}$ until $\mathcal{P}$ emits the next message.

Theorem 1. *A correct and complete prover module in combination with a correct and finitary redundancy module is a correct and refutation complete theorem prover, that is, a theorem prover that halts with output "unsatisfiable" if, and only if, the input set of formulae S_0 is unsatisfiable.*

Proof. From the correctness of both modules it follows that for all i, all sets S_i, U_i, and R_i are logically equivalent to S_0 (by induction on the subindex i: the new formulae coming from a message that is sent are a logical consequence of these sets with smaller index). Since $\mathcal{R}$ is finitary it sends only k messages, for some k. If S_0 is unsatisfiable, the corresponding set U_k is unsatisfiable, and by completeness of $\mathcal{P}$ the prover will halt with output "unsatisfiable". Conversely, if $\mathcal{P}$ halts with output "unsatisfiable" after receiving its j-th message, then, by correctness of $\mathcal{P}$, the set S_j and hence S_0 is unsatisfiable.   □

3   Finitary Redundancy Modules

Here we show how to create a finitary redundancy module, given a well-founded ordering $>$ on formulae. In order to abstract from concrete algorithms and strategies that can be used in a redundancy module, i.e., to be as general as possible, we first model the execution of a redundancy module by the notion of a *redundancy derivation*, a sequence of pairs of sets of formulae. The sets R_i correspond to what we defined in the previous section. The U_i will play the role of sets of formulae that are known to be unsatisfiable as the set U mentioned in the introduction (hence the name U, for unsatisfiable).

Definition 9. *We model the execution of a redundancy module by the notion of a* redundancy derivation, *a sequence* $(U_0, R_0), (U_1, R_1), \ldots$ *where all* U_i *and* R_i *are sets of formulae, and where the transition from a pair* (U_i, R_i) *to a pair* (U_{i+1}, R_{i+1}) *takes place whenever a message is received or sent:*

- $U_0 = R_0 = S_0$, *where* S_0 *is the input set of formulae.*
- *if a message "F has been deduced" is received then* $U_{i+1} = U_i$ *and* $R_{i+1} = R_i \cup \{F\}$
- *if a message "replace F by $G_1 \ldots G_n$" is sent then* $U_{i+1} = U_i \setminus \{F\} \cup \{G_1 \ldots G_n\}$ *and* $R_{i+1} = R_i$.

We now impose conditions ensuring that the redundancy module satisfies the correctness and finiteness requirements:

Definition 10. *Let* $>$ *be a well-founded ordering on formulae. We say that a formula F is* redundant *w.r.t.* $>$ *by* $G_1 \ldots G_n$ *if* $\{G_1 \ldots G_n\} \models F$ *and* $F > G_i$ *for all i in* $1 \ldots n$.

Let $\mathcal{R}_>$ *be a redundancy module modeled by a redundancy derivation* (U_0, R_0), $(U_1, R_1), \ldots$ *such that if* $U_{i+1} = U_i \setminus \{F\} \cup \{G_1 \ldots G_n\}$, *i.e., a message "replace F by $G_1 \ldots G_n$" is sent, then*

- $R_i \models \{G_1 \ldots G_n\}$ *and*
- *F is a formula in* U_i *that is redundant by* $G_1 \ldots G_n$

A practical redundancy module implementing $\mathcal{R}_>$ can be a procedure that updates the set R whenever it receives a message, and keeps looking for formulae $G_1 \ldots G_n$ that can be used for proving the redundancy of some F in the set U, by means of (probably) well-known redundancy elimination techniques, and then replacing F by these $G_1 \ldots G_n$. For example, one can simplify F by rewriting (demodulating) applying some equational formula G_1 of U, and replace F by its simplified version G_2. In practice it will be useful to apply formulae G_k that are in U, since, on the one hand, this helps keeping the set U small, which will be useful, and on the other hand, by construction all formulae in each U_i are logical consequences of R_i, so this does not have to be checked.

Theorem 2. $\mathcal{R}_>$ *is a finitary redundancy module.*

Proof. Let $\gg$ denote the multiset extension of $>$, that is, the smallest ordering such that $S \cup \{F\} \gg S \cup \{F_1, \ldots, F_n\}$ whenever $F > F_i$ for all i in $1 \ldots n$. The ordering $\gg$ is a well-founded ordering on finite (multi)sets of formulae whenever $>$ is a well-founded ordering on formulae (see e.g., [Der87]).

Indeed every U_i is a finite set of formulae (by induction on i: initially it is, and at each step only a finite number of formulae is added). Moreover U decreases wrt. $\gg$ at each message sent. Hence the number of output messages must be finite. □

Theorem 3. $\mathcal{R}_>$ *is a correct redundancy module.*

Proof. ¿From the definition of $\mathcal{R}_>$, whenever a message "replace F by $G_1 \ldots G_n$" is sent then $R_i \models G_1, \ldots, G_n$ and $G_1, \ldots, G_n \models F$. This implies correctness. □

4   How Do We Obtain a Prover Module from a Given Prover

4.1   A Prover Module from Any Prover

It is easy to see that, in a naive way, a complete prover module can be obtained from any correct and refutation complete theorem prover A (with the name A for any): given an unsatisfiable set of formulae S as input, A halts with output "unsatisfiable" after finite time. The main idea is simply to restart A with the new set U_k every time a message "replace F by $G_1, \ldots, G_n$" is received.

Instead of formally defining the communications between A and the "layer" around it, that is, the remaining part of the prover module that uses A, we assume, without going into these details, that the layer knows whether A has inferred a formula, and that it is possible to tell A by means of the operation $Restart(S)$ that it has to restart from scratch with input the set of formulae S.

We now define the prover module in terms of the prover A. In this prover module there will be a set of formulae U that will play the same role as before (see Definition 2) and as the set U in the redundancy module. It will be maintained by this prover module according to this invariant. Hence, initially U is the input set S_0 and each time a received message "replace F by $G_1 \ldots G_n$" is treated, the prover module will update U by doing $U := (U \setminus \{F\}) \cup \{G_1 \ldots G_n\}$.

Summarizing, each time a message "replace F by $G_1 \ldots G_n$" is received we do: $U := (U \setminus \{F\}) \cup \{G_1 \ldots G_n\}$, and $Restart(U)$. Each time the prover A infers F, a message "F has been deduced" is sent.

Definition 11. *Let $\mathcal{P}_A$ denote the prover module as defined above.*

Theorem 4. *If A is a correct and refutationally complete prover, then $\mathcal{P}_A$ is a correct and complete prover module.*

Proof. We clearly have $U = U_k$ after the k-th message in $\mathcal{P}_A$. And moreover $S_k \models U_k$. Then, correctness of $\mathcal{P}_A$ follows from correctness of A and from the fact that A is restarted with input U_k after every received message. For completeness, suppose $\mathcal{P}_A$ receives exactly k messages. We have $U = U_k$ after the k-th message. Then, completeness of $\mathcal{P}_A$ follows from completeness of A, given that $\mathcal{P}_A$ just restarts A with input U_k after the k-th received message. □

It is doubtful that this naive use of A will lead to any practical usefulness, since restarting A from scratch is a very drastic measure that produces large amounts of repeated work. A possible alternative would be not to restart A after each received message, but restart it each n units of time for an arbitrary fixed n. Obviously this does not affect completeness. Another better solution is given in the next subsection, where no work is repeated due to restarts, and an important amount of unnecessary work is avoided.

4.2   A Prover Module Built from a Closure Prover

Here we consider a given prover C (with the name C for closure) that is simply based on some correct and refutation complete inference system, and computes the closure w.r.t. this inference system using some strategy. Furthermore, it is assumed that this is done in such a way that the result is a complete prover: given an unsatisfiable set of formulae S as input, it halts with output "unsatisfiable", since it finds a contradictory formula, denoted by $\square$.

In the following, this kind of prover will be called a *closure prover*. We will show that from C we can build a correct and complete prover module if C fulfils some minimal requirements needed in order to interact with it. The resulting prover module, in combination with a redundancy module, can reasonably well exploit our modular redundancy notions by means of *orphan murder*.

We now define the prover module in terms of the prover C.

In this prover module there will be two disjoint sets of formulae, U and P, that will be maintained by this prover module with the following invariants: the set U will play the same role as before (see Definition 2), and P is the set of remaining formulae stored by the prover C. Hence, initially U is the input set S_0, and P is empty, and each time a received message "replace F by $G_1 \ldots G_n$" is treated, the prover module will update P and U by doing $P := P \setminus \{G_1 \ldots G_n\}$ and $U := (U \setminus \{F\}) \cup \{G_1 \ldots G_n\}$ (note that this may cause F to disappear from $U \cup P$). Similarly, each time the prover C infers a new formula F, the prover module will update P by doing $P := P \cup \{F\}$.

Furthermore, we associate to each F in $U \cup P$ a set of sets of formulae, denoted by $Parents(F)$, with the invariant that the sets $\{F_1, \ldots, F_n\}$ in $Parents(F)$ correspond, one to one, to the inference steps with premises $F_1, \ldots, F_n$ and conclusion F that have taken place in the prover C and none of $F_1, \ldots, F_n$ have been removed after the inference step. In order to maintain these invariants, each time such an inference takes place, $\{F_1, \ldots, F_n\}$ will be added to $Parents(F)$ if F already existed in $U \cup P$, and $Parents(F)$ will be $\{\{F_1, \ldots, F_n\}\}$ if F is new. Similarly, since we require that all formulae occurring in $Parents(F)$ have not been removed afterwards from $U \cup P$, if $\{F_1, \ldots, F_n\} \in Parents(F)$ and some F_i is removed from $U \cup P$ then the set $\{F_1, \ldots, F_n\}$ is removed from $Parents(F)$.

In this context, the idea of orphan murder is the following. If for some formula F in P (and hence $F \notin U$) we have $Parents(F) = \emptyset$ then F appeared because it was inferred, in one or more ways, by the prover C, but, for each one of these ways, the premises do not exist any more. Hence F itself is not needed for the completeness of the prover C and can be removed as well. This triggers new eliminations of other formulae G of P if F occurs in all sets of $Parents(G)$, and so on. We assume that each time a formula is eliminated in this way, this fact can be communicated to the prover C by an operation $Remove(F)$, which has the effect that C simply removes F from its data structures. This operation has its counterpart $Add(S)$, which tells C to consider the formulae of a set of formulae S as additional input formulae.

Expressing this in an algorithmic way, we obtain the following. Initially U is the input set S_0, and P is empty. Each time a message "replace F by $G_1 \ldots G_n$" is received we do:

{ Precondition: $F \in U$ **and** $Parents(F) = \emptyset$ }
$Remove(F)$
$Add(\{G_1, \ldots, G_n\} \setminus (U \cup P))$
$P := P \setminus \{G_1 \ldots G_n\}$
$U := (U \setminus \{F\}) \cup \{G_1 \ldots G_n\}$
While $\exists F' \in (U \cup P)$ **and** $\exists S' \in Parents(F')$ **and** $S' \not\subseteq (U \cup P)$ **Do**
    $Parents(F') := Parents(F') \setminus \{S'\}$
    If $Parents(F') = \emptyset$ **and** $F' \in P$ **Then**
        $P := P \setminus \{F'\}$
        $Remove(F')$
    EndIf
EndWhile

As a precondition we assume that $F \in U$ and $Parents(F) = \emptyset$ since otherwise completeness may be lost. This precondition can be ensured in practice by an additional communication mechanism between both modules, for example, by making the redundancy module ask permission before it sends a message. On the other hand, each time the prover C infers F from premises $G_1 \ldots G_n$, a message "F has been deduced" is sent, and we do:

If $F \in (U \cup P)$ **Then**
    $Parents(F) := Parents(F) \cup \{\{G_1, \ldots, G_n\}\}$
Else
    $P := P \cup \{F\}$
    $Parents(F) := \{\{G_1, \ldots, G_n\}\}$
EndIf

Note that, instead of formally defining the interactions between C and the "layer" around it, that is, the remaining part of the prover module that uses C, we have assumed, without going into these details, that the layer knows whether C has inferred a formula from certain premises, and that it is possible to tell C by means of the operations $Remove(F)$ and $Add(S)$ that a certain formulae can be removed or have to be added.

Definition 12. *Let $\mathcal{P}_C$ denote the prover module as defined above.*

We will now prove that the prover module $\mathcal{P}_C$ is correct and complete. For this purpose we need a reasonable additional assumption on the prover C, namely a notion of fairness:

Definition 13. *A formula F is* persistent *in C if it is stored by C at some point and not removed by the operation remove(F) later on.*
    The prover C is called fair *if, whenever $G_1, \ldots, G_n$ become persistent formulae at some point of the execution, and there is a possible inference with them, then this inference is performed after that point.*

Lemma 1. *Given an execution of $\mathcal{P}_C$, if $G_1, \ldots, G_n$ are persistent formulae in C (and hence in $U \cup P$), and there is a possible inference with them producing F, then F is a persistent formula in C.*

Proof. By fairness, this inference is done at some point of the execution after $G_1, \ldots, G_n$ have become persistent. Furthermore $\{G_1, \ldots, G_n\} \in Parents(F)$ at that point. Moreover $\{G_1, \ldots, G_n\}$ will remain in $Parents(F)$. Then F will not be removed since $Parents(F)$ will never be empty.   □

Theorem 5. $\mathcal{P}_C$ *is a correct prover module.*

Proof. Let S_k be defined as in Definition 2. Then we will show that $S_k \models (U \cup P)$ after the k-th received message and before a new message is received, and the correctness of the procedure will follow trivially. It suffices to show that this property is satisfied initially, and it is preserved during the execution. Initially $U \cup P$ is S_0. When an inference is done, its conclusion F, which is added to P, is a consequence of $(U \cup P)$, and therefore, of the corresponding S_k. When a message is received, $S_k = S_{k-1} \cup \{G_1, \ldots, G_n\}$ and nothing else but $\{G_1, \ldots, G_n\}$ is added to $(U \cup P)$.   □

Lemma 2. *Let U_k be defined as in Definition 2. Then $U_k = U$ after the k-th received message in $\mathcal{P}_C$.*

Theorem 6. $\mathcal{P}_C$ *is a complete prover module.*

Proof. We have to show that, in an execution of our prover module with exactly k input messages, if U_k is unsatisfiable then the prover module halts with output "unsatisfiable". By the previous lemma, $U_k = U$ immediately after the k-th received message. Furthermore, by fairness all inferences with premises in U are performed after the point their premises become persistent. By Lemma 1 the corresponding conclusions are persistent as well. All these descendants will be persistent (by iterating this, i.e., by induction on the number of inference steps required to obtain a descendant F from U). Hence all formulae in the closure of U w.r.t. the given inference system are indeed retained by C, which implies that □ will be inferred since the inference system of C is refutation complete.   □

In this subsection we have developed the prover module in terms of a prover that fulfils a reasonable fairness requirement. We could have followed an alternative approach, without this fairness requirement. Instead of explaining this alternative here for closure prover, we develop it for saturation provers in the next subsection. It will be clear that its application to closure provers will be a particular case.

4.3   A Prover Module from a Saturation Prover

Here we consider a standard saturation-based prover D (with the name D for prover with *deletion* of redundant formulae). Without loss of generality, such a prover D can be assumed to consist of some *strategy* controlling the application of only two basic operations: adding a new formula inferred by some correct deduction system, and removing a formula F whose redundancy follows from other formulae $F_1, \ldots, F_n$ w.r.t. some *internal* redundancy criterion with the minimal requirement that $F_1, \ldots, F_n \models F$ and a kind of transitivity property, namely if

F is redundant by the set of formulae $F_1, \ldots, F_n$ and F_1 is redundant by the set of formulae $G_1, \ldots, G_m$ then F is redundant by the set $G_1, \ldots, G_m, F_2, \ldots, F_n$. Note that redundancy techniques like demodulation of F into F' can be expressed by a combination of these two steps: first add F' (which is a logical consequence), then remove F (which has become redundant).

As usual, a prover like D can be modelled by a derivation, i.e., a sequence of sets of formulae, of the form $D_0, D_1, \ldots$ where at each step the strategy decides by which concrete instance of one of the two basic operations D_{i+1} is obtained from D_i. Of course, this is done in such a way that the result is a complete prover. For technical reasons that will become clear soon, here we will make a reasonable additional assumption, namely that if, for a finite number of basic steps of the derivation (i.e., for a finite number of decisions about which basic step is performed next), the strategy is overruled by an external event, like user interaction, the prover D remains complete. Note that the usual internal fairness requirements of saturation-based provers are not affected by such a finite number of "wrong" decisions.

As in the previous subsection, here we assume that the prover D also provides the operations *Add* and *Remove*, and provides information each time it infers a formula F from a set of premises $F_1, \ldots, F_r$. In addition, here we will assume that each time the prover removes a formula F that is redundant by a set $F_1, \ldots, F_r$, this will be communicated as well.

In a practical prover, its strategy can be based on many different types of information (stored in data structures, like priority queues based on weighting functions) and that may depend on the current state as well as on historical information about previous states. Hence it has to be discussed how to deal with the external operations *Add* and *Remove* that were not foreseen when the strategy was developed. For instance, in a practical prover where a new formula F is added with the operation $Add(F)$, this F has to be inserted in the different data structures on which the strategy is based.

In the most general setting the strategy can be based on the whole *history* of the current derivation. Then the situation is as follows. Suppose that an operation *Add* or *Remove* takes place after step n of a derivation $D_0, \ldots, D_n$, and this operation converts D_n into another set S. Of course the strategy of the prover cannot be expected to behave well any more if it continues based on the history of the derivation $D_0, \ldots, D_n$. But fortunately, our prover module will only perform operations *Add* and *Remove* in such a way that another very similar derivation $D'_0, \ldots, D'_m$ exists where D'_m is S. Hence it is reasonable to expect that the prover D will behave well if it continues with the strategy it would have followed if it had reached S by this alternative similar derivation.

Therefore, we will assume that it is possible to update the internal state of the prover (on which its strategy is based) into the state it would be in after m basic steps of the alternative derivation $D'_0, \ldots, D'_m$. This is how we will proceed.

However, note that, although this alternative sequence consists of j correct steps, it may not be a derivation with the strategy. But since we know that the prover D is complete even if, for a finite number of decisions, the strategy is overruled by an external event, intuitively, the completeness of the prover module follows, since there will only be a finite number of such external events.

We now define the prover module in terms of the prover D. Again there will be sets U and P and $Parents(F)$ for each F in $U \cup P$. As before $U \cup P$ contains the formulae kept by the prover D, but, due to the internal redundancy, U may not be U_k but a subset of it. In addition, each formula in $U \cup P$ can also be *blocked* for external removals, i.e., it will never be removed by a $Remove(F)$ operation, and we denote by $Blocked(U \cup P)$ the set of blocked formulae in $U \cup P$. In this case we only need the following invariant: $U \cup Blocked(U \cup P) \models U_k$. No assumption is needed about the set $Parents(F)$ in order to prove correctness and completeness of the prover module. Nevertheless, as we will see at the end of this section, the way we deal with these sets in combination with the way we construct the alternative derivations will allow us in most of the cases to avoid repeated work. Expressing the previous ideas algorithmically, we obtain the following. Initially U is the input set S_0, and P is empty.

Each time a message "replace F by $G_1 \dots G_r$" is received we do:

> { Precondition: $F \in U$ and $Parents(F) = \emptyset$ and F is not blocked }
> $Remove(F)$
> $Add(\{G_1, \dots, G_r\} \setminus (U \cup P))$
> $P := P \setminus \{G_1 \dots G_r\}$
> $U := (U \setminus \{F\}) \cup \{G_1 \dots G_r\}$
> **While** $\exists F' \in (U \cup P)$ and $\exists S' \in Parents(F')$ and $S' \not\subseteq (U \cup P)$ **Do**
> $\quad Parents(F') := Parents(F') \setminus \{S'\}$
> $\quad$ **If** $Parents(F') = \emptyset$ and $F' \in P$ and F' is not blocked **Then**
> $\quad\quad P := P \setminus \{F'\}$
> $\quad\quad Remove(F')$
> $\quad$ **EndIf**
> **EndWhile**

The precondition is ensured as in the previous subsection. On the other hand, as said at the beginning of this subsection, after each $Remove(F)$ or $Add(F)$ operation, we have to update the internal state of the prover into a new state which can be reached by a correct derivation with the prover D ending with the same final set of formulae. Below, such an alternative derivation is given for each one of these operations. The prover is then updated into the state it would be in after producing the alternative derivation.

The alternative derivation can be built in several different ways. As a general rule, for provers keeping more historical information it pays off to build alternative derivations that mimic the past computations quite accurately in order to avoid repeated computations; for others, similar results will be obtained by shorter derivations leading to the current clause set. In any case, for the main results, it suffices for the moment to assume that the alternative derivation is defined correctly, in the sense that its final state corresponds to the current set of formulae stored in the prover.

Each time the prover D infers F from premises $G_1 \ldots G_r$, a message "F has been deduced" is sent, and we do:

> **If** $F \in (U \cup P)$ **Then**
> $\qquad Parents(F) := Parents(F) \cup \{\{G_1, \ldots, G_r\}\}$
> **Else**
> $\qquad P := P \cup \{F\}$
> $\qquad Parents(F) := \{\{G_1, \ldots, G_r\}\}$
> **EndIf**

Finally, each time the prover D removes a formula F due to its redundancy by a set $F_1 \ldots F_r$, we do:

> **Foreach** F_i in $\{F_1 \ldots F_r\}$ **Do**
> $\qquad Block(F_i)$
> **EndForeach**
> $U := U \setminus F$
> $P := P \setminus F$

Note that, although at this point we do not apply any orphan murder, it will be applied only if, later on, another "replace" message is received by the redundancy module, which triggers the actual *Remove* operations. This is done because we can ensure completeness only if no more than a finite number of *Remove* operations occur, and there may be an infinite number of internal redundancy steps.

Definition 14. *Let $\mathcal{P}_D$ denote the prover module as defined above.*

Theorem 7. *$\mathcal{P}_D$ is a correct prover module.*

Proof. Analogous to the correctness proof of the previous section. $\square$

Theorem 8. *$\mathcal{P}_D$ is a complete prover module.*

Proof. We have to show that, in an execution of our prover module with exactly k input messages, if U_k is unsatisfiable then the prover module halts with output "unsatisfiable".

Since there are finitely many input messages, there are finitely many, say n, operations *Add* or *Remove*. Let $D_0, \ldots, D_m$ be the n-th alternative derivation. By the invariant of the process we have $U \cup Blocked(U \cup P) \models U_k$, therefore $U \cup P$ is unsatisfiable.

Finally, since there are only a finite number of steps (m) contradicting the strategy of the prover D, it remains complete, and hence the prover module halts with output "unsatisfiable". $\square$

Now it remains to define the alternative derivations. Before it was mentioned that for some provers it pays off to build alternative derivations that mimic the past computations quite accurately in order to avoid repeated computations and for others perhaps not. Here we explain a way that attempts to preserve as much as possible the previous history.

Suppose $D_1, \ldots, D_n$ is the current derivation. The following additional invariant, which is necessary to prove the correctness of the alternative derivation, is always satisfied: all formulae in $(\bigcup_{i \leq n} D_i) \setminus (U \cup P)$ are internally redundant by $Blocked(U \cup P)$.

- **an operation $Add(F)$ takes place.** Then, a new derivation, with $D_n \cup F$ as final set, is constructed as follows. There are two cases to be considered. If $F \notin D_i$ for all $i \leq n$ then we consider the derivation $D_0 \cup \{F\}, \ldots, D_n \cup \{F\}$. Otherwise, let D_i be the set with maximal i such that $F \in D_i$. Note that $i < n$. Hence D_{i+1} is obtained by an internal redundancy of F. Then the alternative derivation is $D_0, \ldots, D_i, D_{i+2} \cup \{F\}, \ldots, D_n \cup \{F\}$.
- **an operation $Remove(F)$ takes place.** Then, a new derivation, with $D_n \setminus \{F\}$ as final set, can be constructed. The new derivation $D'_0, \ldots, D'_m$ consists of two parts $D'_0, \ldots, D'_k$ containing only inference steps and $D'_k, \ldots, D'_m$ containing only internal redundancy steps.
  1. We take D'_0 as $(D_0 \setminus \{F\}) \cup Ch(F)$, where $Ch(F)$ denotes the set of children of F in inferences along the sequence $D_0, \ldots, D_n$. Let $D_{i_1}, \ldots, D_{i_k}$ be the sets of the derivation obtained by an inference step where F is neither a premise nor a conclusion. Then $D'_0, \ldots, D'_k$ is the derivation where D'_j is obtained from D'_{j-1} by the same inference step between D_{i_j-1} and D_{i_j}. Note that all steps are correct since we have added $Ch(F)$ to the initial set. Now, we have $D'_k = (\bigcup_{i \leq n} D_i) \setminus \{F\}$, and hence $D_n \setminus \{F\} \subseteq D'_k$.
  2. The second part of the new derivation consists of eliminating the formulae in $D'_k \setminus D_n$ by internal redundancies. This can be done since $D_n \setminus \{F\}$ contains all blocked formulae (F is not blocked since a $Remove(F)$ is done), and, by the invariant above, all formulae in $D'_k \setminus D_n$ are redundant by the blocked formulae.

Regarding the aim of avoiding repeated work, a last invariant, that is also satisfied, shows the reason for mantaining the sets $Parents(F)$: for all inference steps in the derivation with premises $F_1, \ldots, F_r$ and conclusion F, s.t. none of $F_1, \ldots, F_r, F$ are redundant by $Blocked(U \cup P)$, we have $\{F_1, \ldots, F_r\} \in Parents(F)$. This invariant ensures that the only missing information in the sets $Parents(F)$ (about performed inferences) involves either a premise or a conclusion that is redundant by the blocked formulae. Therefore, it is clear that some repeated work might be done, but, in any case, the prover could avoid it by means of internal redundancy steps.

5   Conclusions

A notion of modular redundancy for theorem proving has been introduced, that can be used to exploit redundancy elimination techniques in combination with arbitrary existing theorem provers, in a refutation complete way, even if these provers are not (or not known to be) complete in combination with the redundancy techniques when applied in the usual sense.

We have explained for three cases how to build a prover module in terms of an existing prover: for arbitrary provers, closure provers and saturation-based provers. The latter case seems to be the most general and useful one, since there are several practical applications where the standard general redundancy notions cannot be used, like [Lyn97] [BG98], and [BGNR99].

Especially in the case of [BGNR99], where the completeness of paramodulation w.r.t. non-monotonic orderings is proved, we believe that our modular redundancy results will be the method of choice. The reason is that the standard redundancy notions should be applied w.r.t. an ordering that is built (at the completeness proof level) based on semantic information and hence cannot even be approximated in a strong enough way during the saturation process. Moreover, in many practical applications the existence of a monotonic ordering is too strong a requirement. For example, one may want to apply completion to a terminating set of rules with a given orientation by a reduction ordering that cannot be extended to a total one, like $f(a) \to f(b)$ and $g(b) \to g(a)$, for which a and b must be incomparable in any monotonic extension. Another typical situation is deduction modulo built-in equational theories E, where the existence of a total *E-compatible* reduction ordering is a very strong requirement. We plan to do some experiments in this context with the Saturate prover[NN93,GNN98].

References

BG94. Leo Bachmair and Harald Ganzinger. Rewrite-based equational theorem proving with selection and simplification. *Journal of Logic and Computation*, 4(3):217–247, 1994.

BG98. Leo Bachmair and Harald Ganzinger. Strict basic superposition. In Claude Kirchner and Hélène Kirchner, editors, *Proceedings of the 15th International Conference on Automated Deduction (CADE-98)*, volume 1421 of *LNAI*, pages 160–174, Berlin, July 5–10 1998. Springer.

BGNR99. Miquel Bofill, Guillem Godoy, Robert Nieuwenhuis, and Albert Rubio. Paramodulation with non-monotonic orderings. In *14th IEEE Symposium on Logic in Computer Science (LICS)*, pages 225–233, Trento, Italy, July 2–5, 1999.

Der87. Nachum Dershowitz. Termination of rewriting. *Journal of Symbolic Computation*, 3:69–116, 1987.

GNN98. Harald Ganzinger, Robert Nieuwenhuis, and Pilar Nivela. The Saturate System, 1998. Software and documentation maintained at http://www.mpi-sb.mpg.de.

Lyn97. C. Lynch. Oriented equational logic programming is complete. *Journal of Symbolic Computation*, 23(1):23–46, January 1997.

NN93. Pilar Nivela and Robert Nieuwenhuis. Practical results on the saturation of full first-order clauses: Experiments with the saturate system. (system description). In C. Kirchner, editor, *5th International Conference on Rewriting Techniques and Applications (RTA)*, LNCS 690, pages 436–440, Montreal, Canada, June 16–18, 1993. Springer-Verlag.

Composing and Controlling Search
in Reasoning Theories Using Mappings

Alessandro Coglio[1], Fausto Giunchiglia[2], José Meseguer[3], and
Carolyn L. Talcott[4]

[1] Kestrel Institute, Palo Alto, CA
coglio@kestrel.edu
[2] IRST and Università di Trento, Italy
fausto@irst.itc.it
[3] SRI International, Menlo Park, CA
meseguer@csl.sri.com
[4] Stanford University, CA
clt@cs.stanford.edu

Abstract. Reasoning theories can be used to specify heterogeneous reasoning systems. In this paper we present an equational version of reasoning theories, and we study their structuring and composition, and the use of annotated assertions for the control of search, as mappings between reasoning theories. We define composability and composition using the notion of *faithful inclusion mapping*, we define *annotated reasoning theories* using the notion of *erasing mapping*, and we lift composability and composition to consider also annotations. As an example, we give a modular specification of the top-level control (known as *waterfall*) of NQTHM, the Boyer-Moore theorem prover.

1 Introduction

Reasoning theories are the core of the OMRS (Open Mechanized Reasoning System(s)) framework [14] for the specification of complex, heterogeneous reasoning systems. The long-term goal of OMRS is to provide a technology which will allow the development, integration and interoperation of theorem provers (and similar systems, e.g., computer algebra systems and model checkers) in a principled manner, with minimal changes to the existing modules. Reasoning theories are the *Logic* layer of OMRS, specifying the assertions and the rules used to derive new assertions and to build derivations. There are two other layers: *Control*, specifying the search strategies, and *Interaction*, specifying the primitives which allow an OMRS to interact with its environment. The reasoning theory framework has been used to analyze the NQTHM prover [11] and to develop a modular reconstruction of the high-level structure of the ACL2 prover [3]. The latter work augments ACL2 with the ability to construct proofs. In addition, the reasoning theory framework has been used as the basis of an experiment to re-engineer the integration of the NQTHM linear arithmetic procedure implemented as a re-usable separate process [1]. A formalism for the control component of OMRS

H. Kirchner and C. Ringeissen (Eds.): FroCoS 2000, LNAI 1794, pp. 200–216, 2000.

was developed and used to specify the control component of NQTHM [10]. The OMRS framework has also been used to build provably correct systems [13], and it has been used as the basis for integration of the Isabelle proof system and the Maple symbolic algebra system [4].

In this paper we study structuring and composition mechanisms for reasoning theories. We focus on the special case of equationally presented reasoning theories (ERThs). In passing to ERThs we gain concreteness, executability, and the availability of equational and rewriting tools, while keeping the originally envisioned generality and avoiding making unnecessary ontological commitments. In particular, ERThs have a strong link to equational and rewriting logic [21]. There are increasing levels of *structuring* of an equationally presented reasoning theory: ERths, nested ERThs (NERThs), and also ERThs and NERThs with annotated assertions for the control of search for derivations. Furthermore, there are horizontal *theory composition operations* at each level. The structuring and composition are formalized in terms of mappings between reasoning theories. Therefore, our work is in the spirit of "putting theories together" by means of theory mappings as advocated by Burstall and Goguen [7], that is widely used in algebraic specifications [2]. In order to simplify the presentation, we treat a restricted subcase of horizontal composition that is none-the-less adequate for interesting examples. A more general treatment in terms of pushouts of order-sorted theories [16,22,23] can be given.

In Section 2 we define ERTHs and their composition using the notion of *faithful inclusion mapping*. We then define *annotated reasoning theories* using the notion of *erasing mapping*, and lift composability and composition to consider also annotations. In Section 3 this is extended to NERths. As an example, in Section 4 we give a modular specification of the NQTHM top-level control (known as *waterfall*). In Section 5 we summarize and outline the future work.

Notation

We use standard notation for sets and functions. We let $\bar{\emptyset}$ denote the empty function (the function with empty domain, and whatever codomain). X^* denotes the set of all finite lists of elements in X. $[X_0 \to X_1]$ is the set of total functions, f, with domain, $\mathrm{Dom}(f) = X_0$, and range, $\mathrm{Rng}(f)$, contained in the codomain X_1. Let $(X, \leq)$ be a partial order. If $X' \subseteq X$, then $\leq \lceil X'$ is the restriction of $\leq$ to X', i.e., $\leq \lceil X' = \{(x,y) \mid x \in X', y \in X', x \leq y\}$. X' is *downwards closed* in X if whenever $x \leq y$ and $y \in X'$, then $x \in X'$. X' is *upwards closed* in X if whenever $y \leq x$ and $y \in X'$, then $x \in X'$.

A family Z of sets over a set I is an assignment of a set Z_i for each $i \in I$. This is notated by $Z = \{Z_i\}_{i \in I}$ and we say Z is an I-family. If $I = \{i\}$ is a singleton, then we let Z_i stand for $\{Z_i\}_{i \in I}$. If $I' \subseteq I$, the restriction of Z to I' is defined as $Z \lceil I' = \{Z_i\}_{i \in I'}$. If $I' \subseteq I$, Z is an I-family, Z' is an I'-family, then $Z' \subseteq Z$ means $Z'_i \subseteq Z_i$ for $i \in I'$. If Z_j are I_j-families for $j \in \{1, 2\}$, then we define the $(I_1 \cap I_2)$-family $Z_1 \cap Z_2 = \{(Z_1)_i \cap (Z_2)_i\}_{i \in I_1 \cap I_2}$ and the $(I_1 \cup I_2)$-family $Z_1 \cup Z_2 = \{(Z_1)_i \cup (Z_2)_i\}_{i \in I_1 \cup I_2}$ (where in the last expression we consider $(Z_1)_i = \emptyset$ if $i \in I_2 - I_1$, and $(Z_2)_i = \emptyset$ if $i \in I_1 - I_2$).

2   Equational Reasoning Theories

A reasoning theory specifies a set of *inference rules* of the form

$$l : \{\vec{cn}\} \; \vec{sq} \Rightarrow sq$$

where l is the rule label, sq is the conclusion, $\vec{sq}$ is a list of premisses, and $\vec{cn}$ is a list of side conditions. sq and the elements of $\vec{sq}$ are elements of the set, Sq, of *sequents* specified by the reasoning theory. Sequents represent the assertions to be considered. We use "sequent" in a very broad sense which includes sequent in the Gentzen/Prawitz sense [12,24], simple formulas and judgements in the sense of Martin-Löf [19], as well as the complex data structures (carrying logical information) manipulated by the NQTHM waterfall. Elements of $\vec{cn}$ are elements of the set, Cn, of *constraints* specified by the reasoning theory. If this set is non-empty, then the reasoning theory also provides an inference machinery for deriving constraint assertions. Sequents and constraints can be schematic in the sense that they may contain place holders for missing parts. A reasoning theory also provides a mechanism for filling in missing parts: a set, I, of instantiations and an operation, $\_[\_]$, for applying instantiations to schematic entities. There is an identity instantiation and instantiations can be composed. A reasoning theory determines a set of *derivation structures* (the OMRS notion of proof) formed by instantiating inference rules and linking them by matching premisses to conclusions. We write $ds : \{\vec{cn}\} \; \vec{sq} \Rightarrow \vec{sq}'$ to indicate that ds is a derivation structure with conclusions the list of sequents $\vec{sq}'$, premisses the list $\vec{sq}$, and side conditions the list of constraints $\vec{cn}$. (See [14] for a detailed description of reasoning theories.)

2.1   Sequent Signatures and ERThs

In this paper we focus on *equationally presented reasoning theories* (*ERThs*). An ERTh consists of a sequent signature and a labelled set of rules of inference. A sequent signature consists of an equational signature, Σ, (the choice of equational logic is not so important, here for concreteness we take it to be order-sorted equational logic [15]) together with a sort-indexed family of variables, a set of equations and a distinguished set of sorts – the sequent sorts. Variables are the place holders and instantiations are simply sort preserving maps from variables to terms, applied and composed in the usual way. The equational theory just serves to define the sequents; deduction over such sequents is specified by the rules of inference of the ERTh. Therefore, the expressivity of ERThs is by no means limited to first-order equational logic: other logics can be easily expressed (e.g., higher-order) by using suitable rules of inference (see for example formalizations of Lambda calculus [18], HOL and NuPrl [25], and the Calculus of Constructions [26] in the Maude Rewriting Logic implementation).

Definition: sequent signatures.   A sequent signature is a tuple $eseq = (\Sigma, X, E, Q)$ where $\Sigma = (S, \leq, O)$ is an order-sorted equational signature in which S is the set of sorts, O is an $S^* \times S$-indexed family of operations, and

$\leq$ is a partial order on S; X is an S-indexed family of variables; E is a set of Σ-equations; and $Q \subseteq S$ is the set of sequent sorts. We require that Q be downwards and upwards closed in S. Elements of $O_{\vec{s},s}$ are said to have rank $\vec{s}, s$, where $\vec{s}$ is the list of argument sorts (the arity) and s is the result sort, and we write $o : \vec{s} \to s$ to indicate that o is an element of $O_{\vec{s},s}$. We assume that each set X_s for $s \in S$ is countably infinite. $T_{eseq} = T_{\Sigma}(X)$ is the free order-sorted Σ-algebra over X whose elements are terms formed from elements of X by applying operations of O to lists of terms of the appropriate argument sorts. For $s \in S$, $T_{eseq,s}$ is the set of $eseq$ terms of sort s and for $S' \subseteq S$, $T_{eseq,S'}$ is the set of terms of sort s for $s \in S'$: $T_{eseq,S'} = \bigcup_{s \in S'} T_{eseq,s}$. We further require that Σ be *coherent* [15]. For our purposes this can be taken to mean that every connected component of the sort poset has a top element, and that there is a function $ls : T_{eseq} \to S$ mapping each term to its least sort in the partial order $\leq$ on S.

An equation is a pair of terms of the same sort, thus $E \subseteq \bigcup_{s \in S} T_{eseq,s} \times T_{eseq,s}$. The rules of deduction of order-sorted equational logic [15] then define an order-sorted Σ-congruence $\equiv_E$ on T_{eseq}. We say that t, t' are E-equivalent if $t \equiv_E t'$. For an $eseq$ term t we let $[t]_E$ be the E-equivalence class of t and for any set $T \subseteq T_{eseq,s}$ we let $\{T\}_E$ be the set of E-equivalence classes of terms in T.

The set of sequents, Sq_{eseq}, presented by $eseq$, is the set of equivalence classes of terms with sorts in Q: $Sq_{eseq} = \{T_{eseq,Q}\}_E$. Instantiations are sort-preserving mappings from X to T_{eseq}, with identity and composition defined in the obvious way. Their application to sequents consists in substituting terms for variables in members of the equivalence classes.

Definition: rules and ERThs.   The set of rules $Rules_{eseq}$ over a sequent signature $eseq$ is the set of pairs $(\vec{sq}, sq)$ where $\vec{sq} \in Sq_{eseq}^*$ and $sq \in Sq_{eseq}$. We fix a set of labels, Lab, to use for rule names.

An ERTh, $erth$, is a pair $(eseq, R)$ where $eseq$ is a sequent signature and $R : L \to Rules_{eseq}$ is a function from a set of labels $L \subseteq Lab$ to the set of rules over $eseq$. We write $l : \vec{sq} \Rightarrow sq \in R$ if $l \in L$ and $R(l) = (\vec{sq}, sq)$.

2.2   Composing Sequent Signatures and ERThs

Specifying composition of systems at the logical level amounts to specifying how deductions performed by the different systems relate to each other (e.g., how an inequality proved by a linear arithmetic decider can be used by a rewriter to rewrite a term involving an arithmetic expression). This demands that the ERThs of the systems to be composed share some language, and that the composition preserves the shared language and its meaning. We represent the sharing by theory mappings from the shared part to each component. We restrict the mappings used for composition to guarantee that composition makes sense: there should be a uniquely defined composite with mappings from the component theories that agree on the shared part, and the theory mappings should induce mappings that extend/restrict instantiations in a well-behaved manner, so that the result is a mapping of ERThs. Here we consider a very simple but natural restriction with the desired properties. The general situation will be treated in a forthcoming paper [20].

Definition: faithful inclusions. Let $eseq_j = ((S_j, \leq_j, O_j), X_j, E_j, Q_j)$ be sequent signatures for $j \in \{0, 1\}$. A faithful inclusion arrow exists from $eseq_0$ to $eseq_1$ (written $eseq_0 \hookrightarrow eseq_1$) if

1. $S_0 \subseteq S_1$, $O_0 \subseteq O_1$, $X_0 \subseteq X_1$, and $Q_0 \subseteq Q_1$;
2. $X_1 \lceil S_0 = X_0$, and $Q_1 \cap S_0 = Q_0$;
3. $\leq_1 \lceil S_0 = \leq_0$, and S_0 is downwards and upwards closed in S_1;
4. any operation in O_1 with result sort in S_0, is in O_0;
5. for terms t, t' in T_{eseq_0} having sorts in the same connected component of $\leq_0$, $t' \in [t]_{E_1}$ iff $t' \in [t]_{E_0}$.

Lemma (term.restriction): If $eseq_0 \hookrightarrow eseq_1$, then $T_{eseq_1} \lceil S_0 = T_{eseq_0}$. Therefore any sub-term of a term $t \in T_{eseq_1, S_0}$ has sort in S_0. In fact t is an $eseq_0$-term.

Lemma (incl.transitivity): If $eseq_0 \hookrightarrow eseq_1$ and $eseq_1 \hookrightarrow eseq_2$, then $eseq_0 \hookrightarrow eseq_2$.

Definition: sequent signature composition. Let $eseq_j = ((S_j, \leq_j, O_j), X_j, E_j, Q_j)$ for $j \in \{0, 1, 2\}$. $eseq_1$ is composable with $eseq_2$ sharing $eseq_0$ if $eseq_0 \hookrightarrow eseq_j$ for $j \in \{1, 2\}$ are faithful inclusions such that $S_1 \cap S_2 = S_0$, $O_1 \cap O_2 = O_0$, $X_1 \cap X_2 = X_0$, $Q_1 \cap Q_2 = Q_0$. The composition of $eseq_1$ and $eseq_2$ sharing $eseq_0$ (written $eseq_1 +_{eseq_0} eseq_2$) is the 'least' sequent signature including $eseq_1$ and $eseq_2$. More precisely,

$$eseq_1 +_{eseq_0} eseq_2 = ((S_1 \cup S_2, \leq_1 \cup \leq_2, O_1 \cup O_2), X_1 \cup X_2, Q_1 \cup Q_2).$$

Lemma (incl.composition): If $eseq_j$ for $j \in \{0, 1, 2\}$ satisfy the conditions for composability, then $eseq_1 \hookrightarrow eseq_1 +_{eseq_0} eseq_2$, $eseq_2 \hookrightarrow eseq_1 +_{eseq_0} eseq_2$. Also (by transitivity) $eseq_0 \hookrightarrow eseq_1 +_{eseq_0} eseq_2$.

To avoid the need to make the shared part of a composition explicit we define a function that computes the shared part of a pair of sequent signatures and say that two sequent signatures are composable just if their shared part is faithfully included in both signatures.

Definition: shared part. Let $eseq_j = ((S_j, \leq_j, O_j), X_j, E_j, Q_j)$ be sequent signatures for $j \in \{1, 2\}$. Define $shared(eseq_1, eseq_2)$ by

$$shared(eseq_1, eseq_2) = (\Sigma_0, X_0, E_0, Q_0)$$

where $\Sigma_0 = (S_0, \leq_0, O_0)$, $S_0 = S_1 \cap S_2$, $\leq_0 = \leq_1 \cap \leq_2$, $O_0 = (O_1 \cap O_2) \lceil (S_0^* \times S_0)$, $X_0 = X_1 \cap X_2 \lceil S_0$, $E_0 = \{e \in (E_1 \cup E_2) \mid e \text{ is a } \Sigma_0\text{-equation}\}$, and $Q_0 = Q_1 \cap Q_2$.

Definition: composability and composition. $eseq_1$ is composable with $eseq_2$ (written $eseq_1 \bowtie eseq_2$) if

1. $shared(eseq_1, eseq_2) \hookrightarrow eseq_1$
2. $shared(eseq_1, eseq_2) \hookrightarrow eseq_2$

If $eseq_1 \bowtie eseq_2$, then their composition $eseq_1 + eseq_2$ is defined by

$$eseq_1 + eseq_2 = eseq_1 +_{shared(eseq_1, eseq_2)} eseq_2$$

The two notions of composability and composition are related as follows.

Lemma (composing): $eseq_1$ is composable with $eseq_2$ sharing $eseq_0$ if $eseq_0 = shared(eseq_1, eseq_2)$ and $eseq_1 \bowtie eseq_2$.

Lemma (ACI):   Composition of sequent signatures is associative, commutative and idempotent. (Note that since composition is partial it is necessary to check both definedness and equality.)

(C) if $eseq_1 \bowtie eseq_2$ then $shared(eseq_1, eseq_2) = shared(eseq_2, eseq_1)$ and $eseq_1 + eseq_2 = eseq_2 + eseq_1$
(A) if $eseq_1 \bowtie eseq_2$ and $eseq_0 \bowtie (eseq_1 + eseq_2)$, then $eseq_0 \bowtie eseq_1$ and $(eseq_0 + eseq_1) \bowtie eseq_2$ and $eseq_0 + (eseq_1 + eseq_2) = (eseq_0 + eseq_1) + eseq_2$.
(I) $eseq \bowtie eseq$ and $eseq + eseq = eseq$

Definition: multi-composition.   A special case of multiple composition is when there is a common shared part among several sequent signatures. We say that a set of sequent signatures, $\{eseq_i \mid 1 \leq i \leq k\}$, is composable if there is some $eseq_0$ such that $shared(eseq_i, eseq_j) = eseq_0$ for $1 \leq i \neq j \leq k$ and $eseq_0 \hookrightarrow eseq_i$ for $1 \leq i \leq k$. In this case, the composition of $\{eseq_i \mid 1 \leq i \leq k\}$ is just the repeated binary composition: $eseq_1 + \ldots + eseq_k$. By (**ACI**) the order in which the elements are listed does not matter.

Definition: composing ERThs.   Composability and composition lift naturally to ERThs. Let $erth_j = (eseq_j, R_j)$ for $j \in \{1, 2\}$ be ERThs. Then $erth_1$ is composable with $erth_2$ (written $erth_1 \bowtie erth_2$) if $eseq_1 \bowtie eseq_2$ and the domains of R_1 and R_2 are disjoint. If $erth_1 \bowtie erth_2$, then the composition of $erth_1$ with $erth_2$ is defined by

$$erth_1 + erth_2 = (eseq_1 + eseq_2, R_1 \cup R_2)$$

2.3   Annotated Sequent Signatures and ERThs

Informally an *annotated sequent signature* is a pair of sequent signatures, $(eseq^a, eseq)$ together with an *erasing* mapping $\varepsilon : eseq^a \to eseq$ which "removes annotations", extracting the logical content of annotated assertions. An *annotated ERТh* is a pair of ERThs $(erth^a, erth)$ together with an erasing mapping $\varepsilon : erth^a \to erth$ such that the mapping on the underlying sequent signatures gives an annotated sequent signature and annotated rules map to derivation structures over $erth$.

An erasing map $\varepsilon : eseq^a \to eseq$ is a map of theories that maps each sort of $eseq^a$ (the annotated theory) to either a sort of $eseq$ or to the empty list of sorts (erasing it in that case). The conditions for erasing maps are given below. Erasing maps are more general than the usual maps of theories (that map sorts to sorts). Categorically, an erasing map extends to a functor $\overline{\varepsilon} : \mathcal{L}_{eseq^a} \to \mathcal{L}_{eseq}$ that preserves products and inclusions between the "Lawvere theories" associated to the order-sorted theories $eseq^a$ and $eseq$ [17].

Annotations are very useful for controlling the search for derivations, as will be illustrated in Section 4. The mapping from annotated rules to derivation structures gives for each derivation at the annotated level, a corresponding, possibly more detailed derivation in the original theory.

Definition: annotated sequent signature. Let $eseq^a = (\Sigma^a, X^a, E^a, Q^a)$ and $eseq = (\Sigma, X, E, Q)$ be sequent signatures, where $\Sigma^a = (S^a, \leq^a, O^a)$ and $\Sigma = (S, \leq, O)$. Then $\varepsilon : eseq^a \to eseq$ is an annotated sequent signature if ε, with the action on sorts and terms lifted homomorphically to lists, satisfies the following:

1. $\varepsilon : S^a \to S \cup \{[\,]\}$ such that:
   - if $s_0^a \leq^a s_1^a$, then $\varepsilon(s_0^a) \leq \varepsilon(s_1^a)$
   - if $q^a \in Q^a$, then $\varepsilon(q^a) \in Q$
   - if $s^a \in S^a$ and $\varepsilon(s^a) \in Q$ then $s^a \in Q^a$
2. $\varepsilon : X^a \to X \cup \{[\,]\}$ such that if $x^a \in X^a_{s^a}$ and $\varepsilon(s^a) = s$, then $\varepsilon(x^a) \in X_s$ and if $\varepsilon(s^a) = [\,]$, then $\varepsilon(x^a) = [\,]$ (i.e., the empty list of variables).
3. If $o^a : \vec{s}^a \to s^a$, and $\varepsilon(s^a) = s$ (not $[\,]$), then $\varepsilon(o^a(x_1^a, \ldots, x_n^a))$ is a Σ-term with sort s and free variables list $\varepsilon(x_1^a, \ldots, x_n^a)$. ε is extended homomorphically to $eseq^a$-terms.
4. If $t_0^a \in [t_1^a]_{E^a}$ then $\varepsilon(t_0^a) \in [\varepsilon(t_1^a)]_E$

If $erth^a = (eseq^a, R^a)$ and $erth = (eseq, R)$, then $\varepsilon : erth^a \to erth$ is an annotated ERTh if $\varepsilon : eseq^a \to eseq$ is an annotated sequent signature and for $l : \vec{sq}^a \Rightarrow sq^a \in R^a$ there is a derivation structure $ds : \varepsilon(\vec{sq}^a) \Rightarrow \varepsilon(sq^a)$ in the set of $erth$ derivation structures (such derivation structures are described in detail in [20]).

Now we consider the relation between annotated theories and composition.

Lemma (compose.erase): Let $\varepsilon_j : erth_j^a \to erth_j$ be annotated ERThs for $j \in \{1, 2\}$, where $erth_j^a = (eseq_j^a, R_j^a)$ and $erth_j = (eseq_j, R_j)$. Suppose that $erth_1^a \bowtie erth_2^a$, $erth_1 \bowtie erth_2$, and that $\varepsilon_1, \varepsilon_2$ agree on $shared(eseq_1^a, eseq_2^a)$. Then we have annotated ERThs

$$\varepsilon_1 \cap \varepsilon_2 : shared(eseq_1^a, eseq_2^a) \to shared(eseq_1, eseq_2)$$

$$\varepsilon_1 + \varepsilon_2 : eseq_1^a + eseq_2^a \to eseq_1 + eseq_2$$

where $\varepsilon_1 \cap \varepsilon_2$ is the common action of the two mappings on the shared part, and $\varepsilon_1 + \varepsilon_2$ acts as ε_1 on elements coming from $eseq_1$ and as ε_2 on elements coming from $eseq_2$ (and as $\varepsilon_1 \cap \varepsilon_2$ on the shared elements).

3   Nested Equational Reasoning Theories

To represent proof systems with conditional inference rules and other subsidiary deductions that we may want to treat as side conditions, we represent the side conditions and their rules for verification by a sub-reasoning theory called the constraint theory. This gives rise to a *nested ERTh* (*NERTh* for short). Note that this nesting organization can be further applied to the constraint theory, giving rise to higher levels of nesting. For simplicity we consider only one level here, as all the features and the necessary machinery already appear at this stage.

Definition: NERThs.   A *nested sequent signature* is a tuple $neseq = (eseq, (eseq_c, R_c))$ where $eseq$ is a sequent signature with sequent sorts Q and $(eseq_c, R_c)$ is an ERTh, with sequent sorts Q_c such that $eseq \bowtie eseq_c$ and the shared part has no sequent sorts, i.e., $Q \cap Q_c = \emptyset$.

A NERTh is a pair $nerth = (neseq, R)$ where $neseq$ is a nested sequent signature and, letting $Sq = Sq_{eseq}$ and $Cn = Sq_{eseq_c}$, we have $Rules : L \to Cn^* \times Sq^* \times Sq$. Thus, labelled rules of a nested reasoning structure have the form $l : \vec{cn}, \vec{sq} \Rightarrow sq$ where $\vec{cn}$ are the side conditions or premises to be established by deduction in the constraint theory.

Definition: composition of NERThs.   Let $neseq_j = (eseq_j, (eseq_{j,c}, R_{j,c}))$ be nested sequent signatures for $j \in \{1, 2\}$. $neseq_1$ is composable with $neseq_2$ (written $neseq_0 \bowtie neseq_1$) if:

1. $eseq_1 \bowtie eseq_2$, $eseq_{1,c} \bowtie eseq_{2,c}$, and $eseq_1 + eseq_2 \bowtie eseq_{1,c} + eseq_{2,c}$;
2. $shared(eseq_1, eseq_{2,c})$ has no sequent sorts, and $shared(eseq_2, eseq_{1,c})$ has no sequent sorts.

Note that 2. together with the assumption that $neseq$ is a nested sequent signature implies that there are no sequent sorts in $shared(eseq_1 + eseq_2, eseq_{1,c} + eseq_{2,c})$.

If $neseq_1 \bowtie neseq_2$, then the composition, $neseq_1 + neseq_2$ is defined by

$$neseq_1 + neseq_2 = (eseq_1 + eseq_2, erth_{1,c} + erth_{2,c})$$

Let $nerth_j = (neseq_j, R_j)$ be NERThs for $j \in \{1, 2\}$. $nerth_1$ is composable with $nerth_2$ ($nerth_1 \bowtie nerth_2$) if $neseq_1 \bowtie neseq_2$ and the domains of R_1 and R_2 are disjoint. If $nerth_1 \bowtie nerth_2$, then the composition is defined by

$$nerth_1 + nerth_2 = (neseq_1 + neseq_2, R_1 \cup R_2)$$

Definition: annotated NERThs.   Annotated nested sequent signatures and NERThs are defined analogously to the non-nested case. Let $nerth^a = (neseq^a, R^a)$ and $nerth = (neseq, R)$ where $neseq^a = (eseq^a, erth_c^a)$, $neseq = (eseq, erth_c)$, $erth_c^a = (eseq_c^a, R_c^a)$, and $erth_c = (eseq_c, R_c)$. Then $\varepsilon : nerth^a \to nerth$ is an annotated nested reasoning theory if there are mappings ε' and ε_c such that $\varepsilon' : eseq^a \to eseq$ is an annotated sequent signature, $\varepsilon_c : erth_c^a \to erth_c$

is an annotateed reasoning theory, ε' and ε_c agree on $shared(eseq^a, eseq_c^a)$, and $\varepsilon = \varepsilon' + \varepsilon_c$ is such that $\varepsilon : (eseq^a + eseq_c^a, R^a \cup R_c^a) \rightarrow (eseq + eseq_c, R \cup R_c)$ is an annotated reasoning theory. The final clause is needed to ensure that the annotated rules are mapped to derivations in $nerth$. In fact all that needs to be checked is that the rules in R^a map to derivations, the rest follows by the disjointness of sequents and constraints and the requiement that ε_c is an annotation erasing map on $erth_c^a$.

4   An Annotated Reasoning Theory for the NQTHM Waterfall

We demonstrate the usefulness of the concepts just presented by applying them to a non trivial case study. We describe salient aspects of a specification of the NQTHM [5,6] waterfall as an annotated NERTh $\varepsilon_W : nerWat^a \rightarrow nerWat$. Such an annotated NERTh can be composed with annotated NERThs for the other sub-systems of NQTHM, to produce an annotated NERTh for the whole system. To provide some background, we first describe the features of the waterfall that are relevant to our specification.

4.1   The NQTHM Waterfall

When the user asks NQTHM to prove an assertion, this is first *pre-processed*. While the details of pre-processing are not relevant here, the important fact is that a list of *clauses* (to be considered conjunctively) is produced and is fed into the waterfall. This is shown in the upper part of Fig. 1. The waterfall then tries to prove all the resulting clauses by means of six main inference processes (*simplification, elimination of destructors, cross-fertilization, generalization, elimination of irrelevance,* and *induction*), called upon each clause in turn, until no clauses are left. Each process returns a list of zero or more clauses which replace the input clause; the provability of the input clause is implied by that of the returned ones (i.e., from a logical point of view these inference processes work backwards).

The clauses manipulated by the waterfall are stored in two data structures: the *Top* and the *Pool* (shown in Fig. 1). The Top is a list of pairs where the first element is a clause and the second element is the (current) *history* of the clause. A history of a clause cl_0 is a list of triples $(hid_1, cl_1, hnfo_1), \ldots, (hid_n, cl_n, hnfo_n)$, where each triple is called a *history element*, each $hid_i \in \{SI, ED, CF, GE, EI\}$ and is called *history identifier* (it identifies one of the first five main inference processes) each cl_i is a clause, and each $hnfo_i$ is a *history info*. Such a history expresses that, for $1 \le i \le n$, clause cl_{i-1} has been obtained from cl_i by applying the process identified by hid_i; $hnfo$ contains (process-dependent) details about such backward derivation. Histories are updated by the waterfall as the proof progresses. The Pool is a list, treated as a stack, of pairs whose first component is a clause and second component, called *tag*, is either TBP (To Be Proved) or BP (Being Proved). Just after the pre-processing, the Top is set to the list of resulting clauses, each one equipped with the empty history. The Pool is instead set to the empty list.

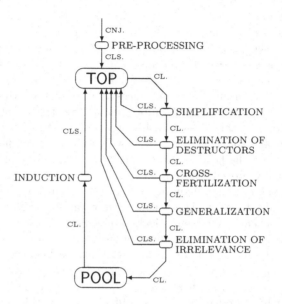

Fig. 1. The NQTHM waterfall.

When the Top is non-empty, its first clause and history are fed into the simplifier, which returns a list of clauses and a history info. If the returned list is equal to the singleton list having as unique element the input clause, then the simplifier has *failed*. Otherwise, it has *succeeded*, and the returned clauses are prepended to the Top, each paired with the history obtained by extending the input history with the history element consisting of SI, the input clause, and the history info returned by the simplifier. If the simplifier fails, the clause and history are fed into the elimination of destructors, cross-fertilization, generalization, and elimination of irrelevance processes (in this order), until one succeeds. As soon as one process applies successfully, the output clauses are added to the Top analogously to the case of the simplifier. If none of the above processes succeeds, then the clause, paired with the tag TBP, is pushed onto the Pool (and the history is discarded), unless it is the empty clause (i.e., with no literals), in which case NQTHM stops with a failure. See the rightmost part of Fig. 1.

When the Top is empty, if also the Pool is empty, then NQTHM terminates with success (the conjecture has been proved). Otherwise, the waterfall performs the following actions. If the clause at the head of the Pool is tagged by BP, then the clause and the tag are popped (that means that the clause has been proved; see below). When the clause at the head is tagged by TBP, another clause in the

Pool is searched, which subsumes it (i.e., the clause at the head is an instance of the subsuming one). In case one clause is found, then if it is tagged by BP, then NQTHM stops with a failure (because a loop in inventing inductions is likely to be taking place); if instead it is tagged by TBP, then the clause and tag at the head are just popped (because if the subsuming clause eventually gets proved, also the subsumed one is provable), and things go on as above (i.e., when the head clause is tagged by BP, it is popped, etc.). If no subsuming clause is found, the tag is changed to BP and the clause is fed into the induction process. In case the induction process cannot invent any induction, NQTHM stops with a failure. Otherwise the induction process produces a list of clauses which are prepended to the Top, each paired with the empty history. See the leftmost part of Fig. 1. In this second case, the clause and history at the head of the Top are fed into the simplifier, and things go on as previously explained. So, note that when the Top is empty and the head clause of the Pool is tagged by BP, this means that it has been really proved (as said above).

We now explain in detail an important feature which we have omitted above for ease of understanding. The feature is motivated as follows: after inventing an induction scheme, it is heuristically convenient to prevent the rewriting of terms appearing in the induction hypothesis, and to force the rewriting of terms appearing in the induction conclusion (so that the hypothesis can be used to prove the conclusion). For this purpose, the induction process also produces two sets of terms as outputs (those appearing in the hypothesis and those in the conclusion), which are fed into the simplifier. Anyway, this special treatment of the terms is only performed the first time the simplifier is called upon a clause produced by induction. To achieve that, as soon as the simplifier fails upon a clause, its history is extended with a history element consisting of a (sixth) *settle-down* history identifier SD, the clause itself, and the empty history info (i.e., carrying no information). SD indicates that the clause has "settled down" with respect to the special treatment of terms appearing in induction hypothesis and conclusion. The clause and the new history are then fed into the simplifier. However, this does not happen if SD is already present in the history; in that case the clause and the history are just fed into the elimination of destructors process. The simplifier ignores the two input sets of terms if and only if SD is present in the input history.

4.2 The NERTh *nerWat*

Clauses are specified by the sequent signature *esCl*, which includes a sort *Cl* for clauses, as well as other sorts and operations to build clauses (e.g., specifying literals and disjunction), which we do not describe here. The waterfall manipulates lists (logically, conjunctions) of clauses. These are specified by the sequent signature *esCls*, with *esCl* ↪ *esCls*. *esCls* includes a sort *Cls* for conjunctions of clauses, with $Cl \leq Cls$, and two operations, $\top : \rightarrow Cls$ and $(\_ \wedge \_) : Cls, Cls \rightarrow Cls$, for the empty clause conjunction and to conjoin clause conjunctions. *esCls* contains equations stating that $(\_ \wedge \_)$ is commutative, associative, idempotent, and has $\top$ as identity. In other words, clause conjunctions are treated as sets.

Although not explicitly mentioned in the previous subsection, all the proving activity performed by NQTHM happens in the context of a database of axioms (of various kinds) with heuristic information (e.g., by which inference techniques a certain axiom should be used). From the logical point of view, this database represents a theory in which theorems can be proved. The sequent signature $esTh$ includes a sort Th for theories, as well as other sorts and operations to build theories (not presented here).

$esTh$ and $esCls$ are composable, and their composition $esThCl = esTh + esCls$ constitutes the base to build sequents formalizing the logical assertions manipulated by the waterfall, by means of the sequent signatures $esWat_0$ and $esWat_1$, with $esThCl \hookrightarrow esWat_0$ and $esThCl \hookrightarrow esWat_1$. $esWat_0$ contains a sequent sort Sq_W (i.e., $Sq_W \in Q$) and an operation $(\_ \vdash_W \_) : Th, Cls \to Sq_W$. A sequent of the form $(th \vdash_W \widehat{cl})$ asserts that all the clauses in $\widehat{cl}$ are provable from axioms in th. $esWat_1$ contains sequent sorts Sq_{SI}, Sq_{ED}, Sq_{CF}, Sq_{GE}, Sq_{EI}, and Sq_{IN}, and operations $(\_ \vdash_{SI} \_ \to \_) : Th, Cls, Cl \to Sq_{SI}$, $(\_ \vdash_{ED} \_ \to \_) : Th, Cls, Cl \to Sq_{ED}$, $(\_ \vdash_{CF} \_ \to \_) : Th, Cls, Cl \to Sq_{CF}$, $(\_ \vdash_{GE} \_ \to \_) : Th, Cls, Cl \to Sq_{GE}$, $(\_ \vdash_{EI} \_ \to \_) : Th, Cls, Cl \to Sq_{EI}$, and $(\_ \vdash_{IN} \_ \to \_) : Th, Cls, Cl \to Sq_{IN}$. A sequent of the form $(th \vdash_{SI} \widehat{cl} \to cl)$ asserts that the provability of the clauses in $\widehat{cl}$ implies that of cl, where $\widehat{cl}$ are obtained by simplification upon cl (in the context of theory th). The other kinds of sequents have analogous meaning (SI, ED, etc. identify the six main inference processes of NQTHM).

The logical inferences performed by the waterfall are specified by suitable rules of inference over the nested sequent systems $nesWat_0 = (esWat_0, erSbsm)$ and $nesWat_1 = (esWat_1, \mathtt{mterth})$. $\mathtt{mtrth}$ is the empty ERTh (i.e., with no sorts, no operations, etc.). $erSbsm$ is an ERTh including a sequent sort Cn_{SBSM} and an operation $Subsumes(\_, \_) : Cl, Cl \to Cn_{SBSM}$. A sequent of the form $Subsumes(cl, cl')$ asserts that clause cl subsumes clause cl', and $erSbsm$ contains rules that specify subsumption (i.e., when $Subsumes(cl, cl')$ can be derived). The NERTh $nerWat_0$ consists of $nesWat_0$ and the following rules:

$$\text{Qed}: \quad \Longrightarrow \quad th \vdash_W \top \, ,$$

$$\text{ElimSubsumed}: \quad \{ Subsumes(cl, cl') \}$$
$$th \vdash_W \widehat{cl} \wedge cl \quad \Longrightarrow \quad th \vdash_W \widehat{cl} \wedge cl \wedge cl' \, .$$

Qed specifies that the empty conjunction of clauses is provable in any theory, and models the end of the proof when there no more clauses to prove in the Top or Pool. ElimSubsumed specifies that to prove a conjunction of clauses it is sufficient to prove all except one, provided this one is subsumed by another one in the conjunction; note the constraint (sequent of $erSbsm$) used as a side condition. This rule models the elimination of a sumsumed clause from the Pool. The NERTh $nerWat_1$ consists of $nesWat_1$ and six rules, one of which is

$$\text{CallSimp}: \quad th \vdash_{SI} \widehat{cl}' \to cl \quad th \vdash_W \widehat{cl} \wedge \widehat{cl}' \quad \Longrightarrow \quad th \vdash_W \widehat{cl} \wedge cl \, .$$

CallSimp specifies that to prove the conjunction of cl and $\widehat{cl}$ it is sufficient to prove the conjunction of $\widehat{cl}'$ and $\widehat{cl}$, where the $\widehat{cl}'$ are the result of simplifying cl. Its backward application models the invocation of the simplification process upon a clause and the replacement of the clause with those resulting from simplification. The other five rules, CallElDes, CallCrFer, CallGen, CallEllrr, and CallInd, are analogous. $nerWat_0$ and $nerWat_1$ are composable, and $nerWat = nerWat_0 + nerWat_1$.

4.3   The NERTh $nerWat^a$

The sequents and rules just presented do not specify anything about clause histories, division between Top and Pool, and so on. That, in fact, constitutes non-logical (i.e., control) information, and is specified, along with its manipulation, by annotated sequents and rules in $nerWat^a$.

Histories are specified by the sequent signature $esHst$, with $esCls \hookrightarrow esHst$. $esHst$ includes a sort Hid and seven constants $\mathtt{SI}, \mathtt{ED}, \mathtt{CF}, \mathtt{GE}, \mathtt{EI}, \mathtt{IN}, \mathtt{SD} : \rightarrow Hid$ for history identifiers, as well as a sort $Hinfo$ and operations for process-specific history information. It also includes a sort $Helem$ and an operation $[\_, \_, \_]$: $Hid, Cl, Hinfo \rightarrow Helem$ for history elements; a sort Hst for histories, with $Helem \le Hst$, and operations $\emptyset : \rightarrow Hst$ and $(\_ \cdot \_) : Hst, Hst \rightarrow Hst$ for the empty history and to concatenate histories, plus equations stating that $(\_ \cdot \_)$ is associative and has identity $\emptyset$ (i.e., histories are treated as lists). Clauses paired with histories are specified by a sort ClH and an operation $(\_/\_) : Cl, Hst \rightarrow ClH$; their conjunctions are specified by a sort $ClHs$, with and $ClH \le ClHs$, and operations $\top : \rightarrow ClHs$ and $(\_ \wedge \_) : ClHs, ClHs \rightarrow ClHs$, which is associative and has identity $\top$ (these conjunctions are thus treated as lists, as opposed to sets; in fact, the ordering of the elements of a list constitutes control information).

Pool tags are formalized by the sequent signature $esTag$, with $esCls \hookrightarrow esTag$, which includes a sort ClT for tagged clauses, and two operations $\mathtt{TBP}(\_) : Cl \rightarrow ClT$ and $\mathtt{BP}(\_) : Cl \rightarrow ClT$ to tag clauses. Conjunctions of tagged clauses are specified by a sort $ClTs$, with $ClT \le ClTs$, and two operations $\top : \rightarrow ClTs$ and $(\_ \wedge \_) : ClTs, ClTs \rightarrow ClTs$; the latter is associative and has identity $\top$.

The sequent signature $esDB$ contains a sort DB for databases of axioms (with heuristic information), as well as sorts and operations to build them. $esDB$, $esHst$, and $esTag$ are composable, and $esDBHstTag = esDB + esHst + esTag$ constitutes the base to build the annotated sequents manipulated by the waterfall. These are specified by sequent signatures $esWat_0^a$ and $esWat_1^a$, with $esDBHstTag \hookrightarrow esWat_0^a$ and $esDBHstTag \hookrightarrow esWat_1^a$. $esWat_0^a$ contains a sequent sort Sq_W^A and an operation $(\_ \vdash_W \_; \_) : DB, ClHs, ClTs \rightarrow Sq_W^A$. An (annotated) sequent of the form $(db \vdash_W \widehat{clh}; \widehat{clt})$ asserts that the clauses in $\widehat{clh}$ and $\widehat{clt}$ are provable from the axioms in db. It also embeds control information: $\widehat{clh}$ and $\widehat{clt}$ constitute the Top and the Pool, with histories and tags; db contains heuristic information for the axioms. $esWat_1^a$ contains sequent sorts $Sq_{SI}^A, Sq_{ED}^A, Sq_{CF}^A, Sq_{GE}^A, Sq_{EI}^A$, and Sq_{IN}^A, and operations $(\_ \vdash_{SI} \_, \_ \rightarrow$

$\_, \_) : DB, Cls, Hinfo, Cl, Hst \to Sq^A_{SI}, (\_ \vdash_{ED} \_, \_ \to \_, \_) : DB, Cls, Hinfo, Cl,$
$Hst \to Sq^A_{ED}, (\_ \vdash_{CF} \_, \_ \to \_, \_) : DB, Cls, Hinfo, Cl, Hst \to Sq^A_{CF}, (\_ \vdash_{GE}$
$\_, \_ \to \_, \_) : DB, Cls, Hinfo, Cl, Hst \to Sq^A_{GE}, (\_ \vdash_{EI} \_, \_ \to \_, \_) : DB, Cls,$
$Hinfo, Cl, Hst \to Sq^A_{EI}, (\_ \vdash_{IN} \_, \_ \to \_) : DB, Cls, Hinfo, Cl \to Sq^A_{IN}$. An (annotated) sequent of the form $(db \vdash_{SI} \widehat{cl}, hnfo \to cl, hst)$ asserts that $\widehat{cl}$ and $hnfo$
are the result of the simplification process when called upon cl with history hst.
The other kinds of (annotated) sequents have analogous meaning.

The nested sequent signature $nesWat^a_0$ consists of $esWat^a_0$ and nested ERThs
for various constraints (obtained by composition) whose details we omit here.
The NERTh $nerWat^a_0$ consists of $nesWat^a_0$ and some (annotated) rules of inference, including

$$\mathsf{Qed}^a : \quad \Longrightarrow \quad db \vdash_W \top; \top ,$$

$$\mathsf{ElimSubsumed}^a : \quad \{Subsumes(cl, cl')\}$$
$$db \vdash_W \widehat{clh}; \widehat{clt} \wedge \mathsf{TBP}(cl') \wedge \widehat{clt}' \quad \Longrightarrow$$
$$db \vdash_W \widehat{clh}; \mathsf{TBP}(cl) \wedge \widehat{clt} \wedge \mathsf{TBP}(cl') \wedge \widehat{clt}' ,$$

which constitute the annotated counterparts of rules Qed and ElimSubsumed
previously shown. Note that the subsuming clause cl' is required to be tagged
by TBP. The elimination of a proved clause from the Pool is formalized by

$$\mathsf{ElimProved}^a : \quad db \vdash_W \top; \widehat{clt} \quad \Longrightarrow \quad db \vdash_W \top; \mathsf{BP}(cl) \wedge \widehat{clt} ,$$

which eliminates the clause at the head of the Pool if it is tagged by BP and
the Top is empty (all these rules must be read backwards). Finally, the pushing
of a clause from the Top onto the Pool, and the settling-down of a clause are
specified by

$$\mathsf{Push}^a : \quad \{NonEmpty(cl)\}$$
$$db \vdash_W \widehat{clh}; \mathsf{TBP}(cl) \wedge \widehat{clt} \quad \Longrightarrow \quad db \vdash_W (cl/hst) \wedge \widehat{clh}; \widehat{clt} ,$$

$$\mathsf{SettleDown}^a : \quad \{NoneSD(hst)\}$$
$$db \vdash_W (cl/([\mathsf{SD}, cl, \emptyset] \cdot hst)) \wedge \widehat{clh}; \widehat{clt} \quad \Longrightarrow$$
$$db \vdash_W (cl/hst) \wedge \widehat{clh}; \widehat{clt} .$$

The side conditions $NonEmpty(cl)$ and $NoneSD(hst)$ respectively require that
cl is non-empty (i.e., it has at least one literal) and that hst does not contain
the history identifier SD. $\emptyset : \to Hinfo$ is the empty history information.

The nested sequents signature $nesWat^a_1$ consists of $esWat^a_1$ and a nested
ERThs for various constraints whose details we omit here. The NERTh $nerWat^a_1$
consists of $nesWat^a_1$ and some (annotated) rules of inference, including

$$\mathsf{CallSimp}^a : \quad \left\{\widehat{cl} \neq cl\right\} \quad db \vdash_{SI} \widehat{cl}, hnfo \to cl, hst$$
$$db \vdash_W atchst(\widehat{cl}, [\mathsf{SI}, cl, hnfo] \cdot hst) \wedge \widehat{clh}; \widehat{clt} \quad \Longrightarrow$$
$$db \vdash_W (cl/hst) \wedge \widehat{clh}; \widehat{clt} ,$$

which constitutes the annotated counterpart of the CallSimp. Read backwards,
it specifies that the clause and history at the head of the Top are fed into the

simplifier and replaced by the clauses returned by it, all paired with the suitably extended history $(atchst(\_,\_) : Cls, Hst \rightarrow ClHs$ pairs each clause of a conjunction with a history, as specified by suitable equations)[1]. The side condition requires that the simplifier does not fail. The rules $\mathsf{CallElDes}^a$, $\mathsf{CallCrFer}^a$, $\mathsf{CallGen}^a$, $\mathsf{CallElIrr}^a$, and $\mathsf{CallInd}^a$ are analogous. $nerWat_0^a$ and $nerWat_1^a$ are composable, and $nerWat^a = nerWat_0^a + nerWat_1^a$.

4.4   The Erasing Mapping

The erasing mapping removes control information. Histories and Pool tags constitute control information. Therefore, we have, for instance, that $\varepsilon_{\mathrm{W}}(Hst) = [\,]$, $\varepsilon_{\mathrm{W}}(ClHs) = \varepsilon_{\mathrm{W}}(ClTs) = Cls$, and $\varepsilon_{\mathrm{W}}(cl/hst) = \varepsilon_{\mathrm{W}}(\mathsf{TBP}(cl)) = cl$. Since clauses tagged by BP are logically redundant, they constitute control information altogether, as specified by $\varepsilon_{\mathrm{W}}(\mathsf{BP}(cl)) = [\,]$. Databases are mapped to their underlying theories, $\varepsilon_{\mathrm{W}}(DB) = Th$. Annotated sequents are mapped to their non-annotated counterparts, e.g., $\varepsilon_{\mathrm{W}}(Sq_{\mathrm{W}}^{\mathrm{A}}) = Sq_{\mathrm{W}}$, $\varepsilon_{\mathrm{W}}(db \vdash_{\mathrm{W}} \widehat{clh}; \widehat{clt}) = (\varepsilon_{\mathrm{W}}(db) \vdash_{\mathrm{W}} \varepsilon_{\mathrm{W}}(\widehat{clh}) \wedge \varepsilon_{\mathrm{W}}(\widehat{clt}))$, $\varepsilon_{\mathrm{W}}(Sq_{\mathrm{SI}}^{\mathrm{A}}) = Sq_{\mathrm{SI}}$, and $\varepsilon_{\mathrm{W}}(db \vdash_{\mathrm{SI}} \widehat{cl}, hnfo \rightarrow cl, hst) = (\varepsilon_{\mathrm{W}}(db) \vdash_{\mathrm{SI}} \widehat{cl} \rightarrow cl)$.

By applying the erasing mapping to the premisses, conclusion, and side condition (which gets erased altogether) of $\mathsf{CallSimp}^a$, we obtain an instance of $\mathsf{CallSimp}$, thereby guaranteeing the existence of a derivation structure as required by the formal definition in Sect. 2.3. If we apply the erasing mapping to the premiss and to the conclusion of Push^a, we respectively obtain $(th \vdash_{\mathrm{W}} \widehat{cl}' \wedge cl \wedge \widehat{cl}'')$ and $(th \vdash_{\mathrm{W}} cl \wedge \widehat{cl}' \wedge \widehat{cl}'')$ (where db, $\widehat{clh}$, and $\widehat{clt}$ are respectively mapped to th, $\widehat{cl}'$, and $\widehat{cl}''$), which are the same sequent. Therefore, this rule maps to a trivial derivation structure (the side condition is just erased). In other words, it only modifies control information, leaving logical content unaltered; the same holds for $\mathsf{ElimProved}^a$ and $\mathsf{SettleDown}^a$.

5   Conclusions and Future Work

In this paper we have formalized the composition and nesting of ERThs by means of faithful inclusions. We have also formalized the concept of annotated (N)ERThs by means of erasing mappings. Working in the special case of ERThs, the mappings we need are well understood notions of mappings of theories in a logic. Mappings for a very general notion of reasoning theory are developed in [20]. The mappings used in this paper provide the starting point for developing

---

[1] In the actual LISP code of NQTHM, the sets of terms appearing in the induction hypothesis and conclusion are passed as separate arguments to the simplifier. However, we believe they can be more nicely and elegantly specified (and implemented as well) as history information produced by the induction process. We have therefore introduced a seventh history identifier for induction, and specified that induction returns a history information together with the clauses constituting the induction schema.

a more general theory of composition and annotation. There is also work in progress to generalize the notion of faithful inclusion to allow a richer collection of mappings to be used directly for composition.

While annotations can be used to specify search information and its manipulation, in order to fully specify an OMRS we also need a formal way to specify the strategy of application of annotated rules. In [9] this is done by means of a tactical language, and a tactical language is also used in [3]. We are working towards the development of a general formalism to capture tactical languages and also more general mechanisms; given the close connections between reasoning theories and rewriting logic studied in [20], such formalisms may perhaps be fruitfully related to the internal strategy languages of rewriting logic [8]. What remains to give a full specification of an OMRS is its interaction level. A formalism for this has still to be developed.

Acknowledgements. The authors would like to thank the anonymous referees for helpful criticisms. This research was partially supported by ONR grants N00014-94-1-0857 and N00014-99-C-0198, NSF grants CRR-9633363 and 9633363, DARPA/Rome Labs grant AF F30602-96-1-0300, and DARPA/NASA contract NAS2-98073.

References

1. A. Armando and S. Ranise. From Integrated Reasoning Specialists to "Plug-and-Play" Reasoning Components. In *Fourth International Conference Artificial Intelligence And Symbolic Computation (AISC'98)*, 1998. also available as DIST Technical Report 97-0049, University of Genova, Italy.
2. E. Astesiano, H.-J. Kreowski, and B. Krieg-Brückner, editors. *Algebraic Foundations of Systems Specifications*. IFIP State-of-the-Art Reports. Springer, 1999.
3. P. G. Bertoli. *Using OMRS in Practice: A Case Study with ACL2*. PhD thesis, University of Rome 3, 1997.
4. P.G. Bertoli, J. Calmet, K. Homann, and F. Giunchiglia. Specification and Integration of Theorem Provers and Computer Algebra Systems. In *Fourth International Conference On Artificial Intelligence and Symbolic Computation (AISC'98)*, Lecture Notes in AI. Springer Verlag, 1998. also Technical Report 9804-03, IRST, Trento, Italy, April 1998.
5. R. S. Boyer and J. S. Moore. *A Computational Logic*. Academic Press, 1979.
6. R. S. Boyer and J. S. Moore. *A Computational Logic Handbook*. Academic Press, 1988.
7. R. M. Burstall and J. A. Goguen. The semantics of Clear, a specification language. In *Proc. 1979 Copenhagen Winter School on Abstract Software Specification*, volume 86 of *Lecture Notes in Computer Science*, pages 292–332. Springer-Verlag, 1980.
8. M. Clavel and J. Meseguer. Reflection and strategies in rewriting logic. In *Rewriting Logic Workshop'96*, number 4 in Electronic Notes in Theoretical Computer Science. Elsevier, 1996. URL: http://www.elsevier.nl/locate/entcs/volume4.html.
9. A. Coglio. The control component of OMRS. Master's thesis, University of Genova, Italy, 1996.

10. A. Coglio. Definizione di un formalismo per la specifica delle strategie di inferenza dei sistemi di ragionamento meccanizzato e sua applicazione ad un sistema allo stato dell'arte. Master's thesis, 1996. Master thesis, DIST - University of Genoa (Italy).

11. A. Coglio, F. Giunchiglia, P. Pecchiari, and C. Talcott. A logic level specification of the nqthm simplification process. Technical report, IRST, University of Genova, Stanford University, 1997.

12. G. Gentzen. *The Collected Papers of Gerhard Gentzen.* North-Holland, 1969. edited by Szabo, M. E.

13. F. Giunchiglia, P. Pecchiari, and A. Armando. Towards provably correct system synthesis and extension. *Future Generation Computer Systems*, 12(458):123–137, 1996.

14. F. Giunchiglia, P. Pecchiari, and C. Talcott. Reasoning theories: Towards an architecture for open mechanized reasoning systems. In *Workshop on Frontiers of Combining Systems FROCOS'96*, 1996.

15. Joseph Goguen and José Meseguer. Order-sorted algebra I: Equational deduction for multiple inheritance, overloading, exceptions and partial operations. *Theoretical Computer Science*, 105:217–273, 1992.

16. A. Haxthausen and F. Nickl. Pushouts of order-sorted algebraic specifications. In *Algebraic Methodology and Software Technology (AMAST'96)*, number 1101 in Lecture Notes in Computer Science, pages 132–148. Springer, 1996.

17. Narciso Martí-Oliet and José Meseguer. Inclusions and subtypes i: First-order case. *J. Logic and Computation*, 6(3):409–438, 1996.

18. Narciso Martí-Oliet and José Meseguer. Rewriting logic as a logical and semantic framework. In D. Gabbay, editor, *Handbook of Philosophical Logic.* Kluwer Academic Publishers, 1997.

19. P. Martin-Löf. *Intuitionistic Type Theory.* Bibliopolis, 1984.

20. J. Meseguer and C. Talcott. Reasoning theories and rewriting logic. (in preparation).

21. J. Meseguer and C. Talcott. Mapping OMRS to Rewriting Logic. In C. Kirchner and H. Kirchner, editors, *2nd International Workshop on Rewriting Logic and its Applications, WRLA'98*, volume 15 of *Electronic Notes in Theoretical Computer Science*, 1998. URL: `http://www.elsevier.nl/locate/entcs/volume15.html`.

22. José Meseguer. Membership algebra as a semantic framework for equational specification. In F. Parisi-Presicce, editor, *12th International Workshop on Abstract Data Types (WADT'97)*, number 1376 in Lecture Notes in Computer Science, pages 18–61. Springer, 1998.

23. T. Mossakowski. Colimits of order-sorted specifications. In F. Parisi-Presicce, editor, *12th International Workshop on Abstract Data Types (WADT'97)*, number 1376 in Lecture Notes in Computer Science, pages 316–322. Springer, 1998.

24. D. Prawitz. *Natural Deduction: A Proof-theoretical Study.* Almquist and Wiksell, 1965.

25. M-O. Stehr and J. Meseguer. The HOL-Nuprl connection from the viewpoint of general logics. manuscript, SRI International, 1999.

26. M-O. Stehr and J. Meseguer. Pure type systems in rewriting logic. In *Workshop on Logical Frameworks and Meta-languages*, 1999.

Why Combined Decision Problems Are Often Intractable*

Klaus U. Schulz

CIS, University of Munich
Oettingenstr. 67, D-80538 München, Germany
`schulz@cis.uni-muenchen.de`

Abstract. Most combination methods for decision procedures known from the area of automated deduction assign to a given "combined" input problem an exponential number of output pairs where the two components represent "pure" decision problems that can be decided with suitable component decision procedures. A mixed input decision problem evaluates to true iff there exists an output pair where both components evaluate to true. We provide a formal framework that explains why in many cases it is impossible to have a polynomial-time optimization of such a combination method, and why often combined decision problems are NP-hard, regardless of the complexity of the component problems. As a first application we consider Oppen's combination method for algorithms that decide satisfiability of quantifier-free formulae in disjunctive normal form w.r.t. a first-order theory. A collection of first-order theories is given where combined problems are always NP-hard and Oppen's method does not have a polynomial-time optimization. As a second example, similar results are given for the problem of combining algorithms that decide whether a unification problem has a solution w.r.t. to a given equational theory.

1 Introduction

The problem of how to combine decision procedures and related algorithms is ubiquitous in many fields. In particular in the area of automated deduction many instances of this problem have been studied [NO79, Op80, Sh84, CLS94, Ri96, TH96, Ki85, He86, Ti86, Ye87, Ni89, SS89, Bo93, BS96, DKR94, BS95a, KR98, Ke98]. Many of these combination methods (e.g., [Op80, BS96, DKR94, BS95a, Ke98]) basically follow a similar strategy: a "mixed" input problem is first decomposed into a pair of two "pure" component problems. Before the two component problems are evaluated in two independent processes, some characteristic restriction is imposed on both components that guarantees that two given solutions of the component problems can safely be combined to a joint solution of the mixed input problem. This restriction is chosen in a non-deterministic way from a set of possible restrictions, which depends on the input problem.

---

* This paper was first presented at the "Workshop on Complexity in Automated Deduction" organized by M. Hermann in Nancy (June 1999).

H. Kirchner and C. Ringeissen (Eds.): FroCoS 2000, LNAI 1794, pp. 217–244, 2000.

Unfortunately, in general (e.g. in [Op80,BS96]) the number of possible restrictions for a given input problem is exponential in the size of the mixed input. Hence the combination algorithm introduces its own NP-complexity, regardless of the complexity of the algorithms that are available for the components. Often it is simple to optimize the naive approach to a certain extent and to eliminate, after a simple inspection, some of the "possible" restrictions (cf. e.g. [NO79,KR98]). The question arises in which cases a polynomial-time procedure can be found that leads to a polynomial number of restrictions that have to be considered. In this paper we show, generalizing previous results in [Sc97,Sc99], that in many cases such a polynomial optimization is impossible (unless P = NP) and the combined decision problems are NP-hard.

The main contribution of the paper is a formal framework that helps to understand why combined problems are often much more difficult to solve than their components, and why this leads to the situation described above. Basically, we first look at a class of simple problems where it is easy to demonstrate that even the combination of two perfectly trivial component problems can become intractable. We then show how to lift impossibility results for polynomial combination and intractability results for a given class of combination problems to another class of combination problems. In order to illustrate this technique we apply it two well-known combination problems from automated deduction.

The structure of the paper is the following. We start with some preliminaries in Section 2. In Section 3 we describe Oppen's procedure for combining algorithms that decide satisfiability of quantifier-free formulae in disjunctive normal form w.r.t. a first-order theory T, and a method for combining algorithms that decide solvablity of unification problems w.r.t. a given equational theory E introduced in [BS96]. In Section 4 we introduce the abstract notion of a so-called protocol-based combination method and show that the two combination methods of Section 3 fall in this class. We also formalize the notion of a polynomial optimization of a given protocol-based combination method. In Section 5 we consider the problem of combining decision procedures for generalized satisfiability problems in the sense of Schaefer [Sh78]. Some trivial examples are given that demonstrate that the combination of decision problems can be NP-hard even if all instances of component problems are always solvable. We also introduce a specific protocol-based combination method and give simple examples where a polynomial optimization of this method is impossible unless P = NP. In Section 6 we introduce the notion of a polynomial interpretation of one protocol-based combination method (source method) in another (target method) and show how impossibility results for polynomial optimization, as well as NP-hardness results for combined input problems, can be lifted from the source method to the target method.

The following two sections are used to illustrate this formal framework. In Section 7 we introduce two abstract classes of first-order theories and show that there is no polynomial optimization of Oppen's combination method if applied to a combination of theories of class 1 and 2 respectively. For both classes, a list of example theories is given. In Section 8 we introduce two classes of equational theories and show that there is no polynomial optimization of the combination method of [BS96] if applied to a combination of equational theories of class 1 and 2 respectively. We prove that each regular equational theory with a commutative (or associative) function symbol belongs to class 1, and each simple equational theory with a non-constant function symbol belongs to class 2. In the conclusion we discuss the relevance of all these observations.

Some remarks on related works are in order. In the area of unification theory, some results are known which show that there cannot be a general polynomial-time algorithm for combining procedures that decide solvability of unification problems, unless P = NP: solvability of associative-commutative-idempotent (*ACI-*) unification problems with free constants is decidable in polynomial time and unifiability in the empty theory can be decided in linear time, but the problem of deciding solvability of *ACI*-unification problems with additional free function symbols (called general *ACI*-unification) is NP-hard (see [KN91]). The latter problem can be considered as a combination of *ACI*-unification with unification in the empty theory. The same increase of complexity has been observed when adding free function symbols to the theory *ACUN* of an associative-commutative and nilpotent function with unit, or to the theory *ACUNh* which contains in addition a homomorphic unary function symbol (see [GN96] for formal definitions). A corresponding impossibility result for the case of algorithms that compute minimal complete sets of unifiers modulo finitary equational theories was obtained by M. Hermann and P.G. Kolaitis [HK96], by proving intractability of the counting problem for general unification in the theory of abelian groups (*AG*-unification). The results on combination of *E*-unification algorithms given in Section 8 generalize previous results of the same author in [Sc97,Sc99].

2   Formal Preliminaries

We assume that the reader is familiar with the basic notions of first-order predicate calculus. For convenience, we recall some more specific concepts from the area of equational unification that are used below.

An *equational theory* is given by a set E of identities $s = t$ between terms s and t. The *signature of E* is the set Σ of all function symbols occurring in E. Let *Var* denote a countably infinite set of variables. With $=_E$ we denote the least congruence relation on the term algebra $\mathcal{T}(\Sigma, Var)$ that is closed under substitution and contains E. An equational theory E is *trivial* iff $x =_E y$ holds for two distinct variables x, y. In the following, we consider only non-trivial equational theories. In this paper, two special types of equational theories will be considered. An equational theory E is *regular* if $Var(s) = Var(t)$ for all identities $s = t$ of E. Here $Var(t)$ denote the set of variables occurring in the term t. An equational theory E is *simple* if no term is equivalent modulo E to a proper subterm. For a detailed explanation of these notions and for an introduction to equational unification we refer to [BS94].

Remark 1. The following fact can easily be proved for regular equational theories E: $\forall s, t \in T(\Sigma, X) : s =_E t$ implies $Var(s) = Var(t)$.

Given an equational theory E with signature Σ, an *elementary E-unification problem* is a finite conjunction of equations between Σ-terms. In E-unification problems *with constants* (*general E*-unification problems), these terms may contain additional free constant (function) symbols, i.e., symbols not contained in the signature Σ of E.

A *solution* (or *E-unifier*) of the E-unification problem φ of the form $s_1 =^? t_1 \wedge \ldots \wedge s_n =^? t_n$ is a substitution σ such that $\sigma(s_i) =_E \sigma(t_i)$ for $i = 1, \ldots, n$. If there exists such a solution, then φ is called *solvable* or *unifiable*.

3   Two Combination Problems for Decision Problems

In this section we briefly introduce the two combination problems from the area of automated deduction that represent the source of motivation and the intended application for the results on intractability of combined decision problems described later.

3.1   Nelson-Oppen Type Combination Problems and Satisfiability of Quantifier-Free Formulae in Disjunctive Normalform

The situation studied by Nelson and Oppen in [NO79] is the following: Let T_1 and T_2 denote two first-order theories, formulated in first-order languages over disjoint signatures Σ_1 and Σ_2 respectively. Assume that for $i = 1, 2$ we have an algorithm for deciding satisfiability of a given quantifier-free Σ_i-formula w.r.t. (i.e., in a model of) theory T_i. Nelson and Oppen show how to obtain an algorithm for deciding satisfiability of quantifier-free $(\Sigma_1 \cup \Sigma_2)$-formulae. A technical assumption is that both theories have to be *stably-infinite*, which means each quantifier-free Σ_i-formula that is satisfiable in some model of T_i is also satisfiable in an infinite model of T_i ($i = 1, 2$).

In the present paper we do not look at the problem treated by Nelson and Oppen, but to a variant studied in [Op80] where input-formulae are restricted to quantifier-free $(\Sigma_1 \cup \Sigma_2)$-formulae in disjunctive normalform. The reason is that we want to characterize situations where the pure component problems are polynomial-time solvable and the combination of both decision problems becomes NP-hard. Quite generally it is simple to encode Boolean satisfiability problems as satisfiability problems of quantifier-free formulae w.r.t. a given component theory T_i. Hence, when admitting arbitrary quantifier-free formulae as input, already the treatment of "pure" input formulae over a single component signature Σ_i is NP-hard. In contrast, satisfiability of quantifier-free formulae in disjunctive normal form (qf-dnf formulae) reduces to satisfiability of conjunctions of literals. In fact a qf-dnf formula φ is satisfiable w.r.t. $T_1 \cup T_2$ iff one disjunct of φ is satisfiable w.r.t. $T_1 \cup T_2$. Since satisfiability of clauses[1] w.r.t. pure component theories T_i is often polynomial-time decidable (see [Op80,NO79] for examples) we have here a situation where combination may in fact cause a complexity gap.

---

[1] With a *clause* we mean a conjunction of atomic formulae.

In the sequel, a $(\Sigma_1 \cup \Sigma_2)$-clause is said to be in *decomposed form* if it has the form $\varphi_1 \wedge \varphi_2$ where the subclauses φ_1 and φ_2 are built over the signatures Σ_1 and Σ_2 respectively. Given a clause φ in the mixed signature $\Sigma_1 \cup \Sigma_2$, which is to be tested for satisfiability w.r.t. $T_1 \cup T_2$, Oppen discusses the following non-deterministic procedure.

Oppen's Procedure. In the *first step* we use "variable abstraction[2]" to transform the input clause φ into a clause in decomposed form $\varphi_1 \wedge \varphi_2$ such that φ and $\varphi_1 \wedge \varphi_2$ are equivalent in terms of satisfiability w.r.t. $T_1 \cup T_2$. In the *second step* we non-deterministically choose a partition $\Pi = \{P_1, \ldots, P_k\}$ of the variables shared by φ_1 and φ_2. For each of the classes P_i, let $x_i \in P_i$ be a representative of this class, and let $X_\Pi := \{x_1, \ldots, x_k\}$ be the set of these representatives. The substitution that replaces, for all $i = 1, \ldots, k$, each element of P_i by its representative x_i is denoted by σ_Π. In the *third step* we test if the formulae $\sigma_\Pi(\varphi_i) \wedge \bigwedge_{x \neq y \in X_\Pi} x \neq y$ are satisfiable in T_i ($i = 1, 2$), in the positive case we return "satisfiable."

Proposition 1. *([Op80]) Assume that T_1 and T_2 have disjoint signatures and both are stably infinite. The input clause φ is satisfiable w.r.t. $T_1 \cup T_2$ iff there exists a choice for partition Π where the algorithm returns "satisfiable".*

The number of partitions of a set of n elements is known as the n-th Bell number and denoted $B(n)$. This function grows faster than 2^n or $(n/2)!$. Hence the practical worst-case behaviour of Oppen's procedure is exponential. The results of this paper will show that there are various component theories T_1 and T_2 where it is impossible to obtain a polynomial-time combination algorithm, in a sense to be made precise.

3.2   Combination of E-Unification Algorithms

In this section we give a brief description of the combination procedure for unification algorithms for equational theories over disjoint signatures given in [BS96]. Before we can describe the algorithm, we have to introduce a generalization of E-unification problems with constants. Let $X \subset Var$ be a finite set of variables, let Π be a partition of X. With $[x]_\Pi$ we denote the equivalence class of $x \in X$ with respect to Π. A partial ordering "$<$" on X is Π-*congruent* if $x < y$ implies $x' < y'$ for all $x, x', y, y' \in X$ such that $[x]_\Pi = [x']_\Pi$ and $[y]_\Pi = [y']_\Pi$. A Π-*total* ordering on X is a Π-congruent ordering "$<$" such that $[x]_\Pi \neq [y]_\Pi$ implies that either $x < y$ or $y < x$, for all $x, y \in X$. Let M be a set. A function $F : X \to M$ is Π-*congruent* if $F(x) = F(x')$ for all $x, x' \in X$ such that $[x]_\Pi = [x']_\Pi$. In the sequel we assume that Var is enumerated in a fixed order. With $rep_\Pi(x)$ we denote the minimal representative of $[x]_\Pi$ w.r.t. this enumeration order.

Definition 1. *Let E_1 be an equational theory with signature Σ_1, let Σ_2 be a second signature. Let φ_1 be an elementary E_1-unification problem and let X be a finite set of variables with $Var(\varphi_1) \subseteq X$. A linear constant restriction for X is*

---

[2] See, e.g., [BS96] for a description of this well-known technique.

a triple $L = (\Pi, \text{Lab}, <_L)$ *where* Π *is a partition of* X, $<_L$ *is a* Π-*total partial ordering on* X *and* $\text{Lab} : X \to \{\Sigma_1, \Sigma_2\}$ *is* Π-*congruent. If* $\text{Lab}(x) = \Sigma_1$ *(resp.* $\text{Lab}(x) = \Sigma_2$) *we say that* x *receives a* domestic *(resp. alien) label, or simply that* x *is a domestic (alien) variable. The pair* (φ_1, L) *is called an* E_1-unification problem with linear constant restriction. A Σ_1-*substitution* σ *solves* (φ_1, L) *if* σ *solves the* E_1-*unification problem* φ_1 *and if the following conditions are satisfied:*

1. *For all* $x \in X$ *we have* $\sigma(x) = \sigma(\text{rep}_\Pi(x))$,
2. $\sigma(y) = \text{rep}_\Pi(y)$ *for each* alien *variable* $y \in X$,
3. *for all* $x, y \in X$: *if* y *is alien,* x *is domestic, and if* $\sigma(y)$ *occurs in* $\sigma(x)$, *then* $y <_L x$.

Let E_1 and E_2 be two consistent equational theories over disjoint signatures Σ_1 and Σ_2 respectively. As in the case of satisfiability problems, an elementary $(E_1 \cup E_2)$-unification problem φ' is in *decomposed form* if φ' has the form $\varphi_1 \wedge \varphi_2$ where the "pure" subproblems φ_1 and φ_2 are built over the signatures Σ_1 and Σ_2 respectively. Suppose that we want to decide solvability of an elementary $(E_1 \cup E_2)$-unification problem φ. The following *Decomposition Algorithm*, described in more detail in [BS96], reduces φ non-deterministically to a finite number of output pairs. Each component of an output pair represents an (E_1- resp. E_2-) unification problem with linear constant restriction.

Decomposition Algorithm. In the *first step*, the input problem φ is transformed via variable abstraction into an elementary $(E_1 \cup E_2)$-unification problem in the decomposed form $\varphi_1 \wedge \varphi_2$ such that φ is solvable iff $\varphi_1 \wedge \varphi_2$ is solvable. Let $X := \text{Var}(\varphi_1 \wedge \varphi_2)$. In the (non-deterministic) *second step*, a linear constant restriction $L = (\Pi, \text{Lab}, <_L)$ for X is chosen. The output pair determined by L is $((\varphi_1, L), (\varphi_2, L))$.

Proposition 2. *([BS96]) Assume that* E_1 *and* E_2 *have disjoint signatures. The input problem,* φ, *has a solution iff there exists an output pair of the Decomposition Algorithm,* $((\varphi_1, L), (\varphi_2, L))$, *such that both the* E_1-*unification problem with linear constant restriction* (φ_1, L) *has a solution and the* E_2-*unification problem with linear constant restriction* (φ_2, L) *has a solution.*

Note that variables with label Σ_1 are domestic in (φ_1, L) but alien in (φ_2, L), the same holds, mutatis mutandis, for variables with label Σ_2. Obviously, if n denotes the number of variables of the the problem $\varphi_1 \wedge \varphi_2$ reached after step 1, the number of output pairs of the Decomposition Algorithm even exceeds the n-th Bell number $B(n)$. Hence the worst-case behaviour of the algorithm is exponential. The results of this paper will show, in a sense to be made precise, that in many cases is is impossible to obtain a polynomial-time combination method.

4   Combined Decision Problems and Protocol-Based Combination of Decision Procedures

In this section we show that Oppen's Procedure and the Decomposition Algorithm for combined E-unification algorithms can be considered as two special instances of a general and abstract notion of combination algorithm. We start with some basic and well-known concepts.

Definition 2. *A decision problem is a pair $\mathcal{D} = (\mathcal{P}, \mathrm{val})$ where $\mathcal{P}$ is a set and* $\mathrm{val} : \mathcal{P} \to \{0, 1\}$ *is a mapping. The elements of $\mathcal{P}$ are called the* instances *of $\mathcal{D}$.*

We assume that each instance $\varphi \in \mathcal{P}$ is of a specific syntactic size. The *complexity of $\mathcal{D}$* is the complexity of deciding, given an instance φ of $\mathcal{D}$, if $\mathrm{val}(\varphi) = 1$. The following examples show how the problems that are treated in Oppen's Procedure resp. in the Decomposition Algorithm can be described as decision procedures in the above sense.

Example 1. Let T be a theory (set of sentences) over a first-order language $\mathcal{L}$, let $\mathcal{P}_T$ denote the set of quantifier-free formulae in disjunctive normal form of $\mathcal{L}$. For $\varphi \in \mathcal{P}_T$ let $\mathrm{var}_T(\varphi) = 1$ iff φ is satisfiable in some model of T. The decision problem $(\mathcal{P}_T, \mathrm{val}_T)$ is called the *qf-dnf satisfiability problem* for T.

Example 2. Let E be an equational theory, let $\mathcal{P}_E$ denote the set of all elementary E-unification problems. For $\varphi \in \mathcal{P}_E$ let $\mathrm{var}_E(\varphi) = 1$ iff φ is unifiable. The decision problem $(\mathcal{P}_E, \mathrm{val}_E)$ is called the *E-unifiability problem*.

Decision problems often come in natural classes. Moreover, given such a class $\mathcal{C}$ there is often a natural notion of *combining* two decision problems in $\mathcal{C}$: given two decision problems $\mathcal{D}_1$ and $\mathcal{D}_2$ in $\mathcal{C}$, there is a unique decision problem in $\mathcal{C}$ that can be considered as "the" natural combination of $\mathcal{D}_1$ and $\mathcal{D}_2$.

Example 3. [Ex. 1 cont.] Let $\mathcal{C}_{qf-dnf}$ denote the class of all qf-dnf satisfiability problems for T, where T denotes any first-order theory. If $\mathcal{D}_{T_i}$ denotes the qf-dnf satisfiability problem for T_i, for $i = 1, 2$ and first-order theories T_1, T_2, then the natural combination is the qf-dnf satisfiability problem for the union $T_1 \cup T_2$.

Example 4. [Ex. 2 cont.] Let $\mathcal{C}_{unif}$ denote the class of all E-unifiability problems, for arbitrary equational theories E. If $\mathcal{D}_{E_i}$ denotes the E_i-unifiability problem, for $i = 1, 2$ and equational theories E_1, E_2, then the natural combination is the problem $\mathcal{D}_{E_1 \cup E_2}$ of $(E_1 \cup E_2)$-unifiability.

For the rest of the section we fix a class $\mathcal{C}$ of decision problems. We assume that there exists a natural notion of combining problems in $\mathcal{C}$. We write $\mathcal{D}_1 \oplus \mathcal{D}_2$ for the combination of the problems $\mathcal{D}_1$ and $\mathcal{D}_2$ of the class $\mathcal{C}$, and $\mathcal{D}_1$ and $\mathcal{D}_2$ will be called the *plain component problems* of $\mathcal{D}_1 \oplus \mathcal{D}_2$. For simplicity we generally assume that the instances of a combined problem $\mathcal{D}_1 \oplus \mathcal{D}_2$ have the form $\varphi_1 \wedge \varphi_2$ where φ_i is an instance of $\mathcal{D}_i$ for $i = 1, 2$. Recall that this assumption is justified in the case of Oppen's Procedure and in the case of the Decomposition Algorithm for combined E-unification algorithms. In both cases the first step is used to reach this form of input. From another perspective, which is used below, this straightforward and deterministic step may be considered as a kind of preprocessing that is not part of the the algorithm. We write $\mathcal{P}_1 \oplus \mathcal{P}_2$ for the set of all conjunctions of the form $\varphi_1 \wedge \varphi_2$ where φ_i is an instance of $\mathcal{D}_i$ for $i = 1, 2$. Hence $\mathcal{D}_1 \oplus \mathcal{D}_2$ has the form $(\mathcal{P}_1 \oplus \mathcal{P}_1, val)$ for suitable val.

Our next aim is to study "combination methods", i.e., methods for deciding combined decision problems by reduction to pure subproblems that can be decided by suitable component decision procedures. We do not require that such a method applies to arbitrary combinations of problems in $\mathcal{C}$, but to some subclass of *admissible* combinations. In the sequel we assume that such a subclass is fixed (later, when studying concrete combination methods, the class of admissible combination problems will be precisely specified.)

Definition 3. *A* protocol-based combination method *(for $\mathcal{C}$ and $\mathcal{C}_{adm}$) is given by a mapping that assigns to each admissible combination $\mathcal{D}_1 \oplus \mathcal{D}_2 = (\mathcal{P}_1 \oplus \mathcal{P}_2, val)$ of plain component problems $\mathcal{D}_1 = (\mathcal{P}_1, val_1)$ and $\mathcal{D}_2 = (\mathcal{P}_2, val_2)$ a quadruple $(\mathcal{S}, Alg, val_1^s, val_2^s)$ where*

1. *$\mathcal{S}$ is a set, called the* protocol. *The elements of $\mathcal{S}$ are called* scenarios,
2. *$Alg : \mathcal{P}_1 \oplus \mathcal{P}_2 \to 2^{\mathcal{S}}$ assigns to each instance $\varphi_1 \wedge \varphi_2$ of $\mathcal{D}_1 \oplus \mathcal{D}_2$ a finite set $Alg(\varphi_1 \wedge \varphi_2) \subseteq \mathcal{S}$ called the set of* output scenarios,
3. *$val_1^s : \mathcal{P}_1 \times \mathcal{S} \to \{0,1\}$ and $val_2^s : \mathcal{P}_2 \times \mathcal{S} \to \{0,1\}$ define decision problems of the form $\mathcal{D}_1^s = (\mathcal{P}_1 \times \mathcal{S}, val_1^s)$ and $\mathcal{D}_2^s = (\mathcal{P}_2 \times \mathcal{S}, val_2^s)$ respectively such that the following condition holds: for each instance $\varphi_1 \wedge \varphi_2$ of $\mathcal{D}_1 \oplus \mathcal{D}_2$ we have $val(\varphi_1 \wedge \varphi_2) = 1$ iff there exists an output scenario $S \in Alg(\varphi_1 \wedge \varphi_2)$ such that $val_1^s(\varphi_1, S) = 1 = val_2^s(\varphi_2, S)$.*

Alg will be called a *combination algorithm* for $\mathcal{D}_1 \oplus \mathcal{D}_2$, the condition given in item 3 will be called *correctness* of Alg. Obviously, correctness of the combination algorithm ensures that the combined decision problem $\mathcal{D}_1 \oplus \mathcal{D}_2$ can be reduced to $\mathcal{D}_1^s$ and $\mathcal{D}_2^s$. Given an instance $\varphi_1 \wedge \varphi_2$ of $\mathcal{D}_1 \oplus \mathcal{D}_2$ we may inspect each of the finitely many output scenarios, and to check if for one scenario S we have $val_1^s(\varphi_1, S) = 1 = val_2^s(\varphi_2, S)$. The situation may be summarized in the following picture.

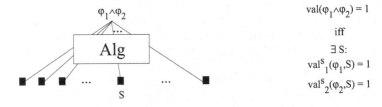

The following examples show that the combination methods introduced in Sections 3.1 and 3.2 are special instances of protocol-based combination algorithms

Example 5. [Ex. 3 cont.] As in Example 3, let $\mathcal{C}_{qf-dnf}$ denote the class of all qf-dnf satisfiability problems for theories T. The combination $\mathcal{D}_{T_1} \oplus \mathcal{D}_{T_2}$ of the problems $\mathcal{D}_{T_1}$ and $\mathcal{D}_{T_2}$ of $\mathcal{C}_{qf-dnf}$ is *admissible* iff the theories T_1 and T_2 are stably-infinite and have disjoint signatures. Oppen's Procedure can be described as a protocol-based combination method for $\mathcal{C}_{qf-dnf}$ in the following way. It assigns to an admissible combination $\mathcal{D}_{T_1} \oplus \mathcal{D}_{T_2}$ of the plain component problems $\mathcal{D}_{T_1}$ and $\mathcal{D}_{T_2}$ of the form $(\mathcal{P}_{T_1}, val_{T_1})$ and $(\mathcal{P}_{T_2}, val_{T_2})$ respectively the quadruple $(\mathcal{S}, Alg, val^s_{T_1}, val^s_{T_2})$ where

1. $\mathcal{S}$ is the set of all partitions Π of finite subsets of Var,
2. $Alg(\varphi_1 \wedge \varphi_2)$ is the set of all partitions Π of the set of shared variables of φ_1 and φ_2, for each $\varphi_1 \wedge \varphi_2 \in \mathcal{P}_{T_1} \oplus \mathcal{P}_{T_2}$,
3. for $\Pi \in \mathcal{S}$ we have $val^s_{T_i}(\varphi_i, \Pi) = 1$ iff there exists a T_i-model $\mathcal{M}$ and a variable assignment ν such that $\mathcal{M}, \nu \models \varphi_i$ and $\nu(x) = \nu(y)$ iff $x =_\Pi y$, for all $x, y \in \bigcup \Pi$.

Proposition 1 expresses correctness of Alg for admissible combinations. In the sequel, if $val^s_{T_i}(\varphi_i, \Pi) = 1$ we simply say that φ_i is *satisfiable w.r.t. T_i under partition Π*.

Example 6. [Ex. 4 cont.] As in Example 4, let $\mathcal{C}_{unif}$ denote the class of all E-unifiability problems. The combination $\mathcal{D}_{E_1} \oplus \mathcal{D}_{E_2}$ of the problems $\mathcal{D}_{E_1}$ and $\mathcal{D}_{E_2}$ of $\mathcal{C}_{unif}$ is *admissible* iff the equational theories E_1 and E_2 have disjoint signatures. The above Decomposition Algorithm can be described as a protocol-based combination method for $\mathcal{C}_{unif}$ in the following way. It assigns to an admissible combination $\mathcal{D}_{E_1} \oplus \mathcal{D}_{E_2}$ of the plain component problems $\mathcal{D}_{E_1}$ and $\mathcal{D}_{E_2}$ of the form $(\mathcal{P}_{E_1}, val_{E_1})$ and $(\mathcal{P}_{E_2}, val_{E_2})$ respectively the quadruple $(\mathcal{S}, Alg, val^s_{E_1}, val^s_{E_2})$ where

1. $\mathcal{S}$ is the set of generalized linear constant restrictions L on a finite subset of Var,
2. $Alg(\varphi_1 \wedge \varphi_2)$ is the set of all generalized linear constant restrictions L of the set of variables occurring in $\varphi_1 \wedge \varphi_2$, for each $\varphi_1 \wedge \varphi_2 \in \mathcal{P}_{E_1} \oplus \mathcal{P}_{E_2}$,
3. for $L \in \mathcal{S}$ we have $val^s_{E_i}(\varphi_i, L) = 1$ iff the E_i-unification problem with generalized linear constant restriction (φ_i, L) is solvable.

Proposition 2 expresses correctness of Alg for admissible combinations. If (φ_i, L) is solvable, we also say that φ_i is *solvable under linear constant restriction L*.

Remark 2. In Example 5 one remarkable phenomenon arises if the input problem $\varphi_1 \wedge \varphi_2$ contains an equation or a disequation α between variables. A formula α of this form can be attached to both sides φ_1 and φ_2, and this choice may affect the cardinality of the set of shared variables. To consider an extreme case, assume that φ_2 is a conjunction of equations and disequations between variables. Assume that we have n shared variables. When treating φ_2 as a part of the subformula for theory T_1, the set of shared variables becomes empty and Oppen's Procedure becomes deterministic. There are variants of the Decomposition Algorithm for combined E-unification problems where the choice of a linear constant restriction is restricted to the set of shared variables. In this situation the same phenomenon occurs.

Due to the large number of output pairs of the two combination algorithms described above, optimization techniques are important. For a systematic study of optimization, the following formal notion is useful.

Definition 4. *Let $\mathcal{C}, \mathcal{C}_{adm}$ and $\mathcal{D}_1 \oplus \mathcal{D}_2$ as in Definition 3. A (polynomial) optimization of Alg for $\mathcal{D}_1 \oplus \mathcal{D}_2$ is an algorithm Alg_{opt} that computes for each instance $\varphi_1 \wedge \varphi_2$ of $\mathcal{D}_1 \oplus \mathcal{D}_2$ (in polynomial time) a set $Alg_{opt}(\varphi_1 \wedge \varphi_2) \subseteq Alg(\varphi_1 \wedge \varphi_2)$ such that $val(\varphi_1 \wedge \varphi_2) = 1$ iff there exists some $S \in Alg_{opt}(\varphi_1 \wedge \varphi_2)$ where $val_1^s(\varphi_1, S) = 1 = val_2^s(\varphi_2, S)$.*

The above notion of optimization, which refers to a specific combined problem $\mathcal{D}_1 \oplus \mathcal{D}_2$, may be distinguished from a "general" optimization technique for a given protocol-based combination method that applies to an arbitrary combined problem. The Nelson-Oppen combination method described in [NO79], if applied to combined T-clause satisfiability problems, is a general optimization technique for Oppen's Procedure in this sense. Kepser and Richts [KR98] describe optimized versions (in the sense of Definition 4) of the Decomposition Algorithm for combined E-unification algorithms, both on the general level and for specific combined problems.

5   Combination of Generalized Satisfiability Problems

In this section we look at a well-known class of decision problems, namely at generalized satisfiability problems as introduced by Schaefer [Sh78]. We shall see that there is a natural notion of combining generalized satisfiability problems. Simple examples are given that demonstrate that combined problems can be NP-hard even if the component problems are decidable in zero time. In the sequel we assume that we are given a fixed set of Boolean variables Var_{bool}. A *truth value assignment* is a partial mapping ν from Var_{bool} to $\{0, 1\}$.

Definition 5. *A logical relation of rank $k \geq 1$ is given by a subset R of $\{0,1\}^k$. Let $\mathcal{R} = \{R_1, \ldots, R_m\}$ be a finite set of logical relations, R_i being of rank k_i. An $\mathcal{R}$-formula is a finite conjunction of clauses of the form $r_i(\boldsymbol{p})$ where r_i is a relation symbol for some $R_i \in \mathcal{R}$ and $\boldsymbol{p} = (p_1, \ldots, p_{k_i})$ is a sequence of k_i (not necessarily distinct) Boolean variables. A truth value assignment ν satisfies a clause $r_i(\boldsymbol{p})$ iff ν is defined for all variables in $\boldsymbol{p}$ and if $(\nu(p_1), \ldots, \nu(p_{k_i})) \in R_i$. A truth value assignment ν satisfies a conjunction φ of clauses (we write $\nu(\varphi) = 1$) iff it satisfies each clause. The $\mathcal{R}$-satisfiability problem is the problem of deciding if a given $\mathcal{R}$-formula is satisfiable.*

The *size* of an $\mathcal{R}$-formula φ is the total number of occurrences of Boolean variables in φ. Obviously each $\mathcal{R}$-satisfiability problem can be described as a decision problem $\mathcal{D}_{\mathcal{R}} = (\mathcal{P}, val)$ in the sense of Definition 2, defining $\mathcal{P}$ as the set of all $\mathcal{R}$-formula and letting $val(\varphi) = 1$ iff there exists a truth value assignment ν that satisfies φ. The class $\mathcal{C}_{gen-sat}$ of all decision problems of the form $\mathcal{D}_{\mathcal{R}}$, for arbitrary finite sets $\mathcal{R}$ of logical relations, is called the class of *generalized satisfiability problems*. It was introduced by Schaefer [Sh78] for studying the complexity of Boolean satisfiability problems in a uniform way. well-known NP-complete satisfiability problems like *3SAT*, *1-in-3 3SAT*, *NOT-ALL-EQUAL 3SAT*, and *1-in-3* problems over positive literals (cf. [GJ79]) can be formalized as generalized satisfiability problems.

Example 7. Let $R_{=1(3)} = \{(1,0,0),(0,1,0),(0,0,1)\}$ and let $\mathcal{R}_{=1(3)}$ denote the singleton set $\{R_{=1(3)}\}$. $R_{=1(3)}$ may be called the *exactly-1-in-3* relation over positive literals, and an $\mathcal{R}_{=1(3)}$-formula is just an *exactly-1-in-3* problem over positive literals[3]. The $\mathcal{R}_{=1(3)}$-satisfiability problem is well-known to be NP-complete (cf. [GJ79]).

Similarly as for the classes $\mathcal{C}_{qf-dnf}$ and $\mathcal{C}_{unif}$, also for the class $\mathcal{C}_{gen-sat}$ of generalized satisfiability problems there is a natural notion of combining two decision problems. Let $\mathcal{R}_1$ and $\mathcal{R}_2$ be two finite sets of logical relations. If $\mathcal{D}_{\mathcal{R}_1}$ (resp. $\mathcal{D}_{\mathcal{R}_2}$) denotes the $\mathcal{R}_1$-satisfiability (resp. $\mathcal{R}_2$-satisfiability) problem, then the natural combination $\mathcal{D}_{\mathcal{R}_1} \oplus \mathcal{D}_{\mathcal{R}_2}$ is the problem $\mathcal{D}_{\mathcal{R}_1 \cup \mathcal{R}_2}$ of $(\mathcal{R}_1 \cup \mathcal{R}_2)$-satisfiability. Our interest on combination of decision problems for fixed sets of logical relations relies on the fact that it is simple to show that often combined problems are NP-hard even for perfectly trivial plain component problems. In the sequel, we say that a problem $\mathcal{D} = (\mathcal{P}, val)$ is an *extension* of $\mathcal{D}' = (\mathcal{P}', val')$ iff $\mathcal{P}' \subseteq \mathcal{P}$ and if *val* and *val'* coincide on $\mathcal{P}'$.

Example 8. As in Example 7, let $\mathcal{R}_{=1(3)} = \{R_{=1(3)}\}$ describe the class of *exactly-1-in-3* problems over positive literals. Let $R_{\geq 1(3)} := \{0,1\}^3 \setminus \{(0,0,0)\}$ denote the *at-least-1-in-3* relation and let $R_{\leq 1(3)} := R_{=1(3)} \cup \{(0,0,0)\}$ denote the *at-most-1-in-3* relation over positive literals. Obviously, given a finite conjunction of clauses, each with three positive literals, there is always a truth value assignment that evaluates at least (at most) one literal of each clause to 1, namely the trivial assignment that maps each atom to 1 (resp. 0). Hence both the $\{R_{\geq 1(3)}\}$-satisfiability problem and the $\{R_{\leq 1(3)}\}$-satisfiability problem are trivial. However, the combination of both problems extends $\mathcal{R}_{=1(3)}$-satisfiability and is NP-complete.

Modifying the above example, we see that this situation is by no means exceptional.

Lemma 1. *For each $\mathcal{R}$-satisfiability decision problem there exist two finite sets of logical relations $\mathcal{R}_1$ and $\mathcal{R}_2$ such that*

1. *the $\mathcal{R}_i$-satisfiability problem is trivial in the sense that each instance of an $\mathcal{R}_i$-satisfiability problem is always solvable $(i = 1, 2)$, and*
2. *the combination of the $\mathcal{R}_1$-satisfiability problem and the $\mathcal{R}_2$-satisfiability problem is an extension of the $\mathcal{R}$-satisfiability problem.*

Proof. Given $\mathcal{R}$, let

$$\mathcal{R}_1 := \{R \cup \{(0)^k\} \mid R \in \mathcal{R} \text{ has arity } k\}$$
$$\mathcal{R}_2 := \{R \cup \{(1)^k\} \mid R \in \mathcal{R} \text{ has arity } k\}$$

---

[3] Usually these problems are just called positive *1-in-3* problems.

where $(0)^k$ denotes a k-tuple $(0, \ldots, 0)$ and similarly for $(1)^k$. Obviously the $\mathcal{R}_1$-satisfiability problem is trivial: each instance φ_1 is satisfiable, by mapping each Boolean variable to 0. Similarly $\mathcal{R}_2$-satisfiability is trivial. Since for $R \in \mathcal{R}$ always $R = (R \cup \{(0)^k\}) \cap (R \cup \{(1)^k\})$ the combination of $\mathcal{R}_1$-satisfiability and $\mathcal{R}_2$-satisfiability is an extension of $\mathcal{R}$-satisfiability. $\square$

Summing up, for the class of generalized satisfiability problems we have seen that combined decision problems are often NP-hard even for trivial plain component problems. Our next aim is to show that this carries over to the other classes of decision problems that we mentioned above. As a first step, we show that combined generalized satisfiability problems can be solved by a simple protocol-based combination algorithm. The idea is to use truth value assignments as scenarios.

Definition 6. *Given $\mathcal{D}_{\mathcal{R}_1} = (\mathcal{P}_1, \mathrm{val}_1)$ and $\mathcal{D}_{\mathcal{R}_2} = (\mathcal{P}_2, \mathrm{val}_2)$ and their combination $\mathcal{D}_{\mathcal{R}_1} \oplus \mathcal{D}_{\mathcal{R}_2} = (\mathcal{P}_1 \oplus \mathcal{P}_2, \mathrm{val})$, let*

1. *$\mathcal{S}$ denote the set of all truth value assignments,*
2. *Alg assign to each instance $\varphi_1 \wedge \varphi_2$ of $\mathcal{D}_{\mathcal{R}_1} \oplus \mathcal{D}_{\mathcal{R}_2}$ the finite set of all truth value assignments that are defined exactly on the Boolean variables occurring in $\varphi_1 \wedge \varphi_2$,*
3. *$\mathrm{val}_i^s(\varphi_i, \nu) = 1$ iff ν satisfies φ_i ($i = 1, 2$).*

Assume that for some $\varphi_1 \wedge \varphi_2 \in \mathcal{P}_1 \oplus \mathcal{P}_2$ we have $\mathrm{val}(\varphi_1 \wedge \varphi_2) = 1$. By definition of val there exists a truth value assignment ν that satisfies all the clauses in $\varphi_1 \wedge \varphi_2$. This implies that there exists a scenario $\nu' \in Alg(\varphi_1 \wedge \varphi_2)$ such that $\nu'(\varphi_1) = \nu'(\varphi_2) = 1$, which means that $\mathrm{val}_1^s(\varphi_1, \nu') = \mathrm{val}_2^s(\varphi_2, \nu') = 1$. Since all implications also hold in the converse direction this shows that Alg is a correct combination algorithm, which proves that Definition 6 describes a protocol-based combination method.

Lemma 2. *Assume that $P \neq NP$. Let $R_{\geq 1(3)}$ (resp. $R_{\leq 1(3)}$) denote the at-least-1-in-3 (resp. at-most-1-in-3) relation over positive literals as defined in Example 8. For the combination method given above, there exists no polynomial optimization of Alg for the combination of the $\{R_{\geq 1(3)}\}$-satisfiability problem with the $\{R_{\leq 1(3)}\}$-satisfiability problem.*

Proof. Given an *exactly-1-in-3* problem over positive literals, φ, and a truth value assignment ν on $\mathrm{var}(\varphi)$ we may check in polynomial-time if ν solves φ. Hence a polynomial optimization of Alg would yield a polynomial-time algorithm for deciding solvability of $\mathcal{R}_{=1(3)}$-satisfiability, a contradiction. $\square$

Lemma 3. *Assume that $P \neq NP$. Let $\mathcal{R}$ be a finite set of logical relations such that $\mathcal{R}$-satisfiability problem is NP-hard, let $\mathcal{R}_1$ and $\mathcal{R}_2$ be as in the proof of Lemma 1. For the combination method given above, there exists no polynomial optimization of Alg for the $\mathcal{R}$-satisfiability problem.*

Proof. Similar as the previous proof. $\square$

6 Polynomial Interpretation of Combination Algorithms

In this section we fix a situation where we have two protocol-based combination methods, for classes $\mathcal{C}$ and $\mathcal{C}'$ respectively, and where for both methods admissible combined decision problems $\mathcal{D}$ and $\mathcal{D}'$ are fixed. We establish a relationship between the two combination algorithms Alg and Alg' and the sets of scenarios computed by the algorithms that may be used

1. to lift complexity bounds for $\mathcal{D}$ to $\mathcal{D}'$, and
2. to obtain impossibility results for polynomial optimization of Alg' on the basis of similar results for Alg.

In the following section this observation will be used to prove complexity bounds for combined qf-dnf satisfiability problems (resp. for combined E-unification decision problems) and impossibility results for polynomial optimization of Oppen's Procedure (resp. the Decomposition Algorithm).

To fix notation, let $\mathcal{C}$ be a class of decision problems with a fixed protocol-based combination method for a subclass of admissible combinations. Let $\mathcal{D}_1 \oplus \mathcal{D}_2 = (\mathcal{P}_1 \oplus \mathcal{P}_2, val)$ be the admissible combination of the plain component problems $\mathcal{D}_1 = (\mathcal{P}_1, val_1)$ and $\mathcal{D}_2 = (\mathcal{P}_2, val_2)$ of $\mathcal{C}$. We assume that application of the combination method to $\mathcal{D}_1 \oplus \mathcal{D}_2$ yields $(\mathcal{S}, Alg, val_1^s, val_2^s)$.

Let $\mathcal{C}'$ denote a second class of decision problems with another protocol-based combination method for a subclass of admissible combinations. Let $\mathcal{D}_1' \oplus \mathcal{D}_2' = (\mathcal{P}_1' \oplus \mathcal{P}_2', val')$ be the admissible combination of the plain component problems $\mathcal{D}_1' = (\mathcal{P}_1', val_1')$ and $\mathcal{D}_2' = (\mathcal{P}_2', val_2')$ of $\mathcal{C}'$. We assume that application of the combination method to $\mathcal{D}_1' \oplus \mathcal{D}_2'$ yields $(\mathcal{S}', Alg', val_1^{s'}, val_2^{s'})$.

Definition 7. *In the context described above, a* polynomial interpretation *of Alg in Alg' is given by three polynomial-time computable functions functions*

$$tr_1 : \mathcal{P}_1 \to \mathcal{P}_1' : \varphi_1 \mapsto \varphi_1^{tr_1}$$
$$tr_2 : \mathcal{P}_2 \to \mathcal{P}_2' : \varphi_2 \mapsto \varphi_2^{tr_2}$$
$$\beta : \mathcal{S}' \to \mathcal{S}$$

such that the following conditions hold for each $\varphi_1 \wedge \varphi_2 \in \mathcal{P}_1 \oplus \mathcal{P}_2$:

1. *for each scenario $S \in Alg(\varphi_1 \wedge \varphi_2)$: if $val_1^s(\varphi_1, S) = val_2^s(\varphi_2, S) = 1$ then there exists $S' \in Alg'(\varphi_1^{tr_1} \wedge \varphi_2^{tr_2})$ such that $val_1^{s'}(\varphi_1^{tr_1}, S') = val_2^{s'}(\varphi_2^{tr_2}, S') = 1$,*
2. *for each $S' \in Alg'(\varphi_1^{tr_1} \wedge \varphi_2^{tr_2})$: if $val_1^{s'}(\varphi_1^{tr_1}, S') = val_2^{s'}(\varphi_2^{tr_2}, S') = 1$, then $\beta(S') \in Alg(\varphi_1 \wedge \varphi_2)$ and $val_1^s(\varphi_1, \beta(S')) = val_2^s(\varphi_2, \beta(S')) = 1$.*

Theorem 1. *In the context given above, assume there exists a polynomial interpretation of Alg in Alg'.*

1. *If $\mathcal{D}_1 \oplus \mathcal{D}_2$ is NP-hard, then also $\mathcal{D}_1' \oplus \mathcal{D}_2'$ is NP-hard. In a similar way, each lower complexity bound for $\mathcal{D}_1 \oplus \mathcal{D}_2$ beyond NP-hardness carries over to $\mathcal{D}_1' \oplus \mathcal{D}_2'$.*

2. *If there is no polynomial optimization of Alg for $\mathcal{D}_1 \oplus \mathcal{D}_2$, then there is no polynomial optimization of Alg' for $\mathcal{D}'_1 \oplus \mathcal{D}'_2$.*

Proof. Assume there exists a polynomial interpretation as described above. We first show that for each $\varphi_1 \wedge \varphi_2 \in \mathcal{P}_1 \oplus \mathcal{P}_2$ we have $val(\varphi_1 \wedge \varphi_2) = 1$ iff $val'(\varphi_1^{tr_1} \wedge \varphi_2^{tr_1}) = 1$. Since tr_1 and tr_2 are polynomial-time this shows that there exists a polynomial transformation of $\mathcal{D}_1 \oplus \mathcal{D}_2$ to $\mathcal{D}'_1 \oplus \mathcal{D}'_2$ which proves the first part of the theorem.

Assume that $val(\varphi_1 \wedge \varphi_2) = 1$. By correctness of Alg there exists a scenario $S \in Alg(\varphi_1 \wedge \varphi_2)$ such that

$$val_1^s(\varphi_1, S) = val_2^s(\varphi_2, S) = 1.$$

Condition 1 of Def. 7 ensures that there exists $S' \in Alg'(\varphi_1^{tr_1} \wedge \varphi_2^{tr_2})$ such that

$$val_1^{s'}(\varphi_1^{tr_1}, S') = val_2^{s'}(\varphi_2^{tr_2}, S') = 1.$$

Now correctness of Alg' shows that $val'(\varphi_1^{tr_1} \wedge \varphi_2^{tr_1}) = 1$.

Conversely assume that $val'(\varphi_1^{tr_1} \wedge \varphi_2^{tr_1}) = 1$. Correctness of Alg' shows that there exists a scenario $S' \in Alg'(\varphi_1^{tr_1} \wedge \varphi_2^{tr_2})$ such that

$$val_1^{s'}(\varphi_1^{tr_1}, S') = val_2^{s'}(\varphi_2^{tr_2}, S') = 1.$$

By Condition 2,
$$val_1^s(\varphi_1, \beta(S')) = val_2^s(\varphi_2, \beta(S')) = 1$$

and $\beta(S') \in Alg(\varphi_1 \wedge \varphi_2)$. Correctness of Alg shows that $val(\varphi_1 \wedge \varphi_2) = 1$.

To prove the second statement, assume that there exists a polynomial optimization of Alg'_{opt}. For an instance $\varphi_1 \wedge \varphi_2$ of $\mathcal{P}$ define

$$Alg_{opt}(\varphi_1 \wedge \varphi_2) := \{\beta(S') \mid S' \in Alg'_{opt}(\varphi_1^{tr} \wedge \varphi_2^{tr})\}.$$

Our assumptions show that Alg_{opt} is polynomial-time computable. Assume that $val(\varphi_1 \wedge \varphi_2) = 1$. As above it follows that $val'(\varphi_1^{tr_1} \wedge \varphi_2^{tr_1}) = 1$. Hence there exists an $S' \in Alg'_{opt}(\varphi_1^{tr} \wedge \varphi_2^{tr})$ such that $val_1^{s'}(\varphi_1^{tr_1}, S') = val_2^{s'}(\varphi_2^{tr_2}, S') = 1$. But then $\beta(S') \in Alg(\varphi_1 \wedge \varphi_2)$ and $val_1^s(\varphi_1, \beta(S')) = val_2^s(\varphi_2, \beta(S')) = 1$. This shows that Alg_{opt} is a polynomial optimization of Alg. $\square$

7   Intractable Combined T-Satisfiability Problems

As a first illustration of the techniques of Section 6 we now characterize first-order theories T_1 and T_2 where satisfiability of qf-dnf formulae w.r.t. $T_1 \cup T_2$ is NP-hard, regardless of the complexity of satisfiability w.r.t. the component theories T_i, and where we cannot obtain a polynomial optimization of Oppen's Procedure. The idea is to use two generalized satisfiability problems whose combination is NP-hard and where the combination algorithm described in Definition 6 does not have a polynomial optimization. By Theorem 1 it then suffices to give a polynomial interpretation between the combination algorithms of the two combination problems. There is an arbitrariness in the selection of the generalized satisfiability problems. We shall use $\mathcal{R}_{\geq 1(3)}$-satisfiability and $\mathcal{R}_{\leq 1(3)}$-satisfiability, but many other combinations could be used as well. Let us fix notation.

Source Problems. As in Example 8, $R_{\geq 1(3)}$ denotes the *at-least-1-in-3* relation over positive literals and $\mathcal{R}_{\geq 1(3)} = \{R_{\geq 1(3)}\}$. With $\mathcal{P}_{\geq 1(3)}$ we denote the set of all instances of $\mathcal{R}_{\geq 1(3)}$-satisfiability problems, and $\mathcal{D}_{\geq 1(3)}$ denotes the $\mathcal{R}_{\geq 1(3)}$-satisfiability problem. In the same way all dual notions are defined, where "$\geq$" is replaced by "$\leq$". With $\mathcal{S}_{1(3)}$ and $Alg_{1(3)}$ we denote the protocol and the combination algorithm introduced in Definition 6.

Target Problems. We fix an admissible combination $\mathcal{D}_{T_1} \oplus \mathcal{D}_{T_2} \in \mathcal{C}_{qf-dnf}$. We assume that $\mathcal{D}_{T_i}$ has the form $(\mathcal{P}_{T_i}, val_i)$, for $i = 1, 2$. With $(\mathcal{S}_T, Alg_T, val_1^s, val_2^s)$ we denote the quadruple assigned to $\mathcal{D}_{T_1} \oplus \mathcal{D}_{T_2}$ by Oppen's Procedure as described in Example 5.

Encoding Principle. Before we describe the technical details of the encoding, two remarks are appropriate.

1. As a matter of fact we cannot expect to obtain a polynomial interpretation between the two combination procedures described above that works for arbitrary combinations $\mathcal{D}_{T_1} \oplus \mathcal{D}_{T_2} \in \mathcal{C}_{qf-dnf}$. Below we give two *sufficient conditions* on T_1 and T_2 respectively that guarantee that a uniform polynomial interpretation exists. Theories that satisfy these two conditions will be called theories of type[4] 1 and 2 respectively (another possible name would be theories of type "$\geq 1(3)$" and "$\leq 1(3)$").
2. From Definition 7 recall that we have to find a translation $\beta : \mathcal{S}_T \to \mathcal{S}_{1(3)}$ that encodes scenarios of target problems (i.e, partitions Π of the set X of "shared" variables of the target problem) into scenarios of the source problems (i.e., truth value assignments ν on the set P of Boolean variables of the source problem). The translations of decision problems (i.e., tr_1 and tr_2) will be organized in such a way that for a fixed pair of source and target problems we always have $P \subseteq X$. To this end we will assume that the set of Boolean variables, Var_{bool}, is a subset of the set Var of variables that is used for formulating instances of $\mathcal{D}_{T_1} \oplus \mathcal{D}_{T_2}$. The latter variables will be called "syntactic" variables, for simplicity. Note that each Boolean variable is also used as a syntactic variable. We assume that $Var \setminus Var_{bool}$ is infinite. One variable $x \in Var \setminus Var_{bool}$ is chosen that plays a special role. The idea for the translation β is to consider a Boolean variable $p \in P$ to be "true" under $\nu = \beta(\Pi)$ iff $p \in [x]_\Pi$, where $[x]_\Pi$ denotes the equivalence class of x w.r.t. Π.

Definition 8. *Theory T_1 is of* type 1 *iff there exists a polynomial-time translation* $tr_1 : \mathcal{P}_{\geq 1(3)} \to \mathcal{P}_{T_1}$ *that assigns to each at-least-1-in-3 problem over positive*

---

[4] The word "type" is not used in a formal sense here, and a theory T may have type 1 and type 2 at the same time.

literals $\varphi_{\geq 1(3)}$ *of the form* $\bigwedge_{i=1}^{k} R_{\geq 1(3)}(p_{i_1}, p_{i_2}, p_{i_3})$ *with set of Boolean variables* P *a conjunction* $\varphi_{\geq 1(3)}^{tr_1}$ *of* T_1*-literals with set of variables* $\{x\} \cup P \cup Y_1$ *such that the following conditions hold:*

"$\rightarrow$" *for each subset* Q *of* P *with* $|Q \cap \{p_{i_1}, p_{i_2}, p_{i_3}\}| = 1$ *for* $i = 1, \ldots, k$[5] *the problem* $\varphi_{\geq 1(3)}^{tr_1}$ *is satisfiable w.r.t.* T_1 *under the partition* $\Pi := \{\{x\} \cup Q, P \setminus Q\}$ *of* $\{x\} \cup P$.

"$\leftarrow$" *If* $\varphi_{\geq 1(3)}^{tr_1}$ *is satisfiable w.r.t.* T_1 *under partition* Π *of* $\{x\} \cup P$, *then* $|[x]_\Pi \cap \{p_{i_1}, p_{i_2}, p_{i_3}\}| \geq 1$, *for all* $1 \leq i \leq k$[6],

Theories of type 2 are defined in the dual way:

Definition 9. *Theory* T_2 *is of* type 2 *iff there exists a polynomial-time translation* $tr_2 : \mathcal{P}_{\leq 1(3)} \rightarrow \mathcal{P}_{T_2}$ *that assigns to each at-most-1-in-3 problem over positive literals* $\varphi_{\leq 1(3)}$ *of the form* $\bigwedge_{i=1}^{k} R_{\leq 1(3)}(p_{i_1}, p_{i_2}, p_{i_3})$ *with set of Boolean variables* P *a conjunction* $\varphi_{\leq 1(3)}^{tr_2}$ *of* T_2*-literals with set of variables* $\{x\} \cup P \cup Y_2$ *such that the following conditions hold:*

"$\rightarrow$" *for each subset* Q *of* P *with* $|Q \cap \{p_{i_1}, p_{i_2}, p_{i_3}\}| = 1$ *for* $i = 1, \ldots, k$ *the problem* $\varphi_{\leq 1(3)}^{tr_2}$ *is satisfiable w.r.t.* T_2 *under the partition* $\Pi := \{\{x\} \cup Q, P \setminus Q\}$ *of* $\{x\} \cup P$.

"$\leftarrow$" *If* $\varphi_{\leq 1(3)}^{tr_2}$ *is satisfiable w.r.t.* T_2 *under partition* Π *of* $\{x\} \cup P$, *then* $|[x]_\Pi \cap \{p_{i_1}, p_{i_2}, p_{i_3}\}| \leq 1$, *for all* $1 \leq i \leq k$,

Theorem 2. *Let* T_1 *and* T_2 *be stably infinite first-order theories with disjoint signatures. If* T_i *is of type* i, *for* $i = 1, 2$, *then* $\mathcal{D}_{T_1} \oplus \mathcal{D}_{T_2}$ *is NP-hard and there is no polynomial optimization of Oppen's Procedure for* $\mathcal{D}_{T_1} \oplus \mathcal{D}_{T_2}$.

Proof. Without loss of generality we may restrict considerations to input formulae $\varphi_{\geq 1(3)}$ and $\varphi_{\leq 1(3)}$ which have the dual structure

$$\bigwedge_{i=1}^{k} R_{\geq 1(3)}(p_{i_1}, p_{i_2}, p_{i_3}) \text{ and } \bigwedge_{i=1}^{k} R_{\leq 1(3)}(p_{i_1}, p_{i_2}, p_{i_3})$$

respectively. With P we denote the common set of variables of $\varphi_{\geq 1(3)}$ and $\varphi_{\leq 1(3)}$. Let tr_1 and tr_2 be translations with the properties described in Definitions 8 and 9 respectively. The sets Y_1 and Y_2 of additional variables used in translated formulae $\varphi_{\geq 1(3)}^{tr_1}$ and $\varphi_{\leq 1(3)}^{tr_2}$ respectively are assumed to be disjoint. Hence $\{x\} \cup P$ is the set of shared variables of $\varphi_{\geq 1(3)}^{tr_1}$ and $\varphi_{\leq 1(3)}^{tr_2}$.

We define the following mapping $\beta : \mathcal{S}_T \rightarrow \mathcal{S}_{1(3)}$. If Π is a partition of $X \subseteq Var$, then $\beta(\Pi)$ is the empty truth value assignment if $x \notin X$. In the other

---

[5] The condition expresses that Q defines a solution of the *exactly-1-in-3* problem $\bigwedge_{i=1}^{k} R_{=1(3)}(p_{i_1}, p_{i_2}, p_{i_3})$.

[6] With comment 2 above, the condition expresses that $[x]_\Pi \cap P$ defines a solution of $\bigwedge_{i=1}^{k} R_{\geq 1(3)}(p_{i_1}, p_{i_2}, p_{i_3})$.

case, $\beta(\Pi)$ is defined on all Boolean variables of X. For a Boolean variable $p \in X$ we define $\beta(\Pi)(p) = 1$ iff $p \in [x]_\Pi$.

We now show that the mappings β, tr_1 and tr_2 satisfy the conditions for a polynomial interpretation of $Alg_{1(3)}$ in Alg_T specified in Definition 7. By choice of tr_1 and tr_2, and by definition of β, these mappings are polynomial-time computable.

We show that Condition 1 of Definition 7 is satisfied. Let ν be a truth value assignment in $Alg_{1(3)}(\varphi_{\geq 1(3)} \wedge \varphi_{\leq 1(3)})$ that satisfies both $\varphi_{\geq 1(3)}$ and $\varphi_{\leq 1(3)}$. We assume that ν has domain P. Let $Q := \{p \in P \mid \nu(p) = 1\}$. Let Π denote the partition $\{\{x\} \cup Q, P \setminus Q\}$ of $\{x\} \cup P$. We have $|Q \cap \{p_{i_1}, p_{i_2}, p_{i_3}\}| = 1$ for $i = 1, \dots, k$. Condition "$\rightarrow$" for theories of type 1 (resp. 2) guarantees that the problem $\varphi_{\geq 1(3)}^{tr_1}$ (resp. $\varphi_{\leq 1(3)}^{tr_2}$) is satisfiable w.r.t. T_1 (resp. T_2) under Π. Since Π is in $Alg_T(\varphi_{\geq 1(3)}^{tr_1} \wedge \varphi_{\leq 1(3)}^{tr_2})$ we are done.

We verify Condition 2 of Definition 7. Let $\Pi \in Alg_T(\varphi_{\geq 1(3)}^{tr_1} \wedge \varphi_{\leq 1(3)}^{tr_2})$ and assume that both $\varphi_{\geq 1(3)}^{tr_1}$ and $\varphi_{\leq 1(3)}^{tr_2}$ are satisfiable (w.r.t. T_1 resp. T_2) under partition Π of $\{x\} \cup P$. Since x belongs to the set of shared variables of $\varphi_{\geq 1(3)}^{tr_1}$ and $\varphi_{\leq 1(3)}^{tr_2}$, and since P is the subset of all Boolean variables among the set of shared variables it follows that $\beta(\Pi)$ is defined exactly on P, hence we have $\beta(\Pi) \in Alg_{1(3)}(\varphi_{\geq 1(3)} \wedge \varphi_{\leq 1(3)})$. Conditions "$\leftarrow$" for theories of type 1 and 2 guarantee that $|[x]_\Pi \cap \{p_{i_1}, p_{i_2}, p_{i_3}\}| = 1$ for all $1 \leq i \leq k$. It follows that $\beta(\Pi)$ solves φ_1 and φ_2.

We have shown that tr_1, tr_2 and β define a polynomial interpretation of $Alg_{1(3)}$ in Alg_T. Using Lemma 2 and Theorem 1 the result follows. $\square$

Some Theories of Type 1 In the sequel we list some examples of theories of type 1. In each case it is simple to see that the theory is stably infinite. In order to show that the theory is of type 1 we generally assume that an *at-least-1-in-3* problem over positive literals, $\varphi_{\geq 1(3)}$, of the form $\bigwedge_{i=1}^{k} R_{\geq 1(3)}(p_{i_1}, p_{i_2}, p_{i_3})$ with set of Boolean variables P is given. In the examples below we only give the polynomial-time translation into a qf-dnf formula $\varphi_{\geq 1(3)}^{tr_1}$. One restriction is imposed on $\varphi_{\geq 1(3)}^{tr_1}$. Remark 2 indicates that in some cases impossibility results for polynomial optimization can be misleading if the input formulae contain equations or disequations between variables. For this reason we demand that $\varphi_{\geq 1(3)}^{tr_1}$ does not contain any equation or disequation between variables.

Example 9. Let T_1 denote the theory of integers with addition $+$, standard ordering $\leq$ and successor s. Let $\varphi_{\geq 1(3)}^{tr_1}$ denote the conjunction of all formulae of the form

$$x = x + x \wedge p_{i_1} + p_{i_2} + p_{i_3} \leq s(s(x + x)) \wedge x + x \leq p_{i_1} \wedge x + x \leq p_{i_2} \wedge x + x \leq p_{i_3}$$

for $1 \leq i \leq k$. *We verify condition "$\rightarrow$" of Definition 8.* If Q is a subset of P with $|Q \cap \{p_{i_1}, p_{i_2}, p_{i_3}\}| = 1$ for $i = 1, \dots, k$ then we may use the standard model of T_1 and map variables in $\{x\} \cup Q$ to 0 and variables in $P \setminus Q$ to 1. This shows that $\varphi_{\geq1(3)}^{tr_1}$ is satisfiable w.r.t. T_1 under the partition $\Pi := \{\{x\} \cup Q, P \setminus Q\}$ of $\{x\} \cup P$. *We verify condition "$\leftarrow$" of Definition 8.* If $\varphi_{\geq1(3)}^{tr_1}$ is satisfiable w.r.t. T_1 under partition Π of $\{x\} \cup P$, then obviously $|[x]_\Pi \cap \{p_{i_1}, p_{i_2}, p_{i_3}\}| \geq 1$, for all $1 \leq i \leq k$.

Example 10. Let T_1 denote the theory of natural numbers with addition $+$ and successor s. We show that T_1 has type 1. Let $\varphi_{\geq1(3)}^{tr_1}$ denote the conjunction of all formulae of the form

$$x = x + x \wedge p_{i_1} + p_{i_2} + p_{i_3} \leq s(s(x + x))$$

for $1 \leq i \leq k$. Conditions "$\rightarrow$" and "$\leftarrow$" of Definition 8 can be verified as in the previous example.

Example 11. Let T_1 denote the theory of any infinite ring with 0, without zero-divisors (e.g. integers, rationals, reals, complex numbers). Let $\varphi_{\geq1(3)}^{tr_1}$ denote the conjunction of all formulae of the form

$$x = 0 \wedge p_{i_1} \cdot p_{i_2} \cdot p_{i_3} = 0$$

for $1 \leq i \leq k$. Conditions "$\rightarrow$" and "$\leftarrow$" of Definition 8 can be verified as in the preceding examples.

Example 12. ([Op80,DS78]) Let T_1 denote the theory of arrays over the signature $\{sel, store\}$ with the axioms

$$\forall v \forall e \forall i : \ sel(store(v, i, e), i) = e,$$
$$\forall v \forall e \forall i \forall j : \ i \neq j \rightarrow sel(store(v, i, e), j) = sel(v, j),$$
$$\forall v \forall i : \ store(v, i, sel(v, i)) = v,$$
$$\forall v \forall i \forall e \forall f : \ store(store(v, i, e), i, f) = store(v, i, f),$$
$$\forall v \forall i \forall j \forall e \forall f : \ i \neq j \rightarrow store(store(v, i, e), j, f) = store(store(v, j, f), i, e).$$

Here $sel(v, i)$ is meant to denote the i-th field $v[i]$ of the array v, and $store(v, i, e)$ is the array whose i-th component is e and whose j-th component for $i \neq j$ is $v[j]$. Downey and Sethi [DS78] have shown that the satisfiability problem for qf-dnf formulae of this theory is NP-complete. Let $\varphi_{\geq1(3)}^{tr_1}$ denote the conjunction of all formulae of the form

$$store(store(store(v, p_{i_1}, e), p_{i_2}, e), p_{i_3}, e)[x] \neq v[x]$$

for $1 \leq i \leq k$. *We verify condition "$\rightarrow$" of Definition 8.* If Q is a subset of P with $|Q \cap \{p_{i_1}, p_{i_2}, p_{i_3}\}| = 1$ for $i = 1, \ldots, k$ it is obviously simple to find an infinite model of T_1 and a satisfying variable assignment where all variables in $Q \cup \{x\}$ (resp. $P \setminus Q$) are identified. *We verify condition "$\leftarrow$" of Definition 8.* A valid sentence of T_1 expresses the following: if we store in array v at positions p_{i_1}, p_{i_2},

p_{i_3} the element e, and if in the new array w that is obtained after these three operations we have a new element stored at position x, then x must coincide with at least one of the positions $p_{i_1}, p_{i_2}, p_{i_3}$. Hence, if $\varphi_{\geq 1(3)}^{tr_1}$ is satisfiable w.r.t. T_1 under partition Π of $P \cup \{x\}$, then we have $|[x]_\Pi \cap \{p_{i_1}, p_{i_2}, p_{i_3}\}| \geq 1$, for all $1 \leq i \leq k$.

Example 13. Another series from examples comes from the area of equational theories. It turns out that for various equational theories E the full first-order theory T_E of the quotient term algebra $\mathcal{T}(\Sigma, \text{Var})/{=_E}$ is of type 1. As one example consider the theory E that expresses commutativity of the function symbol f. Let $\varphi_{\geq 1(3)}^{tr_1}$ denote the conjunction of all formulae of the form

$$f(f(x, y_i), f(u_i, v_i)) = f(f(p_{i_1}, p_{i_2}), f(p_{i_2}, p_{i_3}))$$

for $1 \leq i \leq k$. *We verify condition "$\rightarrow$" of Definition 8.* If Q is a subset of P with $|Q \cap \{p_{i_1}, p_{i_2}, p_{i_3}\}| = 1$ for $i = 1, \ldots, k$ we can give an assignment in the standard model that satisfies $\varphi_{\geq 1(3)}^{tr_1}$ under the partition $\{\{x\} \cup Q, P \setminus Q\}$ of $\{x\} \cup P$. *We verify condition "$\leftarrow$" of Definition 8.* The sentence $\forall x \forall y \forall u \forall v : f(x, y) = f(u, v) \rightarrow (x = y \lor x = v)$ is a valid sentence of T_E. Hence, if $\varphi_{\geq 1(3)}^{tr_1}$ is satisfiable w.r.t. T_E under partition Π of $P \cup \{x\}$, then we have $|[x]_\Pi \cap \{p_{i_1}, p_{i_2}, p_{i_3}\}| \geq 1$, for all $1 \leq i \leq k$.

Example 14. Other examples are provided from formalizations of set theory where elementship and finite set construction is expressible. If we can write a formula that expresses $x \in \{p_{i,1}, p_{i,2}, p_{i,3}\}$, then the conjunction of all such formulae for $1 \leq i \leq k$ can be used to verify that the theory is of type 1.

Some Theories of Type 2 We also list some examples of theories of type 2. All theories below are stably infinite. In order to show that a given theory is of type 2 we generally assume that an *at-most-1-in-3* problem over positive literals, $\varphi_{\leq 1(3)}$, of the form $\bigwedge_{i=1}^{k} R_{\leq 1(3)}(p_{i_1}, p_{i_2}, p_{i_3})$ with set of Boolean variables P is given. We give a polynomial-time translation into a qf-dnf formula $\varphi_{\leq 1(3)}^{tr_2}$. As in the case of theories of type 1 we demand that $\varphi_{\leq 1(3)}^{tr_2}$ does not contain any equation or disequation between variables.

Example 15. Let T_2 denote the theory of equality with function symbols in the signature Σ_2. The theory contains all sentences that are valid in all Σ_2-algebras. Assuming that $f \in \Sigma_2$ is binary, an example is $\forall x \forall y \forall z : x = y \rightarrow f(x, z) = f(y, z)$. Nelson and Oppen [NO79] give an $O(n^2)$ decision procedure for the qf-dnf theory. Let $\varphi_{\leq 1(3)}^{tr_2}$ denote the conjunction of all formulae of the form

$$f(p_{i_1}, p_{i_2}) \neq f(x, x) \land f(p_{i_2}, p_{i_3}) \neq f(x, x) \land f(p_{i_3}, p_{i_1}) \neq f(x, x)$$

for $1 \leq i \leq k$ (a similar encoding is possible as soon as Σ_2 contains any function symbol of arity $k \geq 2$). *We verify condition "$\rightarrow$" of Definition 9.* If Q is a subset of P with $|Q \cap \{p_{i_1}, p_{i_2}, p_{i_3}\}| = 1$ for $i = 1, \ldots, k$, then obviously the problem $\varphi_{\leq 1(3)}^{tr_2}$ is satisfiable w.r.t. T_2 under under the partition $\Pi := \{\{x\} \cup Q, P \setminus Q\}$ of $\{x\} \cup P$. We verify condition "$\leftarrow$" of Definition 9. If $\varphi_{\leq 1(3)}^{tr_2}$ is satisfiable w.r.t. T_2 under partition Π of $\{x\} \cup P$, then obviously $|[x]_\Pi \cap \{p_{i_1}, p_{i_2}, p_{i_3}\}| \leq 1$, for all $1 \leq i \leq k$.

Example 16. Let T_2 denote the theory of arrays as introduced in Example 12. Let $\varphi_{\leq 1(3)}^{tr_2}$ denote the conjunction of all formulae of the form

$$store(store(p_{i_1}, e), p_{i_2}, e) \neq store(v, x, e)$$
$$\wedge \quad store(store(p_{i_2}, e), p_{i_3}, e) \neq store(v, x, e)$$
$$\wedge \quad store(store(p_{i_3}, e), p_{i_1}, e) \neq store(v, x, e)$$

for $1 \leq i \leq k$. It is simple to see that with this translation conditions "$\rightarrow$" and "$\leftarrow$" of Definition 9 are satisfied. Hence, here we have a theory that both has type 1 and type 2.

Example 17. Let T_2 denote the theory of some infinite monoid (e.g., natural, integers, rationals, reals) with multiplication "$\cdot$" and neutral element 1. Let $\varphi_{\leq 1(3)}^{tr_2}$ denote the conjunction of all formulae of the form

$$x = 1 \wedge p_{i_1} \cdot p_{i_1} \neq 1 \wedge p_{i_2} \cdot p_{i_3} \neq 1 \wedge p_{i_3} \cdot p_{i_1} \neq 1$$

for $1 \leq i \leq k$. It is simple to see that with this translation conditions "$\rightarrow$" and "$\leftarrow$" of Definition 9 are satisfied.

Example 18. As in the case of type 1 it turns out that for various equational theories E the full first-order theory T_E of the quotient term algebra $\mathcal{T}(\Sigma, \text{Var})/=_E$ is of type 2. As one example consider the trivial theory E with axiom $\{f(x,y) = f(x,y)\}$. Let $\varphi_{\geq 1(3)}^{tr_1}$ denote the conjunction of all formulae of the form

$$f(p_{i_1}, p_{i_2}) \neq f(x,x) \wedge f(p_{i_2}, p_{i_3}) \neq f(x,x) \wedge f(p_{i_3}, p_{i_1}) \neq f(x,x)$$

for $1 \leq i \leq k$. It is simple to see that with this translation conditions "$\rightarrow$" and "$\leftarrow$" of Definition 9 are satisfied. The same translation works for most other equational theories with a binary symbol f.

8   Intractable Combined E-Unification Problems

As a second and final illustration we now want to use Theorem 1 for obtaining NP-hardness results and impossibility results for polynomial optimizations for admissible combinations of E-unification problems such as considered in Example 6. The results obtained here generalize previous results given in [Sc97,Sc99].

Source problems. Let $R_{\geq 1(3)}$, $\mathcal{R}_{\geq 1(3)}$, $\mathcal{P}_{\geq 1(3)}$, $\mathcal{D}_{\geq 1(3)}$ be defined as before, as well as the dual notions and $\mathcal{S}_{1(3)}$ and $Alg_{1(3)}$.

Target problems. We fix an admissible combination $\mathcal{D}_{E_1} \oplus \mathcal{D}_{E_2} \in \mathcal{C}_{unif}$. We assume that $\mathcal{D}_{E_i}$ has the form $(\mathcal{P}_{E_i}, val_i)$, for $i = 1, 2$. With $(\mathcal{S}_E, Alg_E, val_1^s, val_2^s)$ we denote the quadruple assigned to $\mathcal{D}_{E_1} \oplus \mathcal{D}_{E_2}$ by the combination method for E-unification algorithms described in Example 6.

Encoding principle. The main task, as in the previous section, is to define a translation $\beta : \mathcal{S}_E \to \mathcal{S}_{1(3)}$ that assigns a truth value assignment ν to a given linear constant restriction $L = (\Pi, Lab, <_L)$. Recall that Π represents a partition on the set X of variables of the combined equational unification problem. Hence we may use the same encoding principle as in the previous section: for some selected and fixed non-Boolean variable x, a Boolean variable $p \in P$ is interpreted to be "true" under ν iff $p \in [x]_\Pi$, where $[x]_\Pi$ denotes the equivalence class of x w.r.t. Π. As technical prerequisites, as in the previous section we assume that $Var_{bool} \subseteq Var$ and that $Var \backslash Var_{bool}$ is infinite. Here Var is the set of "syntactic" variables occurring in instances of $\mathcal{D}_{E_1} \oplus \mathcal{D}_{E_2}$. Furthermore we assume that $x \in Var \backslash Var_{bool}$ is the first variable in the given enumeration of Var. Hence, when choosing representatives for equivalence classes of Π as described in Section 3.2, x is always its own representative.

For defining equational theories of type 1, the following notion is needed. Let X be a finite set of variables containing x. A *plain* linear constant restriction for X is a linear constant restriction $(\Pi, Lab, <)$ for X with the following properties:

1. each equivalence class of Π with the possible exception of $[x]_\Pi$ is a singleton,
2. all Boolean variables in X, as well as x, receive label Σ_2, all other variables receive label Σ_1,
3. variables with label Σ_2 are smaller w.r.t. "$<$" than variables with label Σ_1.

Definition 10. *The equational theory E_1 is of type 1 iff there exists a polynomial-time translation*

$$tr_1 : \mathcal{P}_{\geq 1(3)} \to \mathcal{P}_{E_1}$$

that assigns to each at-least-1-in-3 problem over positive literals $\varphi_{\geq 1(3)}$ of the form $\bigwedge_{i=1}^k R_{\geq 1(3)}(p_{i_1}, p_{i_2}, p_{i_3})$ with set of Boolean variables P a conjunction $\varphi_{\geq 1(3)}^{tr_1}$ of elementary E_1-unification problems with set of variables $\{x\} \cup P \cup Y_1$ such that the following conditions hold:

"$\to$" *for each subset Q of P with $|Q \cap \{p_{i_1}, p_{i_2}, p_{i_3}\}| = 1$ for $i = 1, \ldots, k$ there exists a plain linear constant restriction $L_1 = (\Pi_1, Lab_1, <)$ for $\{x\} \cup P \cup Y_1$ such that $[x]_\Pi = Q \cup \{x\}$ and $\varphi_{\geq 1(3)}^{tr_1}$ is solvable under L_1.*

"$\leftarrow$" *If $\varphi_{\geq 1(3)}^{tr_1}$ is solvable under the linear constant restriction L with partition Π where variables in $\{x\} \cup P$ receive alien label Σ_2 and for $i = 1, \ldots, k$ the elements $p_{i_1}, p_{i_2}, p_{i_3}$ belong to distinct equivalence classes, then $|[x]_\Pi \cap \{p_{i_1}, p_{i_2}, p_{i_3}\}| \geq 1$, for all $1 \leq i \leq k$,*

Definition 11. *The equational theory E_2 is of* type 2 *iff there exists a polynomial-time translation*

$$\mathrm{tr}_2 : \mathcal{P}_{\leq 1(3)} \to \mathcal{P}_{E_2}$$

that assigns to each at-most-1-in-3 *problem over positive literals $\varphi_{\leq 1(3)}$ of the form $\bigwedge_{i=1}^{k} R_{\leq 1(3)}(p_{i_1}, p_{i_2}, p_{i_3})$ with set of Boolean variables P a conjunction $\varphi_{\leq 1(3)}^{tr_2}$ of elementary E_2-unification problems with set of variables $\{x\} \cup P \cup Y_2$ such that the following conditions hold:*

"$\rightarrow$" *for each subset Q of P with $|Q \cap \{p_{i_1}, p_{i_2}, p_{i_3}\}| = 1$ for $i = 1, \ldots, k$ there exists a solution σ_2 of $\varphi_{\leq 1(3)}^{tr_2}$ that identifies all elements of $Q \cup \{x\}$.*

"$\leftarrow$" *If $\varphi_{\leq 1(3)}^{tr_2}$ is solvable under linear constant restriction L with partition Π, then variables in $\{x\} \cup P$ receive domestic label Σ_2 and for $1 \leq i \leq k$ the variables $p_{i_1}, p_{i_2}, p_{i_3}$ have distinct equivalence classes w.r.t. Π, in particular $|[x]_\Pi \cap \{p_{i_1}, p_{i_2}, p_{i_3}\}| \leq 1).$*

As to the conditions "$\leftarrow$" there is an unsymmetry between theories of type 1 and 2 respectively. For theories of type 2 (resp. 1) we have refined (relaxed) the basic requirement which says that $|[x]_\Pi \cap \{p_{i_1}, p_{i_2}, p_{i_3}\}| \leq 1$ (resp. $|[x]_\Pi \cap \{p_{i_1}, p_{i_2}, p_{i_3}\}| \geq 1$) for all $1 \leq i \leq k$. This is motivated by the special properties of the classes of equational theories that we treat below.

Theorem 3. *Assume that E_i is of type i, for $i = 1, 2$. Then $\mathcal{D}_{E_1} \oplus \mathcal{D}_{E_2}$ is NP-hard and there is no polynomial optimization of the combination algorithm for $\mathcal{D}_{E_1} \oplus \mathcal{D}_{E_2}$.*

Proof. We define the mapping $\beta : \mathcal{S}_E \to \mathcal{S}_{1(3)}$. If L is a linear constant restriction on $X \subseteq \mathrm{Var}$ with partition Π, then $\beta(L)$ is the empty truth value assignment if $x \notin X$. In the other case, $\beta(L)$ is defined on all Boolean variables of X. For a Boolean variable $p \in X$ we define $\beta(L)(p) = 1$ iff $p \in [x]_\Pi$.

We verify Condition 1 of Definition 7. Let ν be a truth value assignment in $\mathrm{Alg}_{1(3)}(\varphi_{\geq 1(3)} \wedge \varphi_{\leq 1(3)})$ that satisfies both $\varphi_{\geq 1(3)}$ and $\varphi_{\leq 1(3)}$. Let tr_1, tr_2, P, Y_1 and Y_2 be defined as in Definitions 10 and 11. Let $Q := \{p \in P \mid \nu(p) = 1\}$. Since E_1 has type 1 (cf. "$\rightarrow$") there exists a plain linear constant restriction $L_1 = (\Pi_1, \mathrm{Lab}_1, <_1)$ for $\{x\} \cup P \cup Y_1$ such that $[x]_{\Pi_1} = \{x\} \cup Q$ and $\varphi_{\geq 1(3)}^{tr_1}$ is solvable under restriction L_1. Since L_1 is plain, all classes $[y]_{\Pi_1}$ for $y \in Y_1$ are singletons. Let σ_1 be a solution. Since E_2 has type 2 (cf. "$\rightarrow$") there exists a solution σ_2 of $\varphi_{\leq 1(3)}^{tr_2}$ that identifies all elements of $Q \cup \{x\}$.

Let Π be the partition $\Pi_1 \cup \{\{y\} \mid y \in Y_2\}$ that treats the elements of Y_2 as singletons. Let Lab be the extension of Lab_1 where the elements of Y_2 receive label Σ_2. Let $<$ be an extension of $<_1$ where the additional elements in Y_2 are strictly smaller than the elements in $\{x\} \cup P \cup Y_1$. Obviously $L := (\Pi, \mathrm{Lab}, <)$ is in $\mathrm{Alg}_E(\varphi_{\geq 1(3)}^{tr_1} \wedge \varphi_{\leq 1(3)}^{tr_2})$. We claim that both $(\varphi_{\geq 1(3)}^{tr_1}, L)$ and $(\varphi_{\leq 1(3)}^{tr_2}, L)$ are solvable.

We show that σ_1 solves $(\varphi_{\geq 1(3)}^{tr_1}, L)$. W.l.o.g. we may assume that $\sigma_1(y) = y$ for all $y \in Y_2$. Since σ_1 solves $(\varphi_{\geq 1(3)}^{tr_1}, L_1)$ we know that for all $y \in P \cup \{x\} \cup Y_1$

we have $\sigma_1(y) = \sigma_1(rep_{\Pi_1}(y)) = \sigma_1(rep_\Pi(y))$. For $y \in Y_2$ we have $\sigma_1(y) = y = rep_\Pi(y) = \sigma_1(rep_\Pi(y))$. Hence σ_1 satisfies the first condition of Definition 1. For alien variables $y \in Y_2$ we have $\sigma_1(y) = y = rep_\Pi(y)$. For alien variables $y \in P \cup \{x\}$ we have $\sigma_1(y) = rep_{\Pi_1}(y)$ since σ_1 solves $(\varphi^{tr_1}_{\geq 1(3)}, L_1)$. It follows that $\sigma_1(y) = rep_\Pi(y)$ and σ_1 satisfies the second condition of Definition 1. Since all alien variables (i.e., the variables in $P \cup \{x\} \cup Y_2$) are smaller w.r.t. $<$ than domestic variables (i.e., the variables in Y_1) it follows that σ_1 satisfies the third condition of Definition 1. Summing up, we have seen that σ_1 solves $(\varphi^{tr_1}_{\geq 1(3)}, L)$.

W.l.o.g. we may assume that $\sigma_2(y) = y$ for all $y \in Y_1$. Hence σ_2 satisfies the second condition of Definition 1. Because σ_2 identifies the elements of $Q \cup \{x\}$ and all other classes of Π are singletons it follows that σ_2 satisfies the first condition of Definition 1. The formula $\varphi^{tr_2}_{\leq 1(3)}$ does not contain any alien variable in Y_1, hence we may assume that the variables of Y_1 do not occur in any term $\sigma_2(y)$ for $y \in \{x\} \cup P \cup Y_2$. This shows that σ_2 satisfies the third condition of Definition 1. Summing up, we have seen that σ_2 solves $(\varphi^{tr_2}_{\leq 1(3)}, L)$.

We verify Condition 2 of Definition 7. Let $L = (\Pi, Lab, <)$ be a linear constant restriction in $Alg_E(\varphi^{tr_1}_{\geq 1(3)} \wedge \varphi^{tr_2}_{\geq 1(3)})$ and assume that both $\varphi^{tr_1}_{\geq 1(3)}$ and $\varphi^{tr_2}_{\leq 1(3)}$ are solvable under L. Since E_2 is of type 2 (cf. "$\leftarrow$") this implies that variables in $\{x\} \cup P$ receive label Σ_2 and for $1 \leq i \leq k$ the variables $p_{i_1}, p_{i_2}, p_{i_3}$ have distinct equivalence classes w.r.t. Π. In particular $|[x]_\Pi \cap \{p_{i_1}, p_{i_2}, p_{i_3}\}| \leq 1$. Since E_1 is of type 1 (cf. "$\leftarrow$") we conclude that $|[x]_\Pi \cap \{p_{i_1}, p_{i_2}, p_{i_3}\}| \geq 1$ for $1 \leq i \leq k$. It follows that $\beta(L)$ satisfies $\varphi_{\geq 1(3)}$ and $\varphi_{\leq 1(3)}$. Obviously $\beta(L)$ is in $Alg_{1(3)}(\varphi_{\geq 1(3)} \wedge \varphi_{\leq 1(3)})$.

We have shown that tr_1, tr_2 and β define a polynomial interpretation of $Alg_{1(3)}$ in Alg_E. Using Lemma 2 and Theorem 1 the result follows. $\square$

Lemma 4. *Let E_1 be a regular equational theory. If E_1 contains a commutative function symbol[7], or an associative function symbol, then E_1 is of type 1.*

Proof. Let E_1 be regular and assume that "f" is a commutative function symbol of E_1. Let $\varphi_{\geq 1(3)} := \bigwedge_{i=1}^{k} R_{\leq 1(3)}(p_{i_1}, p_{i_2}, p_{i_3})$ be an *at-most-1-in-3* problem over positive literals with set of Boolean variables P. Let $\varphi^{tr_1}_{\geq 1(3)}$ denote the conjunction of all formulae

$$f(f(x, y_i), f(u_i, v_i)) = f(f(p_{i_1}, p_{i_2}), f(p_{i_2}, p_{i_3}))$$

for $i = 1, \ldots, k$ where $x, y_1, \ldots, y_k, u_1, \ldots, u_k, v_1, \ldots, v_k$ are distinct variables. Obviously $tr_1 : \mathcal{P}_{\geq 1(3)} \to \mathcal{P}_{E_1}$ is polynomial-time computable.

We verify condition "$\to$" for theories of type 1. Let Q be a subset of P that contains exactly one literal from each clause. Let $L_1 = (\Pi, Lab, <)$ be a plain linear constant restriction for $Var(\varphi^{tr_1}_{\geq 1(3)})$ such that $[x]_\Pi = \{x\} \cup Q$. Obviously, since f is commutative there is a solution σ_1 of $\varphi^{tr_1}_{\geq 1(3)}$ that maps all

---

[7] Here a binary function symbol "f" is called a commutative (resp. associative) function symbol of E_1 if f belongs to the signature of E_1 and if $E_1 \models \forall x, y : f(x, y) = f(y, x)$ (resp. $E_1 \models \forall x, y, z : f(x, f(y, z)) = f(f(x, y), z)$).

variables in Q to x and the variables y_i, u_i, v_i to elements in $\{p_{i_1}, p_{i_2}, p_{i_3}\}$ for $1 \leq i \leq k$. Elements of $P \setminus Q$ are left fixed under σ_1. We show that σ_1 solves $(\varphi_{\geq 1(3)}^{tr_1}, L_1)$. Since L_1 is plain all equivalence classes of Π with the exception of $[x]_\Pi = Q \cup \{x\}$ are singletons. It follows that σ_1 satisfies the first condition of Definition 1. Since L_1 is plain, $P \cup \{x\}$ is the set of alien variables. For $y \in Q \cup \{x\}$ we have $\sigma_1(y) = x = rep_\Pi(y)$ since by assumption x is the smallest variable of Var. The remaining equivalence classes of $P \cup \{x\}$ are singletons of the form $\{p_{i_j}\}$ and we have $\sigma_1(p_{i_j}) = p_{i_j} = rep_\Pi(p_{i_j})$. This shows that σ_1 satisfies the second condition of Definition 1. Since L_1 is plain, alien variables are smaller w.r.t. $<$ than domestic variables. It follows that σ_1 satisfies the third condition of Definition 1 and solves $(\varphi_{\geq 1(3)}^{tr_1}, L_1)$.

We verify condition "←" for theories of type 1. Assume that $\varphi_{\geq 1(3)}^{tr_1}$ is solvable under the linear constant restriction L with partition Π where variables in $\{x\} \cup P$ receive alien label Σ_2 and for $i = 1, \ldots, k$ the elements $p_{i_1}, p_{i_2}, p_{i_3}$ belong to distinct equivalence classes. Let σ be a solution. Since σ maps each element y of $\{x, p_{i_1}, p_{i_2}, p_{i_3}\}$ to a variable of the form $rep_\Pi(y)$ (cf. Def. 1) it follows from Remark 1 that $\sigma(x) = rep_\Pi(x)$ must coincide with one of the elements $rep_\Pi(p_{i_1})$, $rep_\Pi(p_{i_2})$ or $rep_\Pi(p_{i_3})$. Hence $|[x]_\Pi \cap \{p_{i_1}, p_{i_2}, p_{i_3}\}| \geq 1$.

In the situation where E contains an associative function symbol "∘" we use as $\varphi_{\geq 1(3)}^{tr_1}$ the conjunction of all formulae

$$y_i \circ x \circ u_i = p_{i_1} \circ p_{i_1} \circ p_{i_2} \circ p_{i_3} \circ p_{i_3}.$$

and proceed similarly as above.   □

Lemma 5. *Let E_2 be a simple equational theory. Assume that the signature Σ_2 of E_2 contains at least one function symbol of arity $k \geq 1$. Then E_2 is of type 2.*

Proof. For simplicity we assume that Σ_2 contains a binary function symbol f. (The case where f is unary can be treated in the same way, omitting all arguments u in the terms below. If f has arity $k \geq 3$, use $k - 1$ arguments u instead of a single one.) Let $\varphi_{\leq 1(3)} := \bigwedge_{i=1}^{k} R_{\leq 1(3)}(p_{i_1}, p_{i_2}, p_{i_3})$ be an *at-most-1-in-3* problem over positive literals with set of Boolean variables P. Let $\varphi_{\leq 1(3)}^{tr_2}$ denote the conjunction of all formulae

$$p_{i_1} = f(u, y_i)$$
$$\land \quad p_{i_2} = f(u, f(u, y_i))$$
$$\land \quad p_{i_3} = f(u, f(u, f(u, y_i)))$$
$$\land \quad x = f(u, v_i)$$

for $i = 1, \ldots, k$ where $x, y_1, \ldots, y_k, u, v_1, \ldots, v_k$ are distinct variables and the p_{i_j} are treated as syntactic variables now ($1 \leq i \leq k$, $j = 1, 2, 3$). Obviously $tr_2 : \mathcal{P}_{\leq 1(3)} \to \mathcal{P}_{E_2}$ is polynomial-time computable.

We verify condition "→" for theories of type 2. Let Q be a subset of P that contains exactly one literal from each clause. For $i = 1, \ldots, k$ define $\sigma_2(x) :=$

$f(u, f(u, f(u, z)))$, $\sigma_2(v_i) = f(u, f(u, z))$ and

$$\sigma_2(y_i) := \begin{cases} f(u, f(u, z)) & \text{if } p_{i_1} \in Q, \\ f(u, z) & \text{if } p_{i_2} \in Q, \\ z & \text{if } p_{i_3} \in Q, \end{cases}$$

where z is a new variable and

$$\begin{aligned} \sigma_2(p_{i_1}) &:= & f(u, \sigma_2(y_i)), \\ \sigma_2(p_{i_2}) &:= & f(u, f(u, \sigma_2(y_i))), \\ \sigma_2(p_{i_3}) &:= & f(u, f(u, f(u, \sigma_2(y_i)))). \end{aligned}$$

Obviously all elements of $Q \cup \{x\}$ are mapped to $f(u, f(u, f(u, z)))$ under σ_2. Furthermore, σ_2 is a syntactic unifier of $\varphi^{tr_2}_{\leq 1(3)}$, hence it is a solution of $\varphi^{tr_2}_{\leq 1(3)}$.

We verify condition "←" for theories of type 2. Assume that $\varphi^{tr_2}_{\leq 1(3)}$ is unifiable under the linear constant restriction L with partition Π, say, with unifier σ_2. Obviously, since E_2 is simple the three variables $p_{i_1}, p_{i_2}, p_{i_3}$ cannot be identified by σ_2 modulo E_2 ($1 \leq i \leq k$). Condition 1 of Definition 1 shows that the three variables $p_{i_1}, p_{i_2}, p_{i_3}$ must belong to distinct equivalence classes w.r.t. Π. Furthermore, since E_2 is simple all variables in $P \cup \{x\}$ have to receive domestic label Σ_2 in L: we show this for x, the proof for the other variables is the same. Assume that x is treated as an alien variable. Then, by Definition 1 Nr. 2, we have $\sigma_2(x) = rep_\Pi(x) =_E f(\sigma_2(u), \sigma_2(v_i))$. Since E is simple, the variable $rep_\Pi(x)$ cannot occur in $f(\sigma_2(u), \sigma_2(v_i))$. Let x' be a variable such that $rep_\Pi(x) \neq x'$. Applying the substitution $rep_\Pi(x) \mapsto x'$ we obtain $rep_\Pi(x) =_E f(\sigma_2(u), \sigma_2(v_i)) =_E x'$. This would mean that E_2 is trivial, a contradiction. □

From Theorem 3, Lemma 5 and Lemma 4 we get the following results, which generalizes Theorem 12 of [Sc97].

Corollary 1. *Let E_1 and E_2 be equational theories over disjoint signatures. Assume that E_1 is regular and E_2 is simple. If the signature of E_1 contains a commutative function symbol, or an associative function symbol, and if the signature of E_2 contains at least one function symbol of arity $k \geq 1$, then $\mathcal{D}_{E_1} \oplus \mathcal{D}_{E_2}$ is NP-hard and there is no polynomial optimization of the combination algorithm for $\mathcal{D}_{E_1} \oplus \mathcal{D}_{E_2}$.*

9   Conclusion

We conclude with a brief discussion of three questions that are raised by the results of the preceding section.

1. Is a protocol needed for deciding combined problems? Naively, combination algorithms are often described in an informal way by saying that they allow to combine existing decision procedures (resp. unification algorithms, etc.) for the component theories into a decision procedure (resp. unification algorithm, etc.) for combined input problems. The notion of a protocol-based combination algorithm emphasizes that some kind of additional restriction has to be imposed on the "original" component problems that are obtained via decomposition of combined input problem. Consequently, some form of *adapted* decision procedures are needed for the component problems that can handle these restrictions.

It can be shown that both for Oppen's Procedure and for combination of E-unification algorithms at least some kind of synchronization between the solutions of component problems is needed: in both cases it is not difficult to describe mixed input problems $\varphi_1 \wedge \varphi_2$ where the input component problems φ_1 and φ_2 evaluate to 1 w.r.t. the original (unadapted) decision procedures of the first and second component theory respectively, but $\varphi_1 \wedge \varphi_2$ evaluates to 0 w.r.t. the union of the theories.

2. How significant is the difference between the original input component problems and those with scenarios? The combination method for generalized satisfiability problems introduced in Definition 6 shows that component problems with scenarios can be much simpler than the input component problems. However, the converse is also possible: recently F. Baader [Ba98] has shown that satisfiability of Boolean unification problems with linear constant restriction is PSPACE-complete. It is known that elementary Boolean unification (Boolean unification with constants) is NP-complete (resp. Π_2^p-complete). It is currently unknown if there exist equational theories E where E-unification with constants is decidable and E-unification with linear constant restriction is undecidable.

Even in the case of Oppen's procedure it is remarkable that component problems with scenarios in general contain disequations even if the input component problems do not use negation.

3. Is it possible that there are several protocols that may be used? In this case, how do impossibility results for polynomial optimization depend on the chosen protocol? Since obviously distinct protocols may be equivalent in terms of the solvability conditions which they impose on component problems the answer to the first question is affirmative. It seems interesting to introduce a formal notion of equivalence between protocols and protocol-based combination methods that apply to the same combined decision problems.

The following result seems to indicate that, as a default, impossibility results for polynomial optimization of one protocol-based combination method for a combined decision problem carry over to other protocol-based combination methods for the same combined decision problem. Assume we have two protocol-based methods,

$$(\mathcal{S}, Alg, \mathrm{val}_1^s, \mathrm{val}_2^s) \text{ and } (\mathcal{S}', Alg', \mathrm{val}_1^{s'}, \mathrm{val}_2^{s'})$$

for $\mathcal{D}_1 \oplus \mathcal{D}_2$. Assume that we have a polynomial optimization of Alg', but not for Alg. Then for each combined instance $\varphi_1 \wedge \varphi_2$ the following holds: for each scenario $S' \in Alg'(\varphi_1 \wedge \varphi_2)$ such that $\mathrm{val}_1^{s'}(\varphi_1, S') = 1 = \mathrm{val}_2^{s'}(\varphi_2, S')$ there exists a scenario $S \in Alg(\varphi_1 \wedge \varphi_2)$ such that $\mathrm{val}_1^s(\varphi_1, S) = 1 = \mathrm{val}_2^s(\varphi_2, S)$. However, there is no polynomial-time computable function $\beta : \mathcal{S}' \to \mathcal{S}$ such that always $\mathrm{val}_1^{s'}(\varphi_1, S') = 1 = \mathrm{val}_2^{s'}(\varphi_2, S')$ implies $\mathrm{val}_1^s(\varphi_1, \beta(S')) = 1 = \mathrm{val}_2^s(\varphi_2, \beta(S'))$. Even if we cannot exclude such a situation, it seems to be unusual.

Acknowledgements. The author would like to thank Christophe Ringeissen and Miki Hermann for comments that helped to improve the presentation.

References

Ba98.   F. Baader, "On the Complexity of Boolean Unification," *Information Processing Letters* **67** (4), 1998, pp. 215-220.

BS95a.   F. Baader, K.U. Schulz, "Combination techniques and decision problems for disunification," *Theoretical Computer Science* **142** (1995), pp. 229-255.

BS96.   F. Baader, K.U. Schulz, "Unification in the union of disjoint equational theories: Combining decision procedures," *Journal of Symbolic Computation* **21** (1996), pp. 211-243.

BS94.   F. Baader, J. Siekmann, "Unification Theory," in D.M. Gabbay, C. Hogger, and J. Robinson, Editors, *Handbook of Logic in Artificial Intelligence and Logic Programming*, Oxford University Press, Oxford, UK, 1994, pp. 41-125.

Bo93.   A. Boudet, "Combining Unification Algorithms," *Journal of Symbolic Computation* **16**, (1993), pp. 597-626.

CLS94.   D. Cyrluk, P. Lincoln, N. Shankar, "On Shostak's decision procedure for combinations of theories," in: *Proceedings of the 12th Conference on Automated Deduction*, Springer LNAI 814, 1994, pp. 463-477.

DKR94.   E. Domenjoud, F. Klay, R. Ringeissen, "Combination Techniques for Non-Disjoint Equational Theories," in: *Proceedings of the 12th Conference on Automated Deduction*, Springer LNAI 814, 1994, pp. 267-281.

DS78.   P. Downey, R. Sethi, "Assignment commands with array references," *Journal of the ACM* **25** (4), 1994, pp. 652-666.

GJ79.   M.R. Garey, D.S. Johnson, "Computers and Intractability: A Guide to the Theory of NP-Completeness," W.H. Freeman and Co. San Francisco (1979).

GN96.   Q. Guo, P. Narendran, and D.A. Wolfram, "Unification and Matching modulo Nilpotence," in: *Proceedings of the 13th Conference on Automated Deduction*, Springer LNAI 1104, 1996, pp. 261-274.

HK96.   M. Hermann, P.G. Kolaitis, "Unification Algorithms Cannot be Combined in Polynomial Time," in *Proceedings of the 13th International Conference on Automated Deduction*, M.A. McRobbie and J.K. Slaney (Eds.), Springer LNAI 1104, 1996, pp. 246-260.

He86.   A. Herold, "Combination of Unification Algorithms," *Proceedings of the 8th International Conference on Automated Deduction*, Springer LNCS 230, 1986.

KN91.   D. Kapur, P. Narendran, "Complexity of Unification Problems with Associative-Commutative Operators," *J. Automated Reasoning* **9**, 1992, pp. 261-288.

Ke98.   S. Kepser, "Negation in Combining Constraint Systems," in: *Frontiers of Combining Systems 2, Papers presented at FroCoS'98*, D.M Gabbay, M. de Rijke (Eds.), Research Studies Press/Wiley, Amsterdam 2000, pp. 117–192.

KR98.   S. Kepser, J. Richts, "Optimization Techniques for Combining Constraint Solvers," in: *Frontiers of Combining Systems 2, Papers presented at FroCoS'98*, D.M Gabbay, M. de Rijke (Eds.), Research Studies Press/Wiley, Amsterdam 2000, pp. 193–210.

Ki85.   C. Kirchner, "Méthodes et outils de conception systématique d'algorithmes d'unification dans les théories équationelles," Thèse d'Etat, Université de Nancy 1, France, 1985.

LBB84. D.S. Lankford, G. Butler, and B. Brady, "Abelian Group Unification Algorithms for Elementary Terms," *Contemporary Mathematics* **29**, 1984.

NO79. C.G. Nelson, D.C. Oppen, "Simplification by cooperating decision procedures," *ACM Trans. Programming Languages and Systems* **2** (1), 1979, pp. 245-257.

NO79. C.G. Nelson, D.C. Oppen, "Fast decision algorithms based on congruence closure," *Journal of the ACM* **27** (2), 1980, pp. 356-364.

Ni89. T. Nipkow, "Combining Matching Algorithms: The Regular Case," in: *Proceedings of the 3rd International Conference on Rewriting Techniques and Applications*, N. Dershowitz (Ed.), Springer LNCS 335, 1989, pp. 343-358.

Op80. D.C. Oppen, "Complexity, Convexity and Combination of Theories," *Theoretical Computer Science* **12**, 1980, pp. 291-302.

Pi74. D. Pigozzi, "The Join of Equational Theories," *Colloquium Mathematicum* Vol. XXX (1974) pp. 15-25.

Ri96. C. Ringeissen, "Cooperation of Decision Procedures for the Satisfiability Problem," in: *Frontiers of Combining Systems, Proceedings of the 1st International Workshop, FroCoS'96, Munich, Germany*, F. Baader, K.U. Schulz (Eds.), Applied Logic Series **3**, Kluwer, 1996, pp. 121-141.

Sh78. T. J. Schaefer, "The complexity of satisfiability problems," in: *Proceedings 10th Symposium on Theory of Computing*, San Diego, CA, 1978, pp. 216-226.

SS89. M. Schmidt-Schauß, "Combination of Unification Algorithms," *J. Symbolic Computation* **8**, 1989, pp. 51-99.

Sc97. K.U. Schulz, "A Criterion for Intractability of E-unification with Free Function Symbols and its Relevance for Combination of Unification Algorithms," in: *Proceedings of the 8th International Conference on Rewriting Techniques and Applications*, H. Comon (Ed.), Springer LNCS 1232, 1997, pp. 284-298.

Sc99. K.U. Schulz, "Tractable and Intractable Instances of Combination Problems for Unification and Disunification," to appear in *Journal of Logic and Computation*, Volume 10 (1), 2000.

Sh84. R.E. Shostak, "Deciding Combinations of Theories," *Journal of the ACM* **31**(1), 1984, pp. 1-12.

Ti86. E. Tidén, "Unification in Combinations of Collapse Free Theories with Disjoint Sets of Function Symbols," In: J.H. Siekmann (Ed.), *Proceedings of the 8th International Conference on Automated Deduction*, LNCS 230, 1986, pp. 431-449.

TH96. C. Tinelli, M. Harandi, "A New Correctness Proof of the Nelson-Oppen Combination Procedure," in: *Frontiers of Combining Systems, Proceedings of the 1st International Workshop, FroCoS'96, Munich, Germany*, F. Baader, K.U. Schulz (Eds.), Applied Logic Series **3**, Kluwer, 1996, pp. 103-121.

Ye87. K. Yelick, "Unification in Combinations of Collapse Free Regular Theories," *J. Symbolic Computation* **3**, 1987.

Congruence Closure Modulo Associativity and Commutativity[*]

L. Bachmair[1], I. V. Ramakrishnan[1], A. Tiwari[1], and L. Vigneron[2]

[1] Department of Computer Science
State University of New York
Stony Brook, NY 11794-4400, U.S.A
{leo,ram,astiwari}@cs.sunysb.edu
[2] LORIA – Université Nancy 2
Campus Scientifique, B.P. 239
54506 Vandœuvre-lès-Nancy Cedex, France
vigneron@loria.fr

Abstract. We introduce the notion of an *associative-commutative congruence closure* and show how such closures can be constructed via completion-like transition rules. This method is based on combining completion algorithms for theories over disjoint signatures to produce a convergent rewrite system over an extended signature. This approach can also be used to solve the word problem for ground AC-theories without the need for AC-simplification orderings total on ground terms.

Associative-commutative congruence closure provides a novel way to construct a convergent rewrite system for a ground AC-theory. This is done by transforming an AC-congruence closure, which is described by rewrite rules over an extended signature, to a rewrite system over the original signature. The set of rewrite rules thus obtained is convergent with respect to a new and simpler notion of associative-commutative reduction.

1 Introduction

Congruence closure algorithms have been used to decide if an equality $s \approx t$ logically follows from a set of equalities $E = \{s_1 \approx t_1, s_2 \approx t_2, \cdots, s_k \approx t_k\}$, where all terms are constructed from *uninterpreted or free* function symbols and constants. They also provide a decision procedure for validity problem in the quantifier-free theory of equality (with uninterpreted function symbols) [15].

The Nelson-Oppen method [14,17,19] is a general procedure to decide satisfiability of quantifier-free formulas in combination of certain[1] first-order theories. The crucial steps involved are variable abstraction and equality propagation. The Nelson-Oppen method can be used to obtain a decision procedure for the

---

[*] The research described in this paper was supported in part by the National Science Foundation under grants CCR-9902031, CCR-9711386 and EIA-9705998.

[1] Each individual theory is assumed to be a stably infinite first-order theory (with equality) over a disjoint signature such that satisfiability of quantifier-free formulas in the theory is decidable.

H. Kirchner and C. Ringeissen (Eds.): FroCoS 2000, LNAI 1794, pp. 245–259, 2000.
© Springer-Verlag Berlin Heidelberg 2000

quantifier-free theory of equality (with uninterpreted function symbols Σ)–as the quantifier-free theory of equality over Σ can be viewed as a combination of the quantifier-free theory of equality over the singleton sets $\Sigma_i = \{f\}$, where Σ is a disjoint union $\cup_i \Sigma_i$ of these sets.

Reconciling these two approaches for deciding satisfiability of quantifier-free formulas (in the theory of equality with uninterpreted function symbols) leads to the concept of an *abstract congruence closure*, as described in [3]. In this paper, we extend this notion to include associative-commutative function symbols. In other words, some function symbols in Σ are assumed to satisfy the associativity and commutativity axioms. Thus, if f is such a symbol, the theory over $\Sigma_i = \{f\}$ is that of a commutative semigroup (and not just the pure theory of equality).

We formally define an associative-commutative congruence closure in Section 2 and present a set of completion-like transition rules to construct such closures in Section 3. We prove the correctness of the transition rules using concepts of proof simplification and normalization, and also establish its termination in Section 4. This establishes decidability for ground AC-theories without the use of AC-simplification orderings that are total on ground terms [13,11]. The construction of AC-congruence closure can be optimized. We discuss some optimizations in Section 5.

We also consider the problem of constructing a convergent rewrite system (over the same signature) for a ground AC-theory. However, rather than obtaining a system R of rules such that $\rightarrow_{AC\backslash R^e}$ is convergent, we construct a set R such that a different reduction relation, which we denote by $\rightarrow_{P\backslash R^s}$ is convergent. This new notion is much simpler than usual rewriting modulo AC. This set is constructed by transforming the obtained AC-congruence closure, which is a convergent system over an extended signature, to a set over the original signature, see Section 6.

Preliminaries

Let Σ be a signature consisting of constants and function symbols, and $\mathcal{V}$ be a set of variables. The arity of a symbol f in Σ, denoted by $\alpha(f)$, is a set of natural numbers. The set of ground terms $\mathcal{T}(\Sigma)$ over Σ is the smallest set containing $\{c \in \Sigma : \alpha(c) = \{0\}\}$ and such that $f(t_1, \ldots, t_n) \in \mathcal{T}(\Sigma)$ whenever $f \in \Sigma, n \in \alpha(f)$ and $t_1, \ldots, t_n \in \mathcal{T}(\Sigma)$. The symbols $s, t, u, \ldots$ are used to denote terms; $f, g, \ldots$, function symbols; and $x, y, z, \ldots$, variables. We write $E[t]$ to indicate that an expression E contains t as a subterm and (ambiguously) denote by $E[u]$ the result of replacing a particular occurrence of t by u. (The same notation will be employed if E is a set of expressions.) By $E\sigma$ we denote the result of applying a substitution σ to E.

An *equation* is a pair of terms, written $s \approx t$. We usually identify the two equations $s \approx t$ and $t \approx s$; if the distinction is important, we call the equation a *rewrite rule* and write $s \rightarrow t$. A *rewrite system* is a set of rewrite rules. The *rewrite relation* $\rightarrow_R$ induced by a set of equations R is defined by: $u \rightarrow_R t$ if, and only if, u contains a subterm $l\sigma$ and $t = u[r\sigma]$, for some rewrite rule $l \rightarrow r$ in R and some substitution σ.

A term t is said to be in *normal form* with respect to a rewrite system R, or in *R-normal form*, if there is no term u, such that $t \to_R u$. A rewrite system is said to be (ground) *confluent* if all (ground) terms have a unique normal form. Rewrite systems that are (ground) confluent and terminating are called (ground) *convergent*.

The set of Σ-terms with variables in $\mathcal{V}$ is denoted by $\mathcal{T}(\Sigma, \mathcal{V})$. The *equational theory* induced by a set $\mathcal{E} \subseteq \mathcal{T}(\Sigma, \mathcal{V}) \times \mathcal{T}(\Sigma, \mathcal{V})$ of equations is defined as the reflexive, symmetric, and transitive closure of the rewrite relation induced by $\mathcal{E}$. The pure theory of equality is obtained when the set $\mathcal{E}$ is empty. If $\Sigma_{AC} \subset \Sigma$ is a finite set of function symbols, we denote by P the identitites

$$f(x_1, \dots, x_k, s, y_1, \dots, y_l, t, z_1, \dots, z_m) \approx f(x_1, \dots, x_k, t, y_1, \dots, y_l, s, z_1, \dots, z_m)$$

where $f \in \Sigma_{AC}$ and $k + l + m \in \alpha(f)$; and by F the set of identities

$$f(x_1, \dots, x_m, f(y_1, \dots, y_r), z_1, \dots, z_n) \approx f(x_1, \dots, x_m, y_1, \dots, y_r, z_1, \dots, z_n)$$

where $f \in \Sigma_{AC}$ and $\{m+n+1, m+n+r\} \subset \alpha(f)$. The congruence induced by P is called a *permutation congruence*. Flattening refers to normalizing a term with respect to the set F (considered as rewrite rules). The set $AC = F \cup P$ defines an AC-theory. The symbols in Σ_{AC} are called associative-commutative operators[2]. We require that $\alpha(f)$ be singleton for all $f \in \Sigma - \Sigma_{AC}$ and $\alpha(f) = \{2, 3, 4, \dots\}$ for all $f \in \Sigma_{AC}$.

Rewriting in presence of associative and commutative operators requires *extensions* of rules [16]. Given an *AC-operator* f and a rewrite rule ρ : $f(c_1, \dots, c_k) \to c$, we define its extension ρ^e as $f(c_1, \dots, c_k, x) \to f(c, x)$. Given a set of rewrite rules R, by R^e we denote the set R plus extensions of rules in R. In short, when working with AC-symbols (i) extensions have to be used for rewriting terms and computing critical pairs, and (ii) syntactic matching and unification is replaced by matching and unification modulo AC. Reduction modulo AC is defined by $s \to_{AC \backslash R} t$ iff there exists a rule $l \to r$ in R and a substitution σ such that $s = s[l']$, $l' \leftrightarrow^*_{AC} l\sigma$, and $t = s[r\sigma]$; see [6] for details.

A *multiset* over a set S is a mapping M from S to the natural numbers. Any ordering $\succ$ on a set S can be extended to an ordering $\succ^{mult}$ on multisets over S as follows: $M \succ^{mult} N$ iff $M \neq N$ and whenever $N(x) > M(x)$ then $M(y) > N(y)$, for some $y \succ x$. The multiset ordering $\succ^{mult}$ (on finite multisets) is well founded if the ordering $\succ$ is well founded [7]. Multiset inclusion is defined as $M \subseteq N$ if $M(x) \leq N(x)$ for all $x \in S$.

We consider the problem of deciding satisfiability of quantifier-free formulas in an equational theory presented by $\mathcal{E}$. When $\mathcal{E}$ is empty, congruence closure provides an efficient procedure [15]. The notion of AC-congruence closure handles the case when $\mathcal{E} = AC$.

---

[2] The equations $F \cup P$ define a conservative extension of the theory of associativity and commutativity to varyadic terms. For a fixed arity binary function symbol, the equations $f(x, y) \approx f(y, x)$ and $f(f(x, y), z) \approx f(x, f(y, z))$ define an AC-theory.

2   Associative-Commutative Congruence Closure

Let Σ be a signature, and E a set of ground equations over Σ. Let Σ_{AC} be some subset of Σ, containing all the associative-commutative operators and AC be the set $F \cup P$ as defined above. The set Σ can be written as a disjoint union of singleton sets Σ_i's, where each Σ_i contains exactly one function symbol in Σ. Inspired by the variable abstraction phase in the Nelson-Oppen method [14] and the idea of using constants as "names" of subterms [9], we first attempt to transform the set E of equations into a set of "pure" equations, i.e., each equation is over some signature $\Sigma_i \cup K$, where K is a set of new constants (names) introduced by variable abstraction, and shared between the individual theories. Each pure equation thus obtained is in a special form.

Definition 1. *Let Σ be a signature and K be a set of constants disjoint from Σ. By a D-rule (with respect to Σ and K) we mean a rewrite rule of the form*

$$f(c_1, \ldots, c_k) \to c_0$$

where $f \in \Sigma$ and $c_0, c_1, \ldots, c_k$ are constants in K.

An equation $c \to d$, where c and d are constants in K, is called a C-rule (with respect to K). An undirected equation $c \approx d$ is called a C-equation.

Rewrite rules of the form $f(c_1, \ldots, c_k) \to f(d_1, \ldots, d_l)$, where $f \in \Sigma_{AC}$, and $c_1, \cdots, c_k, d_1, \cdots, d_l \in K$, will be called AC-rules.

Each D- and AC-rule is over exactly one signature $\Sigma_i \cup K$. The C-rules are equations over every signature $\Sigma_i \cup K$.

For example, let Σ consist of function symbols, a, b, c, f and g, (f is AC) and let E_0 be a set of three equations $f(a, c) \approx a$, $f(c, g(f(b, c))) \approx b$ and $g(f(b, c)) \approx f(b, c)$. Viewing Σ as the disjoint union of signatures $\Sigma_1 = \{a\}$, $\Sigma_2 = \{b\}$, $\Sigma_3 = \{c\}$, $\Sigma_4 = \{f\}$ and $\Sigma_5 = \{g\}$, we can introduce new constants $K = \{c_0, c_1, c_2, c_3, c_4\}$ so that each equation (or rule) is over some $\Sigma_i \cup K$. Therefore, if we let

$$D_1 = \{a \to c_1,\ b \to c_2,\ c \to c_3,\ f(c_2, c_3) \to c_4,\ g(c_4) \to c_5\},$$

then, using these rules we can simplify the original equations in E_0 to obtain the new set $E_1 = \{f(c_1, c_3) \approx c_1,\ f(c_3, c_5) \approx c_2,\ c_5 \approx c_4\}$.

Definition 2. *Let R be a set of D-rules, C-rules and AC-rules (with respect to Σ and K). We say that a constant c in K represents a term t in $\mathcal{T}(\Sigma \cup K)$ (via the rewrite system R) if $t \to^*_{AC \backslash R^e} c$. A term t is also said to be represented by R if it is represented by some constant via R.*

For example, the constant c_4 represents the term $f(c, b)$ via D_1.

Definition 3. *Let Σ be a signature and K be a set of constants disjoint from Σ. A ground rewrite system $R = A^e \cup D^e \cup C$ is said to be an associative-commutative congruence closure (with respect to Σ and K) if*

(i) D is a set of D-rules, C is a set of C-rules, A is a set of AC-rules, and if a constant $c \in K$ is in normal form with respect to R, then c represents at least one term $t \in \mathcal{T}(\Sigma)$ via R, and

(ii) AC\R is a ground convergent modulo AC over $\mathcal{T}(\Sigma \cup K)$.

In addition, if E is a set of ground equations over $\mathcal{T}(\Sigma)$ such that,

(iii) If s and t are terms over $\mathcal{T}(\Sigma)$, then $s \leftrightarrow^*_{AC \cup E} t$ if, and only if, $s \rightarrow^*_{AC \setminus R} \circ \leftrightarrow^*_{AC} \circ \leftarrow^*_{AC \setminus R} t$,

then R will be called an associative-commutative congruence closure for E.

When Σ_{AC} is empty this definition specializes to that of an abstract congruence closure [3].

For instance, the rewrite system $D_1 \cup E_1$ above is not a congruence closure for E_0, as it is not a ground convergent rewrite system. But we can transform $D_1 \cup E_1$ into a suitable rewrite system, using a completion-like process described in more detail in the next section, to obtain a congruence closure (Example 4),

$$R' = \{a \rightarrow c_1,\ b \rightarrow c_2,\ c \rightarrow c_3,\ fc_2c_3 \rightarrow c_4,\ fc_3c_4 \rightarrow c_2,\ fc_1c_3 \rightarrow c_1,$$
$$fc_2c_2 \rightarrow fc_4c_4,\ fc_1c_2 \rightarrow fc_1c_4,\ gc_4 \rightarrow c_4\}$$

that provides a more compact representation of E_0. Note that the constant c_1 represents infinitely many terms via R'.

3   Construction of Congruence Closures

The completion procedure is obtained by putting together completion procedures over individual signatures $\Sigma_i \cup K$. Clearly, there are two distinct cases that need to be combined: when $f_i \in \Sigma_i$ is AC, and when it is not. The individual completion procedures working on equations in $\Sigma_i \cup K$ need to exchange equations between constants in K that are generated during the completion procedure. For this purpose, we choose an ordering in which the only terms smaller than a constant in K are other constants in K.

Let U be a set of symbols from which new names (constants) are chosen. We need a (partial) AC-compatible reduction ordering which orients the D-rules in the right way, and orients all the C- and AC-equations. The precedence-based AC-compatible ordering $\succ$ of [18], with any precedence in which $f \succ_{\Sigma \cup U} c$, whenever $f \in \Sigma$ and $c \in U$, serves the purpose. However, much simpler partial orderings would suffice too, but for convenience we use the ordering in [18]. In our case, this simply means that, orientation of D-rules is from left to right. Additionally, the orientation of an AC-rule will be given by: $f(c_1, \ldots, c_i) \succ f(c'_1, \ldots, c'_j)$ iff either $i > j$, or $i = j$ and $\{c_1, \ldots, c_i\} \succ^{mult} \{c'_1, \ldots, c'_j\}$, i.e., if the two terms have the same number of arguments, we compare the multisets of constants using a multiset extension $\succ^{mult}$ of the precedence $\succ_{\Sigma \cup U}$, see [7].

We next present a general method for construction of associative-commutative congruence closures. Our description is fairly abstract, in terms of transition rules that operate on triples (K, E, R), where K is a set of new constants that are introduced (the original signature Σ is fixed); E is a set of ground equations (over $\Sigma \cup K$) yet to be processed; and R is a set of C-rules, D-rules and AC-rules. Triples represent possible *states* in the process of constructing a closure. The initial state is $(\emptyset, E, \emptyset)$, where E is the input set of ground equations.

Abstraction

New constants are introduced by the following transition.

$$\text{Extension:} \qquad \frac{(K, E[f(c_1, \ldots, c_k)], R)}{(K \cup \{c\}, E[c], R \cup \{f(c_1, \cdots, c_k) \to c\})}$$

where $f \in \Sigma$; $c_1, \ldots, c_k$ are constants in K; $f(c_1, \ldots, c_k)$ is a term occurring in (some equation in) E; $c \notin \Sigma \cup K$; and, either $f \notin \Sigma_{AC}$, or, $f(c_1, \ldots, c_k)$ occurs as a proper subterm of a flattened term, or, it occurs as one side of an equation in which the top function symbol at the other side is some $g \neq f$.

Once a D-rule $f(c_1, \ldots, c_k) \to c$ has been introduced by extension, it can be used to eliminate any other occurrence of $f(c_1, \ldots, c_k)$ using the following rule.

$$\text{Simplification:} \qquad \frac{(K, E[s], R)}{(K, E[t], R)}$$

where s occurs in some equation in E, and, $s \to_{AC \setminus R^e} t$.

We will also use flattening to replace a term in E or R by its corresponding flattened form.

Deduction

It is fairly easy to see that any equation in E can be transformed to a C- or a AC-equation by suitable extension, flattening and simplification. Once we have obtained rules that contain terms only from a specific signature $\Sigma_i \cup K$, we can perform deduction steps on the separate parts.

In the case when $f_i \in \Sigma_i$ is an AC-symbol we use AC-superposition wherein we consider overlaps between extensions of AC-rules.

$$\text{ACSuperposition:} \qquad \frac{(K, E, R)}{(K, E \cup \{f(s, x\sigma) \approx f(t, y\sigma)\}, R)}$$

if $f \in \Sigma_{AC}$, there exist D- or AC-rules $f(c_1, \ldots, c_k) \to s$ and $f(d_1, \ldots, d_l) \to t$ in R, substitution σ is a most-general unifier modulo AC of $f(c_1, \ldots, c_k, x)$ and $f(d_1, \ldots, d_l, y)$[3] and $\{c_1, \ldots, c_k\} \cap \{d_1, \ldots, d_l\} \neq \emptyset$.

In the special case when one multiset is contained in the other, we obtain the AC-collapse rule.

$$\text{ACCollapse:} \qquad \frac{(K, E, R \cup \{t \to s\})}{(K, E \cup \{t' \approx s\}, R)}$$

---

[3] For the special case in hand, there is exactly one most general unifier modulo AC, and it is easy to compute.

if for some $u \to v \in R$, $t \to_{AC \setminus \{u \to v\}^e} t'$, and if $t \leftrightarrow^*_{AC} u$ then $s \succ v$.

In case of rules that do not contain AC-symbols at the top, superposition reduces to simplification, and hence AC-collapse also captures such superpositions. Note that we do not explicitly add AC extensions of rules to the set R, and so any rule in R is either a C-rule, or a D-rule, or an AC-rules, and *not* its extension. We implicitly work with extensions in AC-superposition.

Orientation

Equations are moved from the E-component of the state to the R-component by orientation. All rules added to the R-component are either C-rules, D-rules or AC-rules.

$$\text{Orientation:} \qquad \frac{(K, E \cup \{s \approx t\}, R)}{(K, E, R \cup \{s \to t\})}$$

if $s \succ t$, and, $s \to t$ is either a D-rule, or a C-rule, or a AC-rule.

Deletion allows us to delete trivial equations.

$$\text{Deletion:} \qquad \frac{(K, E \cup \{s \approx t\}, R)}{(K, E, R)}$$

if $s \leftrightarrow^*_{AC} t$.

Simplification

We need additional transition rules to incorporate interaction between the individual completion processes. In particular, C-rules can be used to perform simplifications on the left- and right-hand sides of other rules. The use of C-rules to simplify left-hand sides of rules is captured by AC-collapse. The simplification on the right-hand sides is subsumed by the following generalized composition rule.

$$\text{Composition:} \qquad \frac{(K, E, R \cup \{t \to s\})}{(K, E, R \cup \{t \to s'\})}$$

where $s \to_{AC \setminus R^e} s'$.

Example 4. Let $E_0 = \{f(a, c) \approx a,\ f(c, g(f(b, c))) \approx b,\ g(f(b, c)) \approx f(b, c)\}$. We show below some intermediate stages of a derivation.

i	Constants K_i	Equations E_i	Rules R_i	Transitions
0	$\emptyset$	E_0	$\emptyset$	
1	$\{c_1, c_3\}$	$\{fcgfbc \approx b,$ $gfbc \approx fbc\}$	$\{a \to c_1, c \to c_3,$ $fc_1c_3 \to c_1\}$	$\textbf{Ext}^2 \circ \textbf{Sim} \circ$ $\textbf{Ori}$
2	$K_1 \cup \{c_2, c_4\}$	$\{fcgfbc \approx b\}$	$R_1 \cup \{b \to c_2,$ $fc_2c_3 \to c_4, gc_4 \to c_4\}$	$\textbf{Sim}^2 \circ \textbf{Ext}^2 \circ$ $\textbf{Sim} \circ \textbf{Ori}$
3	K_2	$\emptyset$	$R_2 \cup \{fc_3c_4 \to c_2\}$	$\textbf{Sim}^6 \circ \textbf{Ori}$
4	K_2	$\emptyset$	$R_3 \cup \{fc_1c_2 \to fc_1c_4\}$	$\textbf{ACSup} \circ \textbf{Ori}$
5	K_2	$\emptyset$	$R_4 \cup \{fc_2c_2 \to fc_4c_4\}$	$\textbf{ACSup} \circ \textbf{Ori}$

The derivation moves equations, one by one, from the E-component of the state to the R-component through simplification, extension and orientation. It can be verified that the set R_5 is an AC congruence closure for E_0. Note that the side-condition in extension disallows breaking of an AC-rule into two D-rules, which is crucial for termination. We assume that f is AC and $c_i \succ c_j$ if $i < j$.

4   Termination and Correctness

Definition 5. *We use the symbol $\vdash$ to denote the one-step transition relation on states induced by the above transition rules. A derivation is a sequence of states $\xi_0 \vdash \xi_1 \vdash \cdots$. A derivation is said to be fair if any transition rule which is continuously enabled is eventually applied. The set R_∞ of persisting rules is defined as $\cup_i \cap_{j>i} R_j$; and similarly, $K_\infty = \cup_i \cap_{j>i} K_j$.*

We shall prove that any fair derivation will only generate finitely many persisting rewrite rules (in the third component) using Dickson's lemma [5]. Multisets over K_∞ can be compared using the multiset inclusion relation. If K_∞ is finite, this relation defines a Dickson partial order.

Lemma 6. *The set of persisting rules R_∞ in any fair derivation starting from state $(\emptyset, E, \emptyset)$ is finite.*

Proof. We first claim that K_∞ is finite. To see this, note that new constants are created by extension. Using finitely many applications of extension, simplification, flattening and orientation, we can move all rules from the initial second component E of the state tuple to the third component R. Fairness ensures that this will eventually happen. Thereafter, any equations ever added to E can be oriented using flattening and orientation, hence we never apply extension subsequently (see the side condition of the extension rule). Let $K_\infty = \{c_1, \ldots, c_n\}$.

Next we claim that R_∞ is finite. Suppose, R_∞ contains infinitely many rules. No transition rule (except extension) increases the number of D- and C-rules in $E \cup R$. Therefore, R_∞ contains infinitely many AC-rules, and since Σ_{AC} is finite, one AC-operator, say $f \in \Sigma_{AC}$, must occur infinitely often on the left-hand sides of R_∞. By Dickson's lemma, there exists an infinite chain of rules, $f(c_{11}, \ldots, c_{1k_1}) \to s_0, f(c_{21}, \ldots, c_{2k_2}) \to s_1, \ldots$, such that $\{c_{11}, \ldots, c_{1k_1}\} \subseteq \{c_{21}, \ldots, c_{2k_2}\} \subseteq \cdots$, where $\{c_{i1}, \ldots, c_{ik_i}\}$ denotes a multiset and $\subseteq$ denotes multiset inclusion. But, this contradicts fairness (in application of AC-collapse).

Proof Ordering

The correctness of the procedure will be established using proof simplification techniques for associative-commutative completion, as described by Bachmair [1] and Bachmair and Dershowitz [2]. In fact, we can directly use the results and the proof measure from [2]. However, since all rules in R have a special form, we can choose a simpler proof ordering.

Let $s = s[u\sigma] \leftrightarrow s[v\sigma] = t$ be a proof step using the equation (rule) $u \approx v \in AC \cup E \cup R$. The complexity of this proof step is defined by

$$
\begin{array}{ll}
(\{s,t\}, \perp, \perp) & \text{if } u \approx v \in E \qquad\qquad (\{s\}, \perp, t) \text{ if } u \approx v \in AC \\
(\{s\}, u, t) & \text{if } u \to v \in R \qquad\qquad\ (\{t\}, v, s) \text{ if } v \to u \in R \\
(\{s\}, \perp, t) & \text{if } u \to v \in R^e - R \qquad (\{t\}, \perp, s) \text{ if } v \to u \in R^e - R
\end{array}
$$

where $\perp$ is a new symbol. Tuples are compared lexicographically using the multiset extension of the reduction ordering $\succ$ on terms over $\Sigma \cup K_\infty$ in the first component, and the ordering $\succ$ on the second and third component. The constant $\perp$ is assumed to be minimum. The complexity of a proof is the multiset of complexities of its proof steps. The multiset extension of the ordering on tuples yields a proof ordering, denoted by the symbol $\succ_P$. The ordering $\succ_P$ on proofs is well founded as it is a lexicographic combination of well founded orderings.

Lemma 7. *Suppose $(K, E, R) \vdash (K', E', R')$. Then, for any two terms $s, t \in \mathcal{T}(\Sigma)$, it is the case that $s \leftrightarrow^*_{AC \cup E' \cup R'} t$ iff $s \leftrightarrow^*_{AC \cup E \cup R} t$. Further, if π is a ground proof, $s_0 \leftrightarrow s_1 \leftrightarrow \cdots \leftrightarrow s_k$, in $AC \cup E \cup R$, then there is a proof π', $s_0 = s'_0 \leftrightarrow s'_1 \leftrightarrow \cdots \leftrightarrow s'_l = s_k$, in $AC \cup E' \cup R'$ such that $\pi \succeq_P \pi'$.*

The first part of the lemma, which states that the congruence on $\mathcal{T}(\Sigma)$ remains unchanged, is easily verified by exhaustively checking it for each transition rule. For the second part, one needs to check that each equation in $(E - E') \cup (R - R')$ has a simpler proof in $E' \cup R' \cup AC$ for each transition rule application, see [2]. Note that extensions of rules are not added explicitly, and hence, they are never deleted either. Once we converge to R_∞, we introduce extensions to take care of cliffs in proofs.

Lemma 8. *If R_∞ is a set of persisting rules of a fair derivation starting from the state $(\emptyset, E, \emptyset)$, then, R^e_∞ is a ground convergent (modulo AC) rewrite system. Furthermore, $E_\infty = \emptyset$.*

Proof. (Sketch) Fairness implies that all critical pairs (modulo AC) between rules in R^e_∞ are contained in the set $\cup_i E_i$. Since a fair derivation is non failing, $E_\infty = \emptyset$. Since the proof ordering is noetherian, for every proof in $E_i \cup R_i \cup AC$, there exists a minimal proof π in $E_\infty \cup R_\infty \cup AC$. The minimal proof π in $R_\infty \cup AC$ can be further transformed to a minimal proof π' in $R^e_\infty \cup AC$ by eliminating cliffs arising from proper overlaps between AC-rules and R^e_∞-rules.

The proof π' is a rewrite proof. This follows from observing that π' does not contain any non-overlap, variable overlap, cliff or a peak. Any such pattern can be replaced by a smaller proof pattern, thus contradicting the minimality of the proof π. The details are similar, though much simpler in our special case, to those presented in [2].

Using Lemmas 7 and 8, we can easily prove that:

Proposition 9. *Let R_∞ be the set of persisting rules of a fair derivation starting from state $(\emptyset, E, \emptyset)$. Let $s, t \in \mathcal{T}(\Sigma)$. Then $s \leftrightarrow^*_{E \cup AC} t$ iff $s \to^*_{AC \backslash R^e_\infty} \circ \leftrightarrow^*_{AC} \circ \leftarrow^*_{AC \backslash R^e_\infty} t$. Furthermore, the set R^e_∞ is an associative-commutative congruence closure for E.*

Since R_∞ is finite, there exists a k such that $R_\infty \subset R_k$. Though there exist non-terminating fair derivations, we can still identify the finite set of persisting rules if we eagerly apply AC-collapse before any application of AC-superposition at any point after stage k.

Example 10. Let us describe one interesting example. Consider the initial set of equations $E_0 = \{ f(a,b) \approx b, \ g(b) \approx f(a,a), \ f(b, g(f(a,a,b))) \approx c \}$. Let f be an AC-symbol. We show some of the intermediate states of a fair derivation below:

i	Constants K_i	Equations E_i	Rules R_i
0	$\emptyset$	E_0	$\emptyset$
1	$\{c_1, c_2\}$	$\{gb \approx faa, \ fbgfaab \approx c\}$	$\{a \to c_1, \ b \to c_2, \ fc_1c_2 \to c_2\}$
2	$\{c_1, c_2, c_3\}$	$\{fbgfaab \approx c\}$	$R_1 \cup \{gc_2 \to c_3, \ fc_1c_1 \to c_3\}$
3	$\{c_1, c_2, c_3\}$	$\{fbgfaab \approx c\}$	$R_2 \cup \{fc_2c_3 \to c_2\}$
4	$\{c_1, c_2, c_3\}$	$\emptyset$	$R_3 \cup \{c \to c_2\}$

Note that since we applied AC-superposition in step 3 before simplifying the last equation in E, we avoided introducing a lot of new names.

5   Optimizations

Certain optimizations can be incorporated into the basic set of transition rules given above for computation of congruence closures. A lot of inferences can be avoided if we note that we do not need to consider extensions of all rules that have an AC symbol on top of their left-hand sides (for the purpose of critical pair computations). For example, we need not consider extensions of those D-rules that are created by extension to name a *proper subterm* in E. This fact can be easily incorporated using constraints, but we chose not to show it here for the sake of simplicity and clarity. Rather than formalizing this completely here, we illustrate this with an example below.

Example 11. Again suppose that the initial set of equations E_0 is: $\{f(a,b) \approx b, \ g(b) \approx f(a,a), \ f(b, g(f(b, g(f(a,a,b))))) \approx c\}$. If we choose to first apply extension and simplification until all the initial equations can be oriented using orientation, then, we will generate the following set of eleven D-rules:

$$
\begin{array}{llll}
a \to c_1 & b \to c_2 & fc_1c_2 \to c_2 & fc_1c_1 \to c_3 \\
gc_2 \to c_3 & fc_3c_2 \to c_4 & gc_4 \to c_5 & fc_2c_5 \to c_6 \\
gc_6 \to c_7 & fc_2c_7 \to c_8 & c \to c_8 &
\end{array}
$$

Now, note the huge number of possible AC-superpositions. However, we need not consider all of them. In particular, we need to consider extensions only of three rules: $fc_1c_2 \to c_2$, $fc_1c_1 \to c_3$, and $fc_2c_7 \to c_8$, and hence, AC-Superpositions between only these.

Secondly, note that, we can choose the ordering between two constants in K on-the-fly. As an optimization we could always choose it in a way so as to minimize the applications of AC-collapse and composition later. In other words, when we need to choose the orientation for $c \approx d$, we can look at the number of occurrences of c and d in the set of D- and AC-rules (in the R-component of the state), and the constant with fewer occurrences is made larger.

6 Construction of Ground Convergent AC Rewrite Systems

We next discuss the problem of obtaining a ground convergent AC rewrite system for the given ground AC-theory *in the original signature*. Hence, now we focus our attention to the problem of transforming a convergent system over an extended signature to a convergent system in the original signature. If we can successfully achieve this goal, it would mean that we can do ground AC-completion *without* having an AC-compatible ordering.

In the case when there are no AC-symbols in the signature, we can construct ground convergent systems from any given abstract congruence closure. This gives an indirect way to construct ground convergent systems equivalent to a given set of ground equations. However, we run into problems when we use the same method for translation in presence of AC-symbols. Typically, after translating back, the set of rules one gets is usually non-terminating modulo AC (Example 12). But if we suitably define the notion of AC-rewriting, the rules are seen to be convergent in the new definition. This is useful in two ways: (i) the new notion of AC-rewriting seems to be more practical, in the sense that it involves strictly less work than a usual $AC \backslash R^e$ reduction [6]; and, (ii) it helps to clarify the advantage offered by the use of extended signatures when dealing with a set of ground AC-equations.

Transition Rules

We describe the process of transforming a rewrite system R over an extended signature $\Sigma \cup K$ to a rewrite system R over the original signature Σ by transformation rules on states (K, R), where K is the set of constants to be eliminated, and R is a set of rewrite rules over $\Sigma \cup K$ to be transformed.

Constants in K that occur as a left-hand side of some rule in R are called *redundant* in R. Redundant constants can be easily eliminated by the compression rule.

Compression: $$\frac{(K \cup \{c\}, R \cup \{c \to t\})}{(K, R')}$$

if R' is obtained from R by replacing all occurrences of c in R by t.

The basic idea for eliminating a constant c that is not redundant in R involves picking a representative term t (over the signature Σ) in the equivalence class of c, and replacing c by t everywhere in R. The selection rule achieves this.

Selection: $$\frac{(K \cup \{c\}, R \cup \{t \to c\})}{(K, R')}$$

where c is not redundant in R, $t \in \mathcal{T}(\Sigma)$, R' is the set $\{l\sigma \to r\sigma : l \to r \in R\} \cup \{f(t_1, \ldots, t_k, X) \to f(f(t_1, \ldots, t_k), X) : t \equiv f(t_1, \ldots, t_k) \text{ and } f \in \Sigma_{AC}\}$, and σ is the substitution $\{c \mapsto t\}$. After applying the substitution, we do *not* flatten terms. The variable X is a special variable and its role will be be discussed later.

Example 12. Consider the problem of constructing a ground convergent system for the set E_0 of Example 4. A fully-reduced congruence closure for E_0 is given by the set R_0

$$a \to c_1 \qquad b \to c_2 \qquad c \to c_3 \qquad fc_2c_3 \to c_4$$
$$fc_3c_4 \to c_2 \qquad fc_1c_3 \to c_1 \qquad fc_2c_2 \to fc_4c_4 \qquad fc_1c_2 \to fc_1c_4$$
$$gc_4 \to c_4$$

under the ordering $c_2 \succ c_4$ between constants. For the constants c_1, c_2 and c_3 we have no choice but to choose a, b and c as representatives respectively. Thus after three applications of selection, we get

$$fcc_4 \to b \qquad fac \to a \qquad fbb \to fc_4c_4 \qquad fab \to fac_4$$
$$fbc \to c_4 \qquad gc_4 \to c_4$$

Next we are forced to choose fbc as the representative for the class c_4. This gives us the transformed set R_1,

$$fc(fbc) \to b \qquad fac \to a \qquad fbb \to f(fbc)(fbc) \qquad fab \to fa(fbc)$$
$$fbcX \to f(fbc)X \qquad gfbc \to fbc$$

The relation $\to_{AC \setminus R_1^c}$ is clearly non-terminating (with the variable X considered as a regular term variable).

Rewriting with Sequence Extensions Modulo Permutation Congruence

Let X denote a variable ranging over sequences of terms. A sequence substitution σ is a substitution that maps variables to the sequences. If σ is a sequence substitution that maps X to the sequence $\langle s'_1, \ldots, s'_m \rangle$, then $f(s_1, \ldots, s_k, X)\sigma$ is the term $f(t_1, \ldots, s_k, s'_1, \ldots, s'_m)$.

Definition 13. *Let ρ be a ground rule of the form $f(t_1, \ldots, t_k) \to g(s_1, \ldots, s_m)$ where $f \in \Sigma_{AC}$. We define the* sequence extension ρ^s *of ρ as, $f(t_1, \ldots, t_k, X) \to f(s_1, \ldots, s_m, X)$ if $f = g$ and, $f(t_1, \ldots, t_k, X) \to f(g(s_1, \ldots, s_m), X)$ if $f \neq g$.*

Now we are ready to define the notion of rewriting we use. Recall that P denotes the equations defining the permutation congruence, and that $AC = F \cup P$.

Definition 14. *Let R be a set of ground rules. For ground terms $s, t \in \mathcal{T}(\Sigma)$, we say that $s \to_{P \setminus R^s} t$ if there exists a rule $l \to r \in R^s$ and a sequence substitution σ such that $s = C[l']$, $l' \leftrightarrow_P^* l\sigma$ and $r' = r\sigma$.*

Note that the difference with standard rewriting modulo AC is that instead of performing matching modulo AC, we do matching modulo P. For example, if ρ is $fac \to a$, then the term $f(f(a,b),c)$ is not reducible by $\to_{P\backslash\rho^s}$, although it is reducible by $\to_{AC\backslash\rho^e}$. The term $f(f(a,b),c,a)$ can be rewritten by $\to_{P\backslash\rho^s}$ to $f(f(a,b),a)$.

Example 15. Following up on Example 12, we note that the relation $P\backslash R_1^s$ is convergent. For instance, a normalizing rewrite derivation for the term $fabc$ is,

$$fabc \to_{P\backslash R_1^s} fa(fbc)c \to_{P\backslash R_1^s} fab \to_{P\backslash R_1^s} fa(fbc).$$

On closer inspection, we find that we are essentially doing a derivation in the original rewrite system R_0 (over the extended signature).

$$fabc \to_{P\backslash R_0^s} fc_1c_2c_3 \to_{P\backslash R_0^s} fc_1c_4c_3 \to_{P\backslash R_0^s} fc_1c_2 \to_{P\backslash R_0^s} fc_1c_4.$$

There is a one-to-one bijection between a step using $P\backslash R_1^s$ and a step using $P\backslash R_0^s$. This essentially is at the core of the proof of correctness, which we now state without proof.

Theorem 16. *If $(\emptyset, R_\infty)$ is the final state of a maximal derivation starting from state (K, R), where R is a fully-reduced AC-congruence closure, then $\to_{P\backslash R_\infty^s}$ is ground convergent on all fully flattened terms over Σ. The equivalence over flattened $\mathcal{T}(\Sigma)$ terms defined by this relation is the same as the equational theory induced by $R \cup AC$ over flattened $\mathcal{T}(\Sigma)$ terms.*

Note that in the special case when Σ_{AC} is empty, the notion of rewriting corresponds to the standard notion, and hence R_∞ is convergent in the standard sense by this theorem.

7   Conclusion

The fact that we can construct an AC-congruence closure implies that the word problem for finitely presented ground AC-theories is decidable, see [11], [13] and [8]. Note that we arrive at this result *without* assuming the existence of an AC-simplification ordering that is total on ground terms. The existence of such AC-simplification orderings was established in [13], but required a non-trivial proof.

Since we construct a convergent rewrite system, even the problem of determining whether two finitely presented ground AC-theories are equivalent, is decidable. Since commutative semigroups are special kinds of AC-theories, where the signature consists of a single AC-symbol and a finite set of constants, these results carry over to this special case [12], [10].

We have shown that we can construct an AC-congruence closure for a finite set E of ground equations (over a signature Σ containing AC-symbols) using procedures that construct such closures for finitely presented commutative semigroups. In fact the number of invocations of the latter procedure is bounded above by $n|\Sigma|$, where n denotes the size of set E, and $|\Sigma|$ denotes the cardinality of the set Σ. This establishes that the word problem for ground AC-theories is no more difficult than the word problem for finitely presented commutative semigroups (modulo a polynomial time factor).

The idea of using variable abstraction to transform a set equations over several AC-symbols into a set of equations in which each equation contains exactly one AC-symbol appears in [8]. All equations containing the same AC-symbol are separated out, and completed into a canonical rewriting system (modulo AC) using the method proposed in [4]. However, the combination of ground AC-theories with other ground theories is done differently here. In [8], the ground theory (non-AC part) is handled using ground completion (and uses an recursive path ordering during completion). We, on the other hand, use a congruence closure. The usefulness of our approach can also be seen from the simplicity of the correctness proof and the results we obtain for transforming a convergent system over an extended signature to one over the original signature.

The method for completing a finitely presented commutative semigroup (using what we call AC-rules here) has been described in various forms in the literature, e.g. [4])[4]. It is essentially a specialization of Buchberger's algorithm for polynomial ideals to the case of binomial ideals (i.e. when the ideal is defined by polynomials consisting of exactly two monomials with coefficients $+1$ and -1).

Acknowledgements. We wish to thank the anonymous reviewers for helpful comments.

References

1. L. Bachmair. *Canonical Equational Proofs.* Birkhäuser, Boston, 1991.
2. L. Bachmair and N. Dershowitz. Completion for rewriting modulo a congruence. *Theoretical Computer Science,* 67(2 & 3):173–201, Oct. 1989.
3. L. Bachmair, C. Ramakrishnan, I. Ramakrishnan, and A. Tiwari. Normalization via rewrite closures. In P. Narendran and M. Rusinowitch, editors, *Proc. 10th Int. Conf. on Rewriting Techniques and Applications,* volume 1631 of *Lecture Notes in Computer Science,* pages 190–204. Springer-Verlag, Trento, Italy, 1999.
4. A. M. Ballantyne and D. S. Lankford. New decision algorithms for finitely presented commutative semigroups. *Comp. and Maths. with Appls.,* 7:159–165, 1981.
5. T. Becker and V. Weispfenning. *Gröbner bases: a computational approach to commutative algebra.* Springer-Verlag, Berlin, 1993.
6. N. Dershowitz and J. P. Jouannaud. Rewrite systems. In J. van Leeuwen, editor, *Handbook of Theoretical Computer Science (Vol. B: Formal Models and Semantics),* Amsterdam, 1990. North-Holland.

---

[4] Actually there is a subtle difference between the proposed method here and the various other algorithms for deciding the word problem for commutative semigroups too. For example, working with rule extensions is not the same as working with rules on equivalence classes (under AC) of terms. Hence, in our method, we can apply certain optimizations as mentioned in Section 5.

7. N. Dershowitz and Z. Manna. Proving termination with multiset orderings. *Communications of the ACM*, 22(8):465–476, 1979.

8. E. Domenjoud and F. Klay. Shallow AC theories. In *Proceedings of the 2nd CCL Workshop*, La Escala, Spain, Sept. 1993.

9. D. Kapur. Shostak's congruence closure as completion. In H. Comon, editor, *Proceedings of the 8th International Conference on Rewriting Techniques and Applications*, pages 23–37, 1997. Vol. 1232 of *Lecture Notes in Computer Science*, Springer, Berlin.

10. U. Koppenhagen and E. W. Mayr. An optimal algorithm for constructing the reduced Gröbner basis of binomial ideals. In Y. D. Lakshman, editor, *Proceedings of the International Symposium on Symbolic and Algebraic Computation*, pages 55–62, 1996.

11. C. Marche. On ground AC-completion. In R. V. Book, editor, *4th International Conference on Rewriting Techniques and Applications*, pages 411–422, 1991. Vol. 488 of *Lecture Notes in Computer Science*, Springer, Berlin.

12. E. W. Mayr and A. R. Meyer. The complexity of the word problems for commutative semigroups and polynomial ideals. *Advances in Mathematics*, 46:305–329, 1982.

13. P. Narendran and M. Rusinowitch. Any ground associative-commutative theory has a finite canonical system. In R. V. Book, editor, *4th International Conference on Rewriting Techniques and Applications*, pages 423–434, 1991. Vol. 488 of *Lecture Notes in Computer Science*, Springer, Berlin.

14. G. Nelson and D. Oppen. Simplification by cooperating decision procedures. *ACM Transactions on Programming Languages and Systems*, 1(2):245–257, October 1979.

15. G. Nelson and D. Oppen. Fast decision procedures based on congruence closure. *Journal of the Association for Computing Machinery*, 27(2):356–364, Apr. 1980.

16. G. E. Peterson and M. E. Stickel. Complete sets of reductions for some equational theories. *J. ACM*, 28(2):233–264, Apr. 1981.

17. C. Ringeissen. Cooperation of decision procedures for the satisfiability problem. In F. Baader and K. U. Schulz, editors, *Frontiers of Combining Systems*, pages 121–139, 1996. Kluwer Academic Publishers.

18. A. Rubio and R. Nieuwenhuis. A precedence-based total AC-compatible ordering. In C. Kirchner, editor, *Proceedings of the 5 Intl. Conference on Rewriting Techniques and Applications*, pages 374–388, 1993. Vol. 690 of *Lecture Notes in Computer Science*, Springer.

19. C. Tinelli and M. Harandi. A new correctness proof of the Nelson-Oppen combination procedure. In F. Baader and K. U. Schulz, editors, *Frontiers of Combining Systems*, pages 103–119, 1996. Kluwer Academic Publishers.

Combining Equational Theories Sharing
Non-Collapse-Free Constructors

Franz Baader[1]* and Cesare Tinelli[2]**

[1] LuFg Theoretical Computer Science, RWTH Aachen.
baader@informatik.rwth-aachen.de
[2] Department of Computer Science, University of Iowa.
tinelli@cs.uiowa.edu

Abstract. In this paper we extend the applicability of our combination method for decision procedures for the word problem to theories sharing *non-collapse-free* constructors. This extension broadens the scope of the combination procedure considerably, for example in the direction of equational theories axiomatizing the equivalence of modal formulae.

1 Introduction

The word problem for a theory E is concerned with the question of whether two terms are equal in all models of E. In [4] we provided modular decidability results for the word problem in the case of unions of equational theories with possibly non-disjoint signatures, subsuming previous well-known results on the decidability of the word problem for the union of equational theories with disjoint signatures [12,11,13]. Our results were achieved by assuming that the function symbols shared by the component theories were *constructors* in a appropriate sense.

The notion of constructors presented in [4] was modeled after one first introduced in [14], and generalized that in [5]. Its formulation is based on the observation that some equational theories are such that the reducts of their free models to a subset Σ of their signature are themselves free. We would call constructors the symbols in Σ. The actual definition in [4], however, incorporated the restriction that the equational theory of the constructors had to be collapse-free.[1] This restriction was essentially technical, as it was used to provide a syntactic characterizations of the generators of the free Σ-reducts in terms of a certain set G of terms, which was then utilized in various proofs in the paper.

In the present paper, by using a more general way of defining the set G above, we remove the collapse-freeness restriction and show that all the combination results given in [4] continue to hold without it.

---

* Partially supported by the EC Working Group CCL II.
** Partially supported by the National Science Foundation grant no. CCR-99-72311.

[1] In other words, no non-variable term over the constructor symbols could be equivalent in E to one of its variables.

H. Kirchner and C. Ringeissen (Eds.): FroCoS 2000, LNAI 1794, pp. 260–274, 2000.

In [4] we used a rule-based procedure for combining in a modular way a procedure deciding the word problem for a theory E_1 and a procedure deciding the word problem for a theory E_2 into a procedure deciding the word problem for the theory $E_1 \cup E_2$. As mentioned, the main requirement was that the symbols shared by E_1 and E_2 were constructors for each of them. In this paper, we obtain the generalized combination results by using a proper modification of the procedure, which does not rely anymore on the assumption that the constructor theory is collapse-free.

The net effect of lifting the collapse-freeness restriction is a considerable expansion of the scope of our combination results. A lot more equational theories obtained as a conservative extension of a core Σ-theory are now such that Σ is a set of constructors for them. Which means, potentially, that a lot more theories built as conservative extensions of a same Σ-theory can be combined with our method.[2]

One particularly interesting class of such theories includes the equational axiomatizations of some (propositional) modal logics, on which we give more details in Sect. 2.2. A fair amount of research has been done on the combination of modal logics. We believe that our results for the word problems can now be used to contribute to this research by recasting the combination of two modal logics as the union of their corresponding equational theories. However, we have not yet had the time to explore these possibilities in more depth. We are working on this in a joint project with modal logicians.

For now, we present and discuss our generalized notion of constructors, and provide some examples of theories admitting constructors in the new sense but not in the old one, including an equational theory corresponding to a modal logic. Then, we describe the modified version of the combination procedure, provide a sketch of its correctness proof, and show how that leads to exactly the same results given in [4], but of course with the wider scope provided by the new definition of constructors. Because of space limitations, we refer the reader to the longer version of this paper [3] for detailed proofs of our results.

2   Word Problems and Satisfiability Problems

We will use V to denote a countably infinite set of variables, and $T(\Omega, V)$ to denote the set of all Ω-*terms*, that is, terms over the signature Ω with variables in V. An equational theory E over the signature Ω is a set of (implicitly universally quantified) equations between Ω-terms. We use $s \equiv t$ to denote an equation between the terms s, t. For an equational theory E, the *word problem* is concerned with the validity in E of equations between Ω-terms. Equivalently, the word problem asks for the (un)satisfiability of the *disequation* $s \not\equiv t$ in E—where $s \not\equiv t$ is an abbreviation for the formula $\neg(s \equiv t)$. As usual, we write "$s =_E t$" to express that the formula $s \equiv t$ is valid in E. An equational theory E is *collapse-free* iff $x \neq_E t$ for all variables x and non-variable terms t.

---

[2]   The qualification "potentially" is mandatory, of course, because we still need to impose some additional computability requirements on the theories to combine.

Given an Ω-term s, an Ω-algebra $\mathcal{A}$, and a valuation α (of the variables in s by elements of $\mathcal{A}$), we denote by $[\![s]\!]_\alpha^\mathcal{A}$ the interpretation of s in $\mathcal{A}$ under α. Also, if Σ is a subsignature of Ω, we denote by $\mathcal{A}^\Sigma$ the reduct of $\mathcal{A}$ to Σ. An Ω-algebra $\mathcal{A}$ is a *model* of E iff every equation in E is valid in $\mathcal{A}$. The equational theory E over the signature Ω defines an Ω-*variety*, i.e., the class of all models of E. When E is *non-trivial*, i.e., has models of cardinality greater than 1, its variety contains free algebras for any set of (free) generators. If $\mathcal{A}$ is a free algebra in E's Ω-variety with a set X of generators we say that $\mathcal{A}$ is *free in E over X*.

We are interested in *combined* equational theories, that is, equational theories E of the form $E := E_1 \cup E_2$, where E_1 and E_2 are equational theories over two signatures Σ_1 and Σ_2. We call the elements of $\Sigma := \Sigma_1 \cap \Sigma_2$, if any, *shared* symbols. We call 1-*symbols* the elements of Σ_1 and 2-*symbols* the elements of Σ_2. A term $t \in T(\Sigma_1 \cup \Sigma_2, V)$ is an *i-term* iff its *top symbol* $t(\epsilon) \in V \cup \Sigma_i$. Note that variables and terms t with $t(\epsilon) \in \Sigma_1 \cap \Sigma_2$ are both 1- and 2-terms. A subterm s of a 1-term t is an *alien subterm* of t iff it is not a 1-term and every proper superterm of s in t is a 1-term. Alien subterms of 2-terms are defined analogously. A term over the joint signature $\Sigma_1 \cup \Sigma_2$ is called a *shared* term if it is a Σ-term, and a *pure* term if it is a Σ_i-term for $i \in \{1, 2\}$, Similarly, an equation $s \equiv t$ is a *pure* equation if s and t are both Σ_i-terms for $i \in \{1, 2\}$.

A given disequation $s \not\equiv t$ between $(\Sigma_1 \cup \Sigma_2)$-terms s, t can be transformed into an equisatisfiable formula $\varphi_1 \wedge \varphi_2$, where φ_i is a conjunction of pure equations and disequations. This can be achieved by the usual *variable abstraction* process in which alien subterms are replaced by new variables (see the long version of [4] for a detailed description). Obviously, if $\varphi_1 \wedge \varphi_2$ is satisfiable in a model $\mathcal{A}$ of $E_1 \cup E_2$, each φ_i is then satisfiable in the reduct $\mathcal{A}^{\Sigma_i}$, which is a model of E_i ($i = 1, 2$). However, if each φ_i is satisfiable in a model $\mathcal{A}_i$ of E_i, there may be no model of E in which $\varphi_1 \wedge \varphi_2$ is satisfiable. One case in which there always is one is described by the proposition below (see the long version of [4] for a proof).

Proposition 1. *Let $\mathcal{A}_i$ be a model of E_i, φ_i a first-order Σ_i-formula ($i = 1, 2$), and $\Sigma := \Sigma_1 \cap \Sigma_2$. Assume that $\mathcal{A}_1^\Sigma$ and $\mathcal{A}_2^\Sigma$ are both free in the same Σ-variety over respective sets of generators Y_1 and Y_2 with the same cardinality. If φ_i is satisfiable in $\mathcal{A}_i$ with the variables in $Var(\varphi_1) \cap Var(\varphi_2)$ taking distinct values over Y_i for $i = 1, 2$, then $\varphi_1 \wedge \varphi_2$ is satisfiable in a model of $E_1 \cup E_2$.*

As mentioned in the introduction, we will be interested in free models of E_1 and of E_2 whose reducts to their shared signature are themselves free. In general, the property of being a free algebra is not preserved under signature reduction. The problem is that the reduct of an algebra may need more generators than the algebra itself and these generators need not be free. Nonetheless, there are free algebras admitting reducts that are also free, although over a possibly larger set of generators. These algebras are models of equational theories that admit *constructors* in the sense defined in the next subsection.

2.1   Theories Admitting Constructors

In the following, Ω will be an at most countably infinite signature, and Σ a subset of Ω. We will fix a non-trivial equational theory E over Ω and define the Σ-restriction of E as $E^\Sigma := \{s \equiv t \mid s, t \in T(\Sigma, V) \text{ and } s =_E t\}$.

Definition 2 (Constructors). *The subsignature Σ of Ω is a set of constructors for E if for every Ω-algebra $\mathcal{A}$ free in E over a countably infinite set X, $\mathcal{A}^\Sigma$ is free in E^Σ over a set Y including X.*

As we will see, this new definition of constructors is a proper generalization of the definition given in [4], which also requires E^Σ to be collapse-free. Contrary to the one above, that definition does not require the generators of $\mathcal{A}^\Sigma$ to include those of $\mathcal{A}$; but this is always the case when E^Σ is collapse-free.

It is immediate that the whole signature Ω is a set of constructors for the theory E. Similarly, the empty signature is a set of constructors for E, as any model of E is free over its whole carrier in $E^\emptyset$, which is axiomatized by $\{v \equiv v \mid v \in V\}$. If E is the union of two theories over *disjoint* signatures, Σ_1, Σ_2 respectively, then each Σ_i is a set of constructors for E. This is not immediate, but it can be shown as a consequence of some results in [2].

In the following, we provide a more concrete, syntactic characterization of theories admitting constructors. For that we will introduce the concept of a Σ-*base*. But first, some more notation will be needed.

Given a subset G of $T(\Omega, V)$, we denote by $T(\Sigma, G)$ the set of terms over the "variables" G. More precisely, every member t of $T(\Sigma, G)$ is obtained from a term $s \in T(\Sigma, V)$ by replacing the variables of s with terms from G. We will denote such a term t by $s(\bar{r})$ where $\bar{r}$ is the tuple made, without repetitions, of the terms of G that replace the variables of s. Notice that this notation is consistent with the fact that $G \subseteq T(\Sigma, G)$. In fact, every $r \in G$ can be represented as $s(r)$ where s is a variable of V. Also notice that $T(\Sigma, V) \subseteq T(\Sigma, G)$ whenever $V \subseteq G$. In this case, every $s \in T(\Sigma, V)$ can be trivially represented as $s(\bar{v})$ where $\bar{v}$ are the variables of s.

Definition 3 (Σ-base). *A set $G \subseteq T(\Omega, V)$ is a Σ-base of E iff the following holds:*

1. *$V \subseteq G$.*
2. *For all $t \in T(\Omega, V)$, there is an $s(\bar{r}) \in T(\Sigma, G)$ such that $t =_E s(\bar{r})$.*
3. *For all $s_1(\bar{r}_1), s_2(\bar{r}_2) \in T(\Sigma, G)$,*

$$s_1(\bar{r}_1) =_E s_2(\bar{r}_2) \quad \text{iff} \quad s_1(\bar{v}_1) =_E s_2(\bar{v}_2),$$

where $\bar{v}_1, \bar{v}_2$ are fresh variables abstracting $\bar{r}_1, \bar{r}_2$ so that two terms in $\bar{r}_1, \bar{r}_2$ are abstracted by the same variable iff they are equivalent in E.

Theorem 4 (Characterization of constructors). *The signature Σ is a set of constructors for E iff E admits a Σ-base.*

The proof of the theorem—which can be found in [3]—provides a little more information than stated in the theorem.

Corollary 5. *Let G be a Σ-base of E, $\mathcal{A}$ an Ω-algebra free in E over a countably infinite set X, and α a bijective valuation of V onto X. Then $\mathcal{A}^{\Sigma}$ is free in E^{Σ} over the set $Y := \{[\![r]\!]_{\alpha}^{\mathcal{A}} \mid r \in G\}$, and $X \subseteq Y$.*

An interesting question is whether the condition that $X \subseteq Y$ in the definition of constructors is really needed. Does this condition always hold whenever the Σ-reduct of any algebra $\mathcal{A}$ free in E over the countably infinite set X is itself free? It can be easily shown that $\mathcal{A}^{\Sigma}$ can be free in E^{Σ} only over a set Y that is countably infinite. The question is: can Y always be chosen so that it includes X? When E^{Σ} is collapse-free, Y is unique and it does include X, so one needs not worry [4]. When E^{Σ} is not collapse-free, however, $\mathcal{A}^{\Sigma}$ may be free in E^{Σ} over more than one set of generators, not all of which include X.

For instance, consider the Ω-theory $E := \{g(g(x)) = x\}$ and let $\Sigma := \Omega$. Let $\mathcal{A}$ be an Ω-algebra free in E over some set X and α a bijective valuation of V onto X. It is easy to see that $\mathcal{A}$, which is free in E^{Σ} over X of course, is also free over the set $\{[\![g(v)]\!]_{\alpha}^{\mathcal{A}} \mid v \in V\}$, disjoint from X. Now, this example causes no problems because one can always choose $Y := X$ in this case. For the general case, however, the question remains open.

As shown in Cor. 5, a Σ-base of a theory E really denotes the set of free generators of a certain free model of E^{Σ}. Clearly, there may be many Σ-bases for the same theory E. For instance, if G is a Σ-base of E, any set obtained from G by replacing one of its terms by an E-equivalent term is also a Σ-base of E. Also, for any bijective renaming π of V onto itself, the set $\{\pi(r) \mid r \in G\}$ is a Σ-base of E as well.

One may wonder then if it is possible for E to have Σ-bases that are not just syntactic variants of one another. We know that this is not possible when E^{Σ} is collapse-free. In that case in fact, all Σ-bases, if any, denote the unique set Y over which the Σ-reduct of the infinitely generated free model of E is free. As it turns out, then E has a Σ-base $G_E(\Sigma, V)$ which is closed under bijective renaming of variables and under equivalence in E, and as such includes all the Σ-bases of E. In [4] and [14], where the definition of constructors includes the requirement that E^{Σ} be collapse-free, this maximal Σ-base is defined as follows:

$$G_E(\Sigma, V) := \{r \in T(\Omega, V) \mid r \neq_E t \text{ for all } t \in T(\Omega, V) \text{ with } t(\epsilon) \in \Sigma\}.$$

Modulo equivalence in E, $G_E(\Sigma, V)$ is made of Ω-terms whose top symbol is not in Σ, from which it is immediate that $G_E(\Sigma, V)$ is closed under bijective renaming and under equivalence in E. To summarize, the following holds for $G_E(\Sigma, V)$ [4].

Proposition 6. *Whenever E^{Σ} is collapse-free,*

- *every Σ-base of E, if any, is included in $G_E(\Sigma, V)$;*
- *Σ is a set of constructors for E iff $G_E(\Sigma, V)$ is a Σ-base of E.*

Examples of theories admitting (collapse-free) constructors can be found in [4]. We provide below an example of an equational theory admitting non-collapse-free constructors, that is, constructors in the more general sense of Definition 2. But first, it is instructive to look at at least one counterexample.

Let $E := \{f(g(x)) \equiv f(f(g(x)))\}$ and $\Sigma := \{f\}$. Since E^{Σ} is clearly collapse-free, we know that every Σ-base of E, if any, is included in the set $G_E(\Sigma, V)$ defined earlier. It is easy to see that $G_E(\Sigma, V) = V \cup \{g(t) \mid t \in T(\Omega, V)\}$ and that conditions (1) and (2) of Definition 3 hold for $G_E(\Sigma, V)$. However, condition (3) does not since $f(g(x)) =_E f(f(g(x)))$, although $f(y) \neq_E f(f(y))$. In conclusion, Σ is not a set of constructors for E.

Example 7. Consider the signature $\Sigma := \{0, s, p, -\}$ and the equational theory E of the integers with zero, successor, predecessor, and unary minus, axiomatized by the equations:

$$x \equiv s(p(x)), \quad x \equiv p(s(x))$$
$$-0 \equiv 0, \quad -s(x) \equiv p(-x), \quad -p(x) \equiv s(-x), \quad -(-x) \equiv x.$$

The signature $\Sigma := \{0, s, p\}$ is a set of constructors for E. This is proven in [3] by showing that the set $G := V \cup \{-v \mid v \in V\}$ is a Σ-base of E.

Many more examples of theories with constructors can be found in the usual axiomatizations of abstract data types. In the next subsection, we point out another, perhaps less obvious, class of examples for which our combination approach could provide fresh insights and results.

Normal Forms According to Definition 3, if a set G is a Σ-base of E, every Ω-term t is equivalent in E to a term $s(\bar{r}) \in T(\Sigma, G)$. We call $s(\bar{r})$ a *G-normal form of* t *in* E.[3] We say that a term t is *in G-normal form* if it is already of the form $t = s(\bar{r}) \in T(\Sigma, G)$. Because $V \subseteq G$, it is immediate that Σ-terms are in G-normal form, as are terms in G.

We will make use of normal forms in our combination procedure. In particular, we will consider normal forms that are computable in the following sense.

Definition 8 (Computable Normal Forms). *For any Σ-base G of E we say that G-normal forms are computable for Σ and E if there is a computable function $NF_E^{\Sigma}: T(\Omega, V) \longrightarrow T(\Sigma, G)$ such that $NF_E^{\Sigma}(t)$ is a G-normal form of t, i.e., $NF_E^{\Sigma}(t) =_E t$.*

Note that, unless E^{Σ} is collapse-free, the terms of G may as well start with a Σ-symbol themselves. This means that, for any given term t in G-normal form, it may not be possible to effectively identify those terms $\bar{r}$ of G such that $t = s(\bar{r})$ for some Σ-term s. Now, in the combination procedure shown in Sect. 3 sometimes we will need to first compute the normal form of a term and then decompose that into its components s and $\bar{r}$. To be able to do this it will be enough to assume (in addition to the computability of normal forms) that G is a recursive set, thanks to the proposition below.

---

[3] Notice that in general a term may have more than one G-normal form.

Proposition 9. *Let G be a Σ-base of E and $t \in T(\Sigma, G)$. If G is recursive, there is an effective way of computing a term $s(\bar{v}) \in T(\Sigma, V)$ and a sequence $\bar{r}$ of terms in G such that $t = s(\bar{r})$.*

2.2   Constructors and Modal Logics

For all normal modal logics [9], equivalence of formulae is a congruence relation on formulae that is closed under substitution [9]. For example, consider the basic modal logic K. Here, the signature Σ_K contains the Boolean operators ($\wedge$, $\vee$, $\neg$), the Boolean constant $\top$ (for truth), and the unary (modal) operator $\square$. Equivalence of formulae in K can be axiomatized [10] by the equational theory E_K, which consists of the equational axioms for Boolean algebras, and the two additional equational axioms

$$\square(x \wedge y) \equiv \square(x) \wedge \square(y) \quad \text{and} \quad \square(\top) \equiv \top.$$

It is easy to see that satisfiability (and validity) of modal formulae in K is decidable iff the word problem for E_K is decidable. For example, a formula φ is valid iff $\varphi =_{E_K} \top$. Since satisfiability in K is indeed decidable[4] the word problem for E_K is also decidable.

The problem of combining modal logics has been thoroughly investigated (see, e.g., [6,8]). In particular, there are very general results on how decidability of the component logics transfers to their combination (called *fusion* in the literature). We are interested in the question of whether these combination results can also be obtained within our framework for combining decision procedures for the word problem. This line of research appears to be promising for the following two reasons.

First, it follows from results in [9] (Chap. 4.2) that equivalence in the fusion of two modal logics is axiomatized by the union of the equational theories axiomatizing equivalence in the component logics. In this union, the shared symbols are the Boolean symbols, i.e., $\wedge$, $\vee$, $\neg$, and $\top$. Since the axioms for Boolean algebras contain collapse axioms (e.g., $x \wedge x \equiv x$), it is clear that we will really need the generalized version of constructors introduced in this paper.

Second, the requirement that the reduct of the free algebra to the shared symbols be free is always satisfied for modal logics closed under the Boolean operations $\wedge$, $\vee$, and $\neg$. For example, let Σ be the subsignature of Σ_K that consists of $\wedge$, $\vee$, $\neg$, and $\top$. It is easy to show that the Σ-reduct $\mathcal{A}_K{}^\Sigma$ of the E_K-free algebra $\mathcal{A}_K$ over countably infinitely many generators is a countably infinite *atomless* Boolean algebra. Since the free Boolean algebra over countably infinitely many generators is also a countably infinite *atomless* Boolean algebra, and since all countably infinite atomless Boolean algebras are known to be isomorphic [7], we can deduce that the reduct $\mathcal{A}_K{}^\Sigma$ is in fact free. For our combination method to apply, however, this is not sufficient. We need additional conditions; e.g., that normal forms are computable. Unfortunately, it is not even clear how

---

[4]   In fact, it is a well-known PSPACE-complete problem.

a Σ-base could look like in this case. This would depend on an appropriate characterization of the generators of $\mathcal{A}_\mathsf{K}{}^\Sigma$, which appears to be a non-trivial (and, to the best of our knowledge, not yet solved) problem.

For this reason, we restrict our attention in the example below to a certain sublogic of K. Such a sublogic, which is not Boolean closed, is particularly interesting because the current combination results in modal logic are restricted to Boolean closed modal logics.

Example 10. Let us consider just the conjunctive fragment of K. In equational terms, this amounts to restricting the signature Σ_K to the subsignature $\Sigma_\mathsf{K}^0 := \{\wedge, \top, \square\}$ and consider only terms (i.e., modal formulae) built over this signature.

In [1], it is shown[5] that equivalence of such formulae is axiomatized by the theory E_K^0, which consists of the axioms

$$x \wedge (y \wedge z) \equiv (x \wedge y) \wedge z, \quad x \wedge y \equiv y \wedge x, \quad x \wedge x \equiv x, \quad x \wedge \top \equiv \top$$

$$\square(x \wedge y) \equiv \square(x) \wedge \square(y), \quad \square(\top) \equiv \top.$$

We claim that $\Sigma^0 := \{\wedge, \top\}$ is a set of constructors in our sense. In fact, the set

$$G := \{\square^n(v) \mid n \geq 0 \text{ and } v \in V\}$$

can be shown to be a Σ^0-base of E_K^0. This is an easy consequence of the notion of concept-based normal form introduced in [1] and the characterization of equivalence proved in the same paper. The concept-based normal form of a formula is obtained by exhaustively applying the rewrite rules $\square(x \wedge y) \to \square(x) \wedge \square(y)$, $\square(\top) \to \top$, $x \wedge \top \to x$, $\top \wedge x \to x$. It is easy to see that this normal form can be computed in polynomial time, and that any formula in normal form is either $\top$ or a conjunction of elements of G. Thus, the concept-based normal form is also a G-normal form. Since the set G is obviously recursive and closed under variable renaming, the additional prerequisites (see below) for our combination approach to apply to E_K^0 are satisfied as well.

Interestingly, if we consider the conjunctive fragment of the modal logic S_4 in place of K, we obtain a quite different behavior: the reduct of the free algebra to Σ^0 is not free (see [4]). This is surprising as, in the Boolean closed case, S_4 behaves like K in the sense that the reduct of the corresponding free algebra is still free (for the same reasons as for K).

2.3 Combination of Theories Sharing Constructors

To conclude this section we go back to the problem of combining theories and consider two non-trivial equational theories E_1, E_2 with respective signatures Σ_1, Σ_2 such that $\Sigma := \Sigma_1 \cap \Sigma_2$ is a set of constructors for E_1 and for E_2. Moreover, we assume that $E_1{}^\Sigma = E_2{}^\Sigma$.

---

[5] Note, however, that [1] employs description logic syntax rather than modal logic syntax for formulae.

For $i = 1, 2$, let $\mathcal{A}_i$ be a Σ_i-algebra free in E_i over some countably infinite set X_i, and $Y_i := \{\llbracket r \rrbracket_{\alpha_i}^{\mathcal{A}_i} \mid r \in G_i\}$ where G_i is any Σ-base of E_i and α_i any bijection of V onto X_i. From Prop. 1 and Cor. 5, we then have the following:

Proposition 11. *Let φ_1, φ_2 be two first-order formulae of respective signature Σ_1, Σ_2. If φ_i is satisfiable in $\mathcal{A}_i$ with the elements of $Var(\varphi_1) \cap Var(\varphi_2)$ taking distinct values over Y_i for $i = 1, 2$, then $\varphi_1 \wedge \varphi_2$ is satisfiable in $E_1 \cup E_2$.*

An immediate consequence of this result is that the theory $E_1 \cup E_2$ above is non-trivial. To see that, since φ_1 and φ_2 in the proposition are arbitrary formulae, it is enough to take both of them to be the disequation $x \not\equiv y$ between two distinct variables.

In the rest of the paper we show that, under the above assumptions on E_1 and E_2, the combined theory $E_1 \cup E_2$ in fact has much stronger properties: whenever normal forms are computable for Σ and E_i ($i = 1, 2$) with respect to a recursive Σ-base closed under renaming, the decidability of the word problem is a modular property, as is the property of being a set of constructors.

3   A Combination Procedure for the Word Problem

In the following, we present a decision procedure for the word problem in an equational theory of the form $E_1 \cup E_2$ where each E_i is an equational theory with decidable word problem. Such a procedure will be obtained as modular combination of the procedures deciding the word problem for E_1 and for E_2. We will restrict our attention to equational theories E_1, E_2 that satisfy the following conditions for $i = 1, 2$:

1. E_i *is a non-trivial equational theory over the (countable) signature Σ_i;*
2. $\Sigma := \Sigma_1 \cap \Sigma_2$ *is a set of constructors for E_i and $E_1{}^{\Sigma} = E_2{}^{\Sigma}$;*
3. E_i *admits a Σ-base G_i closed under bijective renaming of V;*
4. G_i *is recursive and G_i-normal forms are computable for Σ and E_i;*
5. *the word problem for E_i is decidable.*

As already mentioned, the word problem for $E := E_1 \cup E_2$ can be reduced to the satisfiability problem (in E) for disequations of the form $s_0 \not\equiv t_0$, where s_0, t_0 are $(\Sigma_1 \cup \Sigma_2)$-terms. By variable abstraction it is possible to transform any such disequation into a a set $AS(s_0 \not\equiv t_0)$ consisting of pure equations and a disequation between two variables such that $s_0 \not\equiv t_0$ and $AS(s_0 \not\equiv t_0)$ are "equivalent" in a sense to be made more precise below. The set $AS(s_0 \not\equiv t_0)$ is what we call an *abstraction system*. To define abstraction systems formally we will need some more notation.

To start with, we will use finite sets of formulae in place of conjunctions of such formulae, and say that a set S of formulae is satisfiable in a theory iff the conjunction of its elements is satisfiable in that theory. Now, let T be a set of equations of the form $v \equiv t$ where $v \in V$ and $t \in T(\Sigma_1 \cup \Sigma_2, V) \setminus V$. The relation $\prec$ on T is defined as follows: $(u \equiv s) \prec (v \equiv t)$ iff $v \in Var(s)$.

By $\prec^+$ we denote the *transitive* and by $\prec^*$ the *reflexive-transitive closure* of $\prec$. The relation $\prec$ is *acyclic* if there is no equation $v \equiv t$ in T such that $(v \equiv t) \prec^+ (v \equiv t)$.

Definition 12 (Abstraction System). *A set $\{x \not\equiv y\} \cup T$ is an abstraction system with disequation $x \not\equiv y$ iff $x, y \in V$ and the following holds:*

1. *T is a finite set of equations of the form $v \equiv t$ where $v \in V$ and $t \in (T(\Sigma_1, V) \cup T(\Sigma_2, V)) \setminus V$;*
2. *the relation $\prec$ on T is acyclic;*
3. *for all $(u \equiv s), (v \equiv t) \in T$,*
   a) *if $u = v$ then $s = t$;*
   b) *if $(u \equiv s) \prec (v \equiv t)$ and $s \in T(\Sigma_i, V)$ with $i \in \{1, 2\}$ then $t \notin T(\Sigma_i, V)$.*

The above is a generalization of the definition of abstraction system in [4] in that now the right-hand side of any equation in T can start with a shared symbol. As before, Condition (2) implies that for all $(u \equiv s), (v \equiv t) \in T$, if $(u \equiv s) \prec^* (v \equiv t)$ then $u \notin Var(t)$; Condition (3a) implies that a variable cannot occur as the left-hand side of more than one equation of T; Condition (3b) implies, together with Condition (1), that the elements of every $\prec$-chain of T have *strictly* alternating signatures $(\ldots, \Sigma_1, \Sigma_2, \Sigma_1, \Sigma_2, \ldots)$. In particular, when Σ_1 and Σ_2 have a non-empty intersection Σ, Condition (3b) entails that if $(u \equiv s) \prec (v \equiv t)$ neither s nor t can be a Σ-term: one of the two must contain symbols from $\Sigma_1 \setminus \Sigma$ and the other must contain symbols from $\Sigma_2 \setminus \Sigma$.

Proposition 13. *The set $S := AS(s_0 \not\equiv t_0)$ is an abstraction system. Furthermore, where $\bar{v}$ is the tuple that collects all the variables in the left-hand side of an equations of S, the formula $\exists \bar{v}.S \leftrightarrow (s_0 \not\equiv t_0)$ is logically valid.*

Every abstraction system $\{x \not\equiv y\} \cup T$ induces a finite graph $\mathcal{G}_S := (T, \prec)$ whose set of *edges* consists of all pairs $(n_1, n_2) \in T \times T$ such that $n_1 \prec n_2$. According to Definition 12, $\mathcal{G}_S$ is in fact a directed acyclic graph (or *dag*). Assuming the standard definition of path between two nodes and of length of a path in a dag, the *height* $h(n)$ of the node n is the maximum of the lengths of all the paths in the dag that end with n.[6]

In the previous section, we would have represented the normal form of a term in $T(\Sigma_i, V)$ $(i = 1, 2)$ as $s(\bar{q})$ where s was a term in $T(\Sigma, V)$ and $\bar{q}$ a tuple of terms in G_i. Considering that G_i contains V, we will now use a more descriptive notation. We will distinguish the variables in $\bar{q}$ from the non-variables terms and write $s(\bar{y}, \bar{r})$ instead, where $\bar{y}$ collects the elements of $\bar{q}$ that are in V and $\bar{r}$ those that are in $G_i \setminus V$.

The combination procedure described in Fig. 1 decides the word problem for the theory $E := E_1 \cup E_2$ by deciding the satisfiability in E of disequations of the form $s_0 \not\equiv t_0$ where s_0, t_0 are $(\Sigma_1 \cup \Sigma_2)$-terms. During the execution of the procedure, the set S of formulae on which the procedure works is repeatedly modified by the application of one of the derivation rules defined in Fig. 2. We

---

[6] Since $\mathcal{G}_S$ is acyclic and finite, this maximum exists.

Input: $(s_0, t_0) \in T(\Sigma_1 \cup \Sigma_2, V) \times T(\Sigma_1 \cup \Sigma_2, V)$.

1. Let $S := AS(s_0 \not\equiv t_0)$.
2. Repeatedly apply (in any order) **Coll1**, **Coll2**, **Ident**, **Simpl**, **Shar1**, **Shar2** to S until none of them is applicable.
3. Succeed if S has the form $\{v \not\equiv v\} \cup T$ and fail otherwise.

Fig. 1. The Combination Procedure.

$$\text{Coll1} \quad \frac{T \qquad u \not\equiv v \qquad x \equiv t[y] \quad y \equiv r}{T[x/r] \quad (u \not\equiv v)[x/y] \qquad y \equiv r}$$

$\qquad$ if $t \in T(\Sigma_i, V)$ and $y =_{E_i} t$ for $i = 1$ or $i = 2$.

$$\text{Coll2} \quad \frac{T \qquad x \equiv t[y]}{T[x/y]}$$

$\qquad$ if $t \in T(\Sigma_i, V)$ and $y =_{E_i} t$ for $i = 1$ or $i = 2$, and
$\qquad$ there is no $(y \equiv r) \in T$.

$$\text{Ident} \quad \frac{T \qquad x \equiv s \quad y \equiv t}{T[x/y] \qquad y \equiv t}$$

$\qquad$ if $s, t \in T(\Sigma_i, V)$ and $s =_{E_i} t$ for $i = 1$ or $i = 2$,
$\qquad$ $x \neq y$, and $\mathsf{h}(x \equiv s) \leq \mathsf{h}(y \equiv t)$.

$$\text{Simpl} \quad \frac{T \quad x \equiv t}{T} \quad \text{if } x \notin \mathit{Var}(T).$$

$$\text{Shar1} \quad \frac{T \qquad\qquad u \not\equiv v \quad x \equiv t \qquad \bar{y}_1 \equiv \bar{r}_1}{T[x/s(\bar{y}, \bar{z})][\bar{y}_1/\bar{r}_1]] \quad \bar{z} \equiv \bar{r} \quad u \not\equiv v \quad x \equiv s(\bar{y}, \bar{r}) \quad \bar{y}_1 \equiv \bar{r}_1}$$

$\qquad$ if (a) $x \in \mathit{Var}(T)$,
$\qquad\qquad$ (b) $t \in T(\Sigma_i, V) \setminus G_i$ for $i = 1$ or $i = 2$,
$\qquad\qquad$ (c) $\mathit{NF}^{\Sigma}_{E_i}(t) = s(\bar{y}, \bar{r}) \in T(\Sigma, G_i) \setminus V$,
$\qquad\qquad$ (d) $\bar{r}$ nonempty and $\bar{r} \subseteq G_i \setminus T(\Sigma, V)$,
$\qquad\qquad$ (e) $\bar{z}$ fresh variables with no repetitions,
$\qquad\qquad$ (f) $\bar{y}_1 \subseteq \mathit{Var}(s(\bar{y}, \bar{r}))$ and
$\qquad\qquad\qquad$ $(x \equiv s(\bar{y}, \bar{r})) \prec (y \equiv r)$ for no $(y \equiv r) \in T$.

$$\text{Shar2} \quad \frac{T \qquad\qquad u \not\equiv v \quad x \equiv t \qquad \bar{y}_1 \equiv \bar{r}_1}{T[x/s[\bar{y}_1/\bar{r}_1]] \quad u \not\equiv v \quad x \equiv s[\bar{y}_1/\bar{r}_1] \quad \bar{y}_1 \equiv \bar{r}_1}$$

$\qquad$ if (a) $x \in \mathit{Var}(T)$,
$\qquad\qquad$ (b) $t \in T(\Sigma_i, V) \setminus G_i$ for $i = 1$ or $i = 2$,
$\qquad\qquad$ (c) $\mathit{NF}^{\Sigma}_{E_i}(t) = s \in T(\Sigma, V) \setminus V$,
$\qquad\qquad$ (d) $\bar{y}_1 \subseteq \mathit{Var}(s)$,
$\qquad\qquad$ (e) $\bar{r}_1 \subseteq G_\iota$ with $\iota \in \{1, 2\} \setminus \{i\}$, and
$\qquad\qquad\qquad$ $(x \equiv s) \prec (y \equiv r)$ for no $(y \equiv r) \in T$.

Fig. 2. The Derivation Rules.

describe these rules in the style of a sequent calculus. The premise of each rule lists all the formulae in S before the application of the rule, where T stands for all the formulae not explicitly listed. The conclusion of the rule lists all the formulae in S after the application of the rule.

The procedure and the rules are almost identical to those given in [4]. The only difference is that the rules **Shar1** and **Shar2** have a different set of preconditions to account for the generalization of the notion of normal form caused by the new definition of constructors.

As before, **Coll1** and **Coll2** remove from S collapse equations that are valid in E_1 or E_2, while **Ident** identifies any two variables equated to equivalent Σ_i-terms and then discards one of the corresponding equations. The ordering restriction in the precondition of **Ident** is on the heights that the two equations involved have in the dag induced by S. It is there to prevent the creation of cycles in the relation $\prec$ over S. We have used the notation $t[y]$ to express that the variable y occurs in the term t, and the notation $T[x/t]$ to denote the set of formulae obtained by substituting every occurrence of the variable x by the term t in the set T.

The rule **Simpl** reduces clutter in S by eliminating those equations that have become unreachable along a $\prec$-path from the initial disequation because of the application of previous rules.

The rules **Shar1** and **Shar2**, which only apply when Σ_1 and Σ_2 are non-disjoint, are used in essence to propagate the constraint information represented by shared terms. To do that, the rules replace the right-hand side t of an equation $x \equiv t$ by its normal form, and then plug the "shared part" of the normal form in all those equations whose right-hand side contains x. In the description of the rules, an expression like $\bar{z} \equiv \bar{r}$ denotes the set $\{z_1 \equiv r_1, \ldots, z_n \equiv r_n\}$ where $\bar{z} = (z_1, \ldots, z_n)$ and $\bar{r} = (r_1, \ldots, r_n)$, and $s(\bar{y}, \bar{z})$ denotes the term obtained from $s(\bar{y}, \bar{r})$ by replacing the subterm r_j with z_j for each $j \in \{1, \ldots, n\}$. This notation also accounts for the possibility that t reduces to a non-variable term of G_i. In that case, s will be a variable, $\bar{y}$ will be empty, and $\bar{r}$ will be a tuple of length 1. Substitution expressions containing tuples are to be interpreted accordingly; also, tuples are sometimes used to denote the set of their components.

We make one assumption on **Shar1** and **Shar2** which is not explicitly listed in their preconditions. We assume that $NF_{E_i}^{\Sigma}$ ($i = 1, 2$) is such that, whenever the set $V_0 := Var(NF_{E_i}^{\Sigma}(t)) \setminus Var(t)$ is not empty,[7] each variable in V_0 is fresh with respect to the current set S. As explained in [3], this assumption can be made without loss of generality whenever G_i is closed under renaming of variables.

By requiring that $\bar{r}$ be non-empty, **Shar1** excludes the possibility that the normal form of the term t is a shared term. It is **Shar2** that deals with this case. The reason for a separate case is that we want to preserve the property that every $\prec$-chain is made of equations with alternating signatures (cf. Definition 12(3b)). When the equation $x \equiv t$ has immediate $\prec$-successors, the replacement of t by the Σ-term s may destroy the alternating signatures property because $x \equiv s$,

---

[7] This might happen if E_i is non-regular because then Definition 8 does not necessarily entail that all the variables of $NF_{E_i}^{\Sigma}(t)$ occur in t.

which is both a Σ_1- and a Σ_2-equation, may inherit some of these successors from $x \equiv t$.[8] **Shar2** restores this property by merging into $x \equiv s$ all of its immediate successors, if any. Condition (e) in **Shar2** makes sure that the tuple $\bar{y}_1 \equiv \bar{r}_1$ collects all these successors. The replacement of $\bar{y}_1$ by $\bar{r}_1$ in **Shar1** is done for similar reasons. In **Shar2**, the restriction that the terms in $\bar{r}_1$ be elements of G_i is necessary to ensure termination, as is the condition $x \in Var(T)$ in both rules.

A Sketch of the Correctness Proof

The total correctness of the combination procedure can be proven more or less in the same way as in [4]. We can first show that an application of one of the rules of Fig. 2 transforms abstraction systems into abstraction systems, preserves satisfiability, and leads to a decrease with respect to a certain well-founded ordering. This ordering can be obtained as follows: every node in the dag induced by the abstraction system S is associated with a pair (h, r), where h is the height of the node, r is 1 if the right-hand side of the node is neither in G_1 nor in G_2, and 0 otherwise. The abstraction system S is associated with the multiset $M(S)$ consisting of all these pairs. Let $\sqsupset$ be the multiset ordering induced by the lexicographic ordering on pairs.

Lemma 14. *Assume that S' is obtained from the abstraction system S by an application of one of the rules of Fig. 2. Then the following holds:*

1. $M(S) \sqsupset M(S')$.
2. S' is an abstraction system.
3. $\exists \bar{v}.S \leftrightarrow \exists \bar{v}'.S'$ is valid in E, where $\bar{v}$ lists all the left-hand side variables of S and $\bar{v}'$ the left-hand side variables of S'.

Since the multiset ordering $\sqsupset$ is well-founded, the first point of the lemma implies that the derivation rules can be applied only finitely many times. Given that the preconditions of each rule are effective because of the computability assumptions on the component theories and of Prop. 9, we can then conclude that the combination procedure halts on all inputs. The last point of the lemma together with Prop. 13 implies that the procedure is sound, that is, if it succeeds on an input (s_0, t_0), then $s_0 =_E t_0$. The second point implies that the final system obtained after the termination of the procedure is an abstraction system, which plays an important rôle in the completeness proof. We can prove that the combination procedure is complete, that is, succeeds on an input (s_0, t_0) whenever $s_0 =_E t_0$, by showing that Prop. 11 can be applied (see [3] for details). To sum up, this shows the overall correctness of the procedure, and thus:

Lemma 15. *Under the given assumptions, the word problem for $E := E_1 \cup E_2$ is decidable.*

---

[8] As explained above, we assume that the variables in $Var(s) \setminus Var(t)$ do not occur in the abstraction system. Thus, the equations in $\bar{y}_1 \equiv \bar{r}_1$ are in fact successors of $x \equiv t$.

The corresponding result in [4] is indeed a corollary of this one. The difference there is that we have the additional restriction that $E_i{}^\Sigma$ is collapse-free and we use the largest Σ-base of E_i, namely $G_{E_i}(\Sigma, V)$, instead of an arbitrary one ($i = 1, 2$). In [4], we do not *explicitly* assume that $G_{E_i}(\Sigma, V)$ is closed under renaming. But this is always the case, as we mentioned earlier. Also, we do not postulate that $G_{E_i}(\Sigma, V)$ is recursive because that is always the case whenever $G_{E_i}(\Sigma, V)$-normal forms are computable for Σ and E_i.

Similarly to [4], the decidability result of Lemma 15 is actually extensible to the union of any (finite) number of theories, all (pairwise) sharing the same signature Σ and satisfying the same properties as E_1 and E_2 above. The reason is that, again, all needed properties are modular with respect to theory union.

Theorem 16. *For all theories E_1, E_2 satisfying assumptions(1)-(5) stated at the beginning of Sect. 3, the following holds:*

1. *Σ is a set of constructors for $E := E_1 \cup E_2$ and $E^\Sigma = E_1{}^\Sigma = E_2{}^\Sigma$.*
2. *E admits a Σ-base G^* closed under bijective renaming of V;*
3. *G^* is recursive and G^*-normal forms are computable for Σ and E;*
4. *the word problem for E is decidable.*

The proof of the first three points is still quite involved (see [3] for details) but somewhat simpler than the corresponding proof in [4]. It depends on the explicit construction of the set G^*, given below. There, for all terms $r \in T(\Sigma_1 \cup \Sigma_2, V)$, we denote by $\hat{r}$ the pure term obtained from r by abstracting its alien subterms.

Definition 17. *The set G^* is inductively defined as follows:*

1. *Every variable is an element of G^*, that is, $V \subseteq G^*$.*
2. *Assume that $r(\bar{v}) \in G_i \setminus V$ for $i \in \{1, 2\}$ and $\bar{r}$ is a tuple of elements of G^* with the same length as $\bar{v}$ such that the following holds:*
   a) *$r(\bar{v}) \neq_E v$ for all variables $v \in V$;*
   b) *$\hat{r}_k \notin T(\Sigma_i, V)$ for all non-variable components r_k of $\bar{r}$;*
   c) *$r_k \neq_E r_\ell$ if r_k, r_ℓ occur at different positions in the tuple $\bar{r}$.*
   Then $r(\bar{r}) \in G^*$.

Notice that every non-collapsing term of G_i is in G^* for $i = 1, 2$ because the components of $\bar{r}$ in the definition above can also be variables. Every non-variable element r of G^* then has a stratified structure. Each layer of r is made of terms all belonging to G_1 or to G_2. Moreover, if the terms in one layer are in G_i, then the terms in the layer above and below, if any, are not in G_i.

4   Future Work

The results presented in this paper are preliminary in two respects. First, they depend on two *technical restrictions* for which we do not yet know whether they are necessary. One is the requirement in the definition of constructors that $X \subseteq Y$ (used to prove the completeness of the combination procedure); the other is the requirement that the Σ-bases employed in the combination procedure be closed under renaming (used in the derivation rules and to prove the modularity results). We are trying to find out whether these restrictions can be removed, or there is a deeper, non-technical, reason for them.

Second, in order to demonstrate the power of our new combination procedure, we intend to investigate more thoroughly its applicability to the combination of decision procedures for modal logics. This probably depends on a deep understanding of the structure of the free algebras corresponding to the modal logics in question.

References

1. F. Baader and P. Narendran. Unification of concept terms in description logics. In H. Prade, editor, *Proc. of ECAI-98*, pages 331–335. John Wiley & Sons Ltd, 1998.
2. F. Baader and K. U. Schulz. Combination of constraint solvers for free and quasi-free structures. *Theoretical Computer Science*, 192:107–161, 1998.
3. F. Baader and C. Tinelli. Combining equational theories sharing non-collapse-free constructors. TR 99-13, Dept. of Computer Science, University of Iowa, 1999.
4. F. Baader and C. Tinelli. Deciding the word problem in the union of equational theories sharing constructors. In P. Narendran and M. Rusinowitch, editors, *Proc. of RTA'99*, volume 1631 of *LNCS*, pages 175–189. Springer-Verlag, 1999. Longer version available as TR UIUCDCS-R-98-2073, Dept. of Computer Science, University of Illinois at Urbana-Champaign, 1998.
5. E. Domenjoud, F. Klay, and Ch. Ringeissen. Combination techniques for non-disjoint equational theories. In A. Bundy, editor, *Proc. of CADE-12*, volume 814 of *LNAI*, pages 267–281. Springer-Verlag, 1994.
6. E. Hemaspaandra. Complexity transfer for modal logic (extended abstract). In *Proc, of LICS'94*, pages 164–173, 1994. IEEE Computer Society Press.
7. S. Koppelberg. General theory of Boolean algebras. In J. Monk, editor, *Handbook of Boolean Algebras*. North-Holland, Amsterdam, The Netherlands, 1988.
8. M. Kracht and F. Wolter. Simulation and transfer results in modal logic: A survey. *Studia Logica*, 59:149–177, 1997.
9. M. Kracht. *Tools and Techniques in Modal Logic*. Elsevier, Amsterdam, The Netherlands, 1999.
10. E. J. Lemmon. Algebraic semantics for modal logics. *J. of Symbolic Logic*, 31(1):46–65, 1966.
11. T. Nipkow. Combining matching algorithms: The regular case. In N. Dershowitz, editor, *Proc. of RTA'89*, volume 335 of *LNCS*, pages 343–358. Springer-Verlag, 1989.
12. Don Pigozzi. The join of equational theories. *Colloquium Mathematicum*, 30(1):15–25, 1974.
13. Manfred Schmidt-Schauß. Combination of unification algorithms. *J. of Symbolic Computation*, 8(1–2):51–100, 1989.
14. C. Tinelli and Ch. Ringeissen. Non-disjoint unions of theories and combinations of satisfiability procedures: First results. TR UIUCDCS-R-98-2044, Dept. of Computer Science, University of Illinois at Urbana-Champaign, 1998.

Comparing Expressiveness of Set Constructor Symbols

Agostino Dovier[1], Carla Piazza[2], and Alberto Policriti[2]

[1] Dip. Scientifico-Tecnologico, Univ. di Verona.
Strada Le Grazie, 37134 Verona (Italy). dovier@sci.univr.it
[2] Dip. di Matematica e Informatica, Univ. di Udine.
Via Le Scienze 206, 33100 Udine (Italy). (piazza|policrit)@dimi.uniud.it

Abstract. In this paper we consider the relative expressive power of two very common operators applicable to sets and multisets: the *with* and the *union* operators. For such operators we prove that they are not mutually expressible by means of existentially quantified formulae. In order to prove our results, canonical forms for set-theoretic and multiset-theoretic formulae are established and a particularly natural axiomatization of multisets is given and studied.

Keywords. *Sets and Multisets Theory, Unification, Constraints.*

Introduction

In the practice of programming, as well in axiomatic study of set (and multiset) theory, two families of constructors are usually employed:

- Constructors of the form (*with*) that can build a (possibly) new set by adding one element to a given set. A typical element of this family is the *cons*-constructor of lists;
- Constructors (*union*) that can build a set made by all the elements of two given sets. A typical element of this family is the *append*-constructor of lists.

Bernays in [5] was the first to use the *with* constructor symbol in his axiomatization of Set Theory. Vaught in [22] proved, by giving an essential undecidability result, that theories involving such a constructor are extremely expressive and powerful. On the other hand, classical set theories (e.g., Zermelo-Fraenkel [16]) are more often based on union-like symbols (either unary or binary) and are sufficiently powerful to define both *with* and *union*.

In this paper we analyze the relationships between these two (kinds of) operators. We show that in both a set and a multiset setting, it is impossible to express the union-like operators with existentially quantified formulae based on with-like operators and vice versa. These results hold in any set theory sufficiently expressive to introduce the above operators.

With-like constructors can be associated with an equational theory containing two axioms (left commutativity (C_ℓ) and absorption (A_b)—cf., e.g., [13]) while union-like symbols are associated to $ACI1$ equational theories (see, e.g., [4]).

H. Kirchner and C. Ringeissen (Eds.): FroCoS 2000, LNAI 1794, pp. 275–289, 2000.

A by-product of the results of this paper is a systematic proof of the fact that the classes of formulae and problems that can be expressed with an $(A_b)(C_\ell)$ unification (constraint) problem and those concerned with $ACI(1)$ unification (constraint) problems with constants can not be (trivially) reduced to each other.

Other consequences of the results of this paper are criteria for choosing the classes of constraints that can be managed by a constraint solver (e.g., for programming with constraints, or when analyzing programs by making use of constraints), or for choosing the basic operators for dealing with sets in programming languages.

In Section 1 we formally introduce the problem faced in this paper and we fix the notation. In Section 2 we show that it is impossible to express the *with* using union-like constraints in a set framework. In Section 3 we show the vice versa. Section 4 shows how to apply the results obtained for expressiveness of unification problems and constraints. In Section 5 the results are repeated in the context of multisets. Finally, some conclusions are drawn.

1   Preliminaries

We assume standard notions and notation of first-order logic (cf. [20]). We use $\mathcal{L}$ as a meta-variable for first-order languages with equality whose variables are denoted by capital letters. We write $\varphi(X_1, \ldots, X_n)$ for a formula of $\mathcal{L}$ with $X_1, \ldots, X_n$ as free variables; when the context is clear, we denote a list $Z_1, \ldots, Z_n$ of variables by $\boldsymbol{Z}$. The symbol false stands for a generic unsatisfiable formula, such as $X \neq X$. A formula without quantifiers is said *open* and FV is a function returning the set of free variables of a formula.

Definition of the problem. Let $\mathbb{T}$ be a first-order theory over the language $\mathcal{L}$ and Φ a class of (generally open) $\mathcal{L}$-formulae. Let f be a function symbol such that $\mathbb{T} \models \forall \boldsymbol{X} \exists Y \, (Y = f(\boldsymbol{X}))$. We say that f can be *existentially expressed* by Φ in $\mathbb{T}$ if there is $\psi \in \Phi$ such that

$$\mathbb{T} \models \forall \boldsymbol{X} Y \, (Y = f(\boldsymbol{X}) \leftrightarrow \exists \boldsymbol{Z} \psi(\boldsymbol{X}, Y, \boldsymbol{Z}))$$

Assume $\mathcal{L}$ contains equality '$=$' and membership '$\in$' as binary predicate symbols, the constant symbol $\emptyset$ for the empty set, and the binary function symbols $\{\cdot \mid \cdot\}$ for set construction *with*,[1] and $\cup$ for set union. Assume that $\mathbb{T}$ is any set theory such that these three symbols are governed by the following axioms:[2]

$$
\begin{array}{lll}
(N) & \forall X & (X \notin \emptyset) \\
(W) & \forall Y V X & (X \in \{Y \mid V\} \leftrightarrow (X \in V \vee X = Y)) \\
(U) & \forall Y V X & (X \in Y \cup V \leftrightarrow (X \in Y \vee X \in V))
\end{array}
$$

---

[1] The interpretation of $\{x \mid y\}$ is $\{x\} \cup y$, hence $\{\cdot \mid \cdot\}$ is unnecessary whenever $\cup$ and the singleton set operator are present. Notice that $\{\cdot \mid \cdot\}$ is a list-like symbol and not the intensional set former $\{x : \varphi(x)\}$.

[2] These axioms are explicitly introduced in minimal set theories (e.g. [5,22,17]) while they are consequence of other axioms in stronger ones (e.g. ZF, cf. [16]).

and also assume that the extensionality principle (E) and the regularity axiom (R) hold in $\mathbb{T}$:

$$(E) \qquad \forall X\,Y \qquad (\forall Z\,(Z \in X \leftrightarrow Z \in Y) \rightarrow X = Y)$$
$$(R) \qquad \forall X\,\exists Z\,\forall Y\,(Y \in X \rightarrow (Z \in X \wedge Y \notin Z)).$$

Let $\Phi_\cup$ and $\Phi_{\texttt{with}}$ be the classes of open formulae built using $\emptyset, \cup, \in, =$ and $\emptyset, \{\,\cdot\,|\,\cdot\,\}, \in, =$, respectively. We will prove that for any $\mathbb{T}$ satisfying the above axioms, the symbol $\{\,\cdot\,|\,\cdot\,\}$ can not be existentially expressed by $\Phi_\cup$ in $\mathbb{T}$ and $\cup$ can not be existentially expressed by $\Phi_{\texttt{with}}$ in $\mathbb{T}$.

Let $\mathbb{HF}$ (the class of hereditarily finite well-founded sets) be the model whose domain $\mathcal{U}$ can be inductively defined as $\mathcal{U} = \bigcup_{i \geq 0} U_i$ with $U_0 = \emptyset, U_{i+1} = \wp(U_i)$, (where $\wp$ stands for the power-set operator) and whose interpretation for $\emptyset, \{\,\cdot\,|\,\cdot\,\}, \cup, \in, =$ is the natural one. $\mathbb{HF}$ is a submodel of any model of a set theory satisfying the above axioms and any formula of $\Phi_\cup$ and $\Phi_{\texttt{with}}$ is satisfiable if and only if it is satisfiable over $\mathbb{HF}$ (cf. [6]). Elements of $\mathcal{U}$ can be represented by directed acyclic graphs where edges represent the membership relation.

Technical notations. We use the following syntactic convention for the set constructor symbol $\{\,\cdot\,|\,\cdot\,\}$: the term $\{\,s_1\,|\,\{\,s_2\,|\,\cdots\,\{\,s_n\,|\,t\,\}\cdots\}\}$ will be denoted by $\{s_1, \ldots, s_n\,|\,t\}$ or simply by $\{s_1, \ldots, s_n\}$ when t is $\emptyset$. The function *rank*, defined as: $rank(\emptyset) = 0, rank(\{t\,|\,s\}) = \max\{rank(s), 1 + rank(t)\}$ returns the maximum 'depth' of a ground set, while the function *find*:

$$find(x, t) = \begin{cases} \emptyset & \text{if } t = \emptyset,\ x \neq \emptyset \\ \{0\} & \text{if } t = x \\ \{1 + n : n \in find(x, y)\} & \text{if } t = \{y\,|\,\emptyset\} \\ \{1 + n : n \in find(x, y)\} \cup find(x, s) & \text{if } t = \{y\,|\,s\},\ s \neq \emptyset \end{cases}$$

returns the set of 'depths' at which a given element x occurs in the set t (there is an exception for the unique case $find(\emptyset, \emptyset)$). For instance, if t is $\{\{\emptyset\}, \{\emptyset, \{\emptyset\}\}\}$, then $rank(t) = 3$ (it is sufficient to compute the maximum nesting of braces), and $find(\emptyset, t) = \{2, 3\}, find(\{\emptyset\}, t) = \{1, 2\}, find(\{\emptyset, \{\emptyset\}\}, t) = \{1\}$. The two above-defined functions will be used in Lemma 3 to build a suitable truth valuation for (canonical) formulae over $\mathbb{HF}$. We also denote by $\{\emptyset\}^n$ the (simplest) singleton set of rank n, that is the set inductively defined as: $\{\emptyset\}^0 = \emptyset, \{\emptyset\}^{n+1} = \{\,\{\emptyset\}^n\}$.

2   Union vs *with*

In this section we show show that the function symbol $\{\,\cdot\,|\,\cdot\,\}$ can not be existentially expressed by the class of formulae $\Phi_\cup$. To prove this fact it is sufficient to verify that it is impossible to express the singleton operator, namely the formula $X = \{Y\}$, i.e., $X = \{Y\,|\,\emptyset\}$:

Lemma 1. *Let φ be a satisfiable formula of $\Phi_\cup$. Then there is a satisfying valuation σ over $\mathbb{HF}$ for φ s.t. for all $X \in FV(\varphi)$ either $\sigma(X) = \emptyset$ or $|\sigma(X)| \geq 2$.*

Proof. Without loss of generality, we can restrict our attention to flat formulae in DNF, namely disjunctions of conjunctions of literals of the form: $X \ op \ Y, X = Y \cup Z, X = \emptyset$, where op can be $=, \neq, \in, \notin$, and X, Y, Z are variables. If φ is satisfiable, then it is satisfiable over $\mathbb{HF}$; let σ be a valuation which satisfies φ over $\mathbb{HF}$ and consider a disjunct ψ satisfied by σ. Assume σ does not fulfill the requirements (otherwise we have directly the thesis) and consider the term substitution $\theta = [X/\emptyset : \sigma(X) = \emptyset]$; let ψ' be the formula obtained by

1. removing from $\psi\theta$ all the literals of the form $\emptyset = \emptyset, \emptyset = \emptyset \cup \emptyset, X \notin \emptyset, \emptyset \notin \emptyset$,
2. replacing the literals of the form $X = Y \cup \emptyset, X = \emptyset \cup Y$ with $X = Y$, and
3. replacing all literals of the form $\emptyset \neq X$ with $X \neq \emptyset$.

All the literals of ψ' are of the form $X = Y \cup Z, X = Y, X \neq Y, X \neq \emptyset, X \in Y, \emptyset \in Y, X \notin Y, \emptyset \notin Y$ and $\sigma(X) \neq \emptyset$ for all $X \in FV(\psi')$.[3]

Let $\bar{n} = \max_{X \in FV(\psi')} rank(\sigma(X))$ and let c be a set of rank $\bar{n}$. We obtain a satisfying valuation σ' such that $\sigma'(X)$ is not a singleton for all $X \in FV(\psi')$, by adding c to all the non-empty elements in the transitive closure of the $\sigma(X)$'s. More in detail, given the membership graphs denoting the sets associated by σ to the variables in ψ, we can build the (unique) global minimum graph $\mathcal{G}_\sigma$. Now, let $\mathcal{G}_c$ be the minimum (rooted) graph denoting the set c. Add an edge from all the nodes of $\mathcal{G}_\sigma$, save the unique leaf denoting $\emptyset$ to (the root of) $\mathcal{G}_c$, obtaining a new graph $\mathcal{G}_{\sigma'}$ (the entry points of the variables remain the same). We prove that all literals of ψ' are satisfied by $\mathcal{G}_{\sigma'}$ (i.e. by σ'):

$X \neq \emptyset, \emptyset \notin Y, \emptyset \in Y$: These literals are fulfilled by $\mathcal{G}_\sigma$ and we have added no edges to the leaves.

$X = Y$: X and Y have the same entry point in the graph $\mathcal{G}_\sigma$, and thus in $\mathcal{G}_{\sigma'}$.

$X = Y \cup Z$: We have added the same entity to all sets. Equality remains true.

$X \in Y$: This means that there is an edge from the node associated to Y to that associated to X in $\mathcal{G}_\sigma$. The edge remains in $\mathcal{G}_{\sigma'}$.

$X \neq Y$: By contradiction. Let ν be a minimal (w.r.t. the membership relation) node of the graph $\mathcal{G}_\sigma$ such that there is another node ν' in $\mathcal{G}_\sigma$ bisimilar to ν in $\mathcal{G}_{\sigma'}$. This means that:
   - a node μ $\in$-successor of ν is different from a a node μ', $\in$-successor of ν' in $\mathcal{G}_\sigma$ and they are equal in $\mathcal{G}_{\sigma'}$: this is absurdum since ν is a $\in$-minimal node with this property; or
   - a node μ $\in$-successor of ν or a node μ' $\in$-successor of ν' are equal in $\mathcal{G}_{\sigma'}$ to the root of $\mathcal{G}_c$: this is absurdum since either they represent $\emptyset$ or $\mathcal{G}_c$ is also an element of the set represented by them and the overall graph is acyclic.

$X \notin Y$: In $\mathcal{G}_\sigma$ there is no edge from the node associated to Y to that associated to X. Using the same considerations of the previous point, it is impossible that now the former collapse with a node reached by an outgoing edge from the node associated to Y.

---

[3] Observe that, for instance, a literal $X = \emptyset$ cannot be in ψ' since σ satisfies ψ and, by hypothesis, X is mapped to a non-empty set.

To complete the valuation, map to $\emptyset$ all variables occurring only in the other disjuncts. $\square$

Theorem 1. *Let $\mathbb{T}$ be any set theory implying $NW \cup ER$. Then the functional symbol $\{\cdot\,|\,\cdot\}$ can not be existentially expressed by $\Phi_\cup$ in $\mathbb{T}$.*

Proof. By contradiction, assume that ψ in $\Phi_\cup$ existentially expresses $\{\cdot\,|\,\cdot\}$ in $\mathbb{T}$. Then ψ it is satisfiable over $\mathbb{HF}$. The result follows from Lemma 1 and the fact that $\mathbb{HF}$ is a submodel of any model of $NW \cup ER$. $\square$

3   *with* vs *union*

In this section we show that it is impossible to existentially express the function symbol $\cup$ using the class of formulae Φ_{with}. We make use of the following notion: a conjunction φ of Φ_{with}-literals is said to be in *canonical form* if $\varphi \equiv \mathsf{false}$ or each literal is either of the form $A = t, r \notin B, C \neq s$ where A, B, C are variables and A does not occur neither in t nor elsewhere in φ, B does not occur in r, C does not occur in s.

Lemma 2 ([8,9]). *Let $\varphi(\boldsymbol{X})$ be a formula in Φ_{with}, then there is a formula $\varphi'(\boldsymbol{X}, \boldsymbol{Y}) = \bigvee_i \varphi_i(\boldsymbol{X}_i, \boldsymbol{Y}_i)$ s.t. $\{\boldsymbol{X}_i\} \subseteq \{\boldsymbol{X}\}$, $\{\boldsymbol{Y}\} = \bigcup_i \{\boldsymbol{Y}_i\}$, and each φ_i is (a conjunction of literals) in canonical form. Moreover, $\mathbb{HF} \models \forall \boldsymbol{X}\ (\varphi \leftrightarrow \exists \boldsymbol{Y} \bigvee_i \varphi_i)$.*

The above lemma, guaranteeing that any formula can be rewritten as a disjunction of canonical formulae, goes a long way towards providing a satisfiability test. In fact, consider the following

Lemma 3. *1. If $\varphi(\boldsymbol{X})$ is in canonical form and different from the formula false, then $\mathbb{HF} \models \exists \boldsymbol{X}\, \varphi$.*
  2. If $\varphi(X, Y, \boldsymbol{Z})$ is in canonical form, different from false, and there are neither atoms of the form $X = s$ nor of the form $Y = t$ in φ, then there is a valuation γ such that $\mathbb{HF} \models \varphi\gamma$, but $\mathbb{HF} \models (X \not\subseteq Y)\gamma$.
  3. Let φ be a formula in canonical form, different from false, in which there is at least one of the atoms:

$$X = \{s_1, \ldots, s_m \,|\, A\}, Y = \{t_1, \ldots, t_n \,|\, B\}$$

  where A and B are different variables and $B \not\equiv X, A \not\equiv Y$. Then there is a valuation γ such that $\mathbb{HF} \models \varphi\gamma$ and $\mathbb{HF} \models (X \not\subseteq Y)\gamma$.

Proof. We prove (1) finding a particular valuation that satisfies the condition (2) and helps us in proving (3).
  1) We split φ into $\varphi^=$, $\varphi^{\neq}$, and $\varphi^{\notin}$, containing $=, \neq$, and $\notin$ literals, respectively.
  $\varphi^=$ has the form $X_1 = t_1 \wedge \cdots \wedge X_m = t_m$ and for all $i = 1, \ldots, m$, X_i appears uniquely in $X_i = t_i$ and $X_i \notin FV(t_i)$. We define the mapping $\theta_1 = [X_1/t_1, \ldots, X_m/t_m]$.

$\varphi^{\notin}$ has the form $r_1 \notin Y_1 \wedge \cdots \wedge r_n \notin Y_n$ (Y_i does not occur in r_i) and $\varphi^{\neq}$ has the form $Z_1 \neq s_1 \wedge \cdots \wedge Z_p \neq s_p$ (Z_i does not occur in s_i). Let $W_1, \ldots, W_h$ be the variables occurring in φ other than $X_1, \ldots, X_m, Y_1, \ldots, Y_n, Z_1, \ldots, Z_p$; we define $\theta_2 = [W_1/\{\emptyset\}^1, \ldots, W_h/\{\emptyset\}^h]$.

Let $\bar{s} = \max\{rank(t) : t \text{ occurs in } \varphi\theta_1\theta_2\} + 1$ and $R_1, \ldots, R_j$ be the variables occurring in $(\varphi^{\notin} \wedge \varphi^{\neq})\theta_2$ (actually, the variables Y and Z) and $n_1, \ldots, n_j$ auxiliary variables ranging over $\mathbb{N}$. We build an integer disequation system E in the following way:

1. $E = \{n_i > \bar{s} : \forall i \in \{1, \ldots, j\}\} \cup \{n_{i_1} \neq n_{i_2} : \forall i_1, i_2 \in \{1, \ldots, j\}, i_1 \neq i_2\}$.
2. For each literal $(R_{i_1} \neq t)$ in $\varphi^{\neq}\theta_2$

$$E = E \cup \{n_{i_1} \neq n_{i_2} + c : \forall i_2 \neq i_1, \forall c \in find(R_{i_2}, t)\}$$

3. For each literal $(t \notin R_{i_1})$ in $\varphi^{\notin}\theta_2$

$$E = E \cup \{n_{i_1} \neq n_{i_2} + c + 1 : \forall i_2 \neq i_1, \forall c \in find(R_{i_2}, t)\}$$

A system of this form admits always integer solutions. Let $\{n_1 = \bar{n}_1, \ldots, n_j = \bar{n}_j\}$ be a solution, define $\theta_3 = [R_i/\{\emptyset\}^{\bar{n}_i} : \forall i \in \{1, \ldots, j\}]$.

Let $\gamma = \theta_1\theta_2\theta_3$, and observe that $\varphi\gamma$ is a conjunction of ground literals. We show that $\mathbb{HF} \models \varphi\gamma$. We analize each literal of φ.

$X = t$: $X\theta_1$ coincides syntactically with $t\theta_1 = t$. Thus, $X\gamma = X\theta_1\theta_2\theta_3 = t\theta_2\theta_3 = t\gamma$. Hence, a literal of this form is true in any model of the equality.

$r \notin Y$: Two cases are possible:

1. if $r = \emptyset$ or r is one of the variables W_i, then $r\gamma = r\theta_2 = \{\emptyset\}^{i_1}$, with $i_1 < \bar{s}$. Thus $r\gamma$ can not belong to $Y\gamma = \{\emptyset\}^{i_2}$ since $i_2 > \bar{s} \geq i_1 + 1$;
2. Otherwise, r is a term containing at least one of the variables Y_i or Z_i. From the solution to the integer system E, we obtain $rank(r\gamma) \neq rank(Y\gamma) - 1$. Since $Y\gamma$ is a term denoting a singleton set, this is sufficient to force this literal to be true in each well-founded model of membership.

$Z \neq s$: Similar to the case above.

2) Consider the formula φ as in point (2) of the statement. Since X and Y are different variables and are not among the $X_1, \ldots, X_m$, then $X\gamma$ and $Y\gamma$ are two different singleton sets. Thus, it can not be that $X\gamma \subseteq Y\gamma$.

3) Consider the formula φ as in point (3) of the statement.

If only the atom $X = \{s_1, \ldots, s_m \mid A\}$ is in φ, then also $\varphi' \equiv \varphi \wedge A \neq Y$ is in canonical form, since $A \not\equiv Y$. Thus, by (1), there is a valuation γ which satisfies φ' and $A\gamma = \{\emptyset\}^a$ and $Y\gamma = \{\emptyset\}^y$ with $a \neq y$. Hence, $\{\emptyset\}^{a-1} \in X\gamma$ and $\{\emptyset\}^{a-1} \notin Y\gamma$: $\mathbb{HF} \models \neg(X \subseteq Y)\gamma$.

If only the atom $Y = \{t_1, \ldots, t_n \mid B\}$ is in φ, then consider

$$\varphi' \equiv \varphi \wedge t_1 \notin X \wedge \cdots \wedge t_n \notin X \wedge B \neq X$$

Remove all the literals $t_i \notin X$ from φ' when $X \in FV(t_i)$ and obtain φ'', equivalent to φ' and in canonical form, since $X \not\equiv B$. Thus, by (1), there is a valuation γ

which satisfies φ' and $X\gamma = \{\emptyset\}^x$ and $B\gamma = \{\emptyset\}^b$ with $x \neq b$. Since γ is a solution, then $t_i\gamma \notin X\gamma$. Therefore $\{\emptyset\}^{x-1} \in X\gamma$ and $\{\emptyset\}^{x-1} \notin Y\gamma$: $\mathbb{HF} \models \neg(X \subseteq Y)\gamma$.

If the two atoms $X = \{s_1, \ldots, s_m \mid A\}, Y = \{t_1, \ldots, t_n \mid B\}$ are in φ and φ is in canonical form, then A and B are different variables not in those occurring as left hand side of an atom in $\varphi^=$. Let $X_1 = t_1, \ldots, X_m = t_m$ be the remaining atoms of $\varphi^=$ and consider

$$\varphi' \equiv \varphi \wedge t_1 \notin A \wedge \cdots \wedge t_n \notin A \wedge A \neq B.$$

Remove all the literals $t_i \notin A$ from φ' when $A \in FV(t_i)$ and obtain φ'', equivalent to φ' and in canonical form. Thus, by (1), there is a valuation γ which satisfies φ' and $A\gamma = \{\emptyset\}^a$ and $B\gamma = \{\emptyset\}^b$ with $a \neq b$. Since γ is a solution, then $t_i\gamma \notin A\gamma$. Therefore $\{\emptyset\}^{a-1} \in X\gamma$ and $\{\emptyset\}^{a-1} \notin Y\gamma$: $\mathbb{HF} \models \neg(X \subseteq Y)\gamma$.   □

Lemma 4. *Let $\varphi_i(X, Y, \mathbf{Z}_i)$ $i \in \{1, \ldots, c\}$ be canonical formulae containing:*

- *an atom of the form $X = s_i$, where s_i is neither a variable of $X, \mathbf{Z}_i$, nor $\emptyset$,*
- *or an atom of the form $Y = \{t_1^i, \ldots, t_{m_i}^i \mid X\}$, or $Y = \{t_1^i, \ldots, t_{m_i}^i\}$, $m_i \geq 0$ (when $m_i = 0$ the equations become $Y = X$, and $Y = \emptyset$, respectively).*

Then, there are hereditarily finite sets x and y such that for all hereditarily finite sets $z_1, \ldots, z_n$, it holds that $\gamma = [X/x, Y/y, Z_1/z_1, \ldots, Z_n/z_n]$ implies:

$$\mathbb{HF} \models (X \subseteq Y)\gamma \text{ and } \mathbb{HF} \models ((\neg\varphi_1) \wedge \ldots \wedge (\neg\varphi_c))\gamma$$

Proof. Let $\hat{m} = \max\{m_1, \ldots, m_c\}$. We prove that $\gamma = [X/\emptyset, Y/\{\{\emptyset\}^0, \ldots, \{\emptyset\}^{\hat{m}}\}]$ fulfills the requirement.

Clearly $(X \subseteq Y)\gamma$ holds. Since they differ by $\hat{m} + 1$ elements, every atom of the form $Y = \{t_1^i, \ldots, t_{m_i}^i \mid X\}$ can not be true, thus if φ_i contains one of these atoms, it is all right. Otherwise, we need to prove that $(X = s_i)\gamma$ is false. If s_i is Y then it is the same as above. Otherwise, $s_i = \{s \mid t\}$ for some s and t. For any valuation γ of the variables in it, $\{s \mid t\}\gamma$ is different from $\emptyset$.   □

Theorem 2. *Let $\mathbb{T}$ be any set theory implying $NW \cup ER$. There is no formula $\varphi(X, Y, \mathbf{Z})$ in $\Phi_{\texttt{with}}$ such that: $\mathbb{HF} \models \forall XY (X \subseteq Y \leftrightarrow \exists \mathbf{Z}\varphi)$.*

Proof. $X \subseteq Y$ is equivalent to $(X = \emptyset) \vee (X \neq \emptyset \wedge X \subseteq Y)$. Thus, if there is a φ equivalent to $X \neq \emptyset \wedge X \subseteq Y$, then $X = \emptyset \vee \varphi$ is equivalent to $X \subseteq Y$.

Without loss of generality (cf. Lemma 2), we can assume that φ to be a disjunction of canonical formulae $\bigvee_{i=1}^{c} \varphi_i$, each of them different from false. Moreover, we can assume that no atom of the form $X = \emptyset$ is in φ_i (since φ_i should imply $X \neq \emptyset$), and that also atoms of the form $X = Z, Z = X, Y = Z, Z = Y$ with Z in $\mathbf{Z}$ is not in φ_i, since Z is a new variable and therefore, such a conjunct can be eliminated after the application of the substitutions Z/X (Z/Y) to the remaining part of φ_i.

Assume first $c = 1$, i.e., φ be a canonical formula. We prove that one of the following two sentences holds:

$$(a) \; \mathcal{D} \models \exists XY\mathbf{Z} (\varphi(X, Y, \mathbf{Z}) \wedge X \not\subseteq Y)$$
$$(b) \; \mathcal{D} \models \exists XY (X \subseteq Y \wedge \forall \mathbf{Z}\neg\varphi(X, Y, \mathbf{Z}))$$

1. If X and Y occur only in negative literals or in the r.h.s. of some equality atom of φ, then by Lemma 3(2) we are in case (a).
2. If $Y = \emptyset$ or $X = Y$ or $Y = X$ or $X = \{s_1, \ldots, s_m\}$ or $Y = \{t_1, \ldots, t_n\}$ or $X = \{s_1, \ldots, s_m \mid Y\}$ are in φ, then it is easy to find two values for X and Y fulfilling the inclusion but invalidating φ for any possible evaluation for $\mathbf{Z}$. Thus, we are in case (b).
3. If $Y = \{t_1, \ldots, t_n \mid X\}$ is in φ, then, by Lemma 4 (with a unique disjunct, i.e. $c = 1$) we are in case (b).
4. If $X = \{s_1, \ldots, s_m \mid Z_i\}$ and Y occurs only in negative literals or in the r.h.s. of some equality atom of φ, then by Lemma 3(3) we are in case (a).
5. If $Y = \{t_1, \ldots, t_n \mid Z_j\}$ and X occurs only in negative literals or in the r.h.s. of some equality atom of φ, then by Lemma 3(3) we are in case (a).
6. $X = \{s_1, \ldots, s_m \mid Z_i\}$ and $Y = \{t_1, \ldots, t_n \mid Z_j\}$, $m, n > 0$ are in φ.
   If Z_i and Z_j are the same variable, then we are in case (b) since we can find two sets s and t, $s \subseteq t$, differing for more than $n - m$ elements.
   When Z_i and Z_j are different variables, by Lemma 3(3) we are in case (a).

Assume now that φ be $\varphi_1 \vee \cdots \vee \varphi_c$. We prove again that one of the two sentences holds:

$$(a)\ \mathcal{D} \models \exists XYZ\,((\varphi_1 \vee \cdots \vee \varphi_n) \wedge X \not\subseteq Y)$$
$$(b)\ \mathcal{D} \models \exists XY\,(X \subseteq Y \wedge \forall \mathbf{Z}(\neg\varphi_1 \wedge \cdots \wedge \neg\varphi_c))$$

If there is one of the φ_i that fulfills case (a), then the result holds for the disjunction. Otherwise, assume that all the φ_is fulfill case (b). By the case analysis above, all these cases are those dealt by Lemma 4. Thus, we are globally in the case (b). □

Theorem 3. *Let $\mathbb{T}$ be any set theory implying $NW \cup ER$. Then $\{\cdot \mid \cdot\}$ can not be existentially expressed by $\Phi_{\texttt{with}}$ in $\mathbb{T}$.*

Proof. If a formula $\varphi(X, Y, Z, \mathbf{W})$ in $\Phi_{\texttt{with}}$ s.t. $\mathbb{HF} \models \forall XYZ\,(X = Y \cup Z \leftrightarrow \exists \mathbf{W}\varphi)$ exists, then also $\varphi \wedge X = Z$ is in $\Phi_{\texttt{with}}$ and it is equivalent to $Y \subseteq X$. A contradiction to Theorem 2. □

A similar result can be obtained for $\cap$ since $X = Y \cap X \leftrightarrow X \subseteq Y$.

4 Independence Results for Equational Theories

The two set constructor symbols analyzed in this paper have been studied in the context of unification theory [4] and constraints. In this contexts the properties of the constructors are usually given by equational axioms. $\{\cdot \mid \cdot\}$, $\cup$, and $\emptyset$ are governed by the axioms of Fig. 1.

As far as the theory $(A_b)(C_\ell)$ is concerned, in [10] it is presented the first unification algorithm for the *general* case (other free function symbols are admitted [4]). The algorithms in [2,21] reduces redundancies of unifiers, while those

(A_b)	$\{X \mid X \mid Y\}\} = \{X \mid Y\}$
(C_ℓ)	$\{X \mid \{Y \mid Z\}\} = \{Y \mid \{X \mid Z\}\}$
(A)	$(X \cup Y) \cup Z = X \cup (Y \cup Z)$
(C)	$X \cup Y = Y \cup X$
(I)	$X \cup X = X$
(1)	$\emptyset \cup X = X$

Fig. 1. Equational axioms for $\emptyset, \{\cdot \mid \cdot\}$, and $\cup$

in [9,1,7] are ensured to remain in the class NP. In [8,9,11] positive and negative constraints of this theory has been studied and solved.

In [3] a unification algorithm for the $ACI1$ unification *with constants* (no other free function symbols, save constants, are admitted [4]) is presented.

The expressiveness results of the previous sections allows us to prove the independence of the class of formulae that can be expressed using these two families of problems:

1. From Theorem 3 we know that $X = Y \cup Z$ can not be expressed by an open *with*-based formula. Thus, in particular for simple formulae made only by disjunctions of conjunctions of equations.
2. From Theorem 1 we know that it is impossible to express $X = \{Y \mid \emptyset\}$ using *union*-like formulae. Again, this implies the result.

A similar result holds for constraints problem involving positive and negative equality literals.

Notice that in the framework of *general $ACI1$* unification, we can encode all $(A_b)(C_\ell)$ unification problems, since, if $f(\cdot)$ is a unary (free) function symbol, then $\{X \mid Y\}$ is equivalent to $f(X) \cup Y$ (cf. also Footnote 1 and [19]). In [12] $ACI1$ constraints (including unification) for the general case are studied and solved.

5   Multisets

Expressiveness results similar to those of the previous sections can be obtained in a multiset framework. However, while the meaning of set operators is a common and unambiguous knowledge, there is no uniform and universally accepted view of multisets. Thus, we begin by recalling an existing axiomatizazion; then we discuss the introduction of new axioms that help us to formalize multisets in a reasonable way.

Let $\mathcal{L}^m$ be a first-order language having $=$ and $\in$ as binary predicate symbols, $\emptyset$ as constant, and $\{\![\cdot \mid \cdot]\!\}$ as binary function symbols, whose behavior is regulated by axioms $(N), (R)$, and (cf. [13]):

$$(W^m) \ \forall Y V X \ (X \in \{\![\,Y \mid V\,]\!\} \leftrightarrow X \in V \vee X = Y)$$
$$(E_p^m) \ \forall XYZ \ \{\![X, Y \mid Z]\!\} = \{\![Y, X \mid Z]\!\}.$$

The extensionality axiom (E) for sets has been replaced by (E_p^m), which does not imply the absorption property, i.e.: $NW^m E_p^m R \not\models \{\!|\, X, X \,|\, Y \,|\!\} = \{\!|\, X \,|\, Y \,|\!\}$. The elements of the models of $NW^m E_p^m R$ are called *multisets*.

The usual definition for inclusion between sets $X \subseteq Y \leftrightarrow \forall Z (Z \in X \to Z \in Y)$ does not capture the intended meaning of inclusion between multisets, since, for instance, $\{\!|\, \emptyset, \emptyset \,|\!\} \subseteq \{\!|\, \emptyset \,|\!\}$ oppositely to intuition. A similar consideration can be done for the union symbol if interpreted in the same way as for sets. A more tuned definition for multiset-inclusion can be recursively described by:

$$X \sqsubseteq Y \leftrightarrow X = \emptyset \vee$$
$$\exists VWZ \, (X = \{\!|\, V \,|\, W \,|\!\} \wedge Y = \{\!|\, V \,|\, Z \,|\!\} \wedge W \sqsubseteq Z)$$

Multisets are often called bags in literature. A bag is the intuitive way to imagine a multiset: consider two bags, one containing 1 apple and 2 oranges and another containing 2 apples and 3 oranges. The former is a sub-bag $\sqsubseteq$ of the latter. This second definition of inclusion can be given in a non-recursive way using a language $\mathcal{L}^{mn}$, more natural to deal with multisets, in which $\in$ is replaced by an infinite set of predicate symbols: $\in^0, \in^1, \in^2, \ldots$ Intuitively, the meaning of $X \in^i Y$, with $i \geq 0$, is that there are at least i occurrences of X in the multiset Y. Thus, $X \in^0 Y$ is always satisfied.

The axioms to model multisets in this language are ($i \in \mathbb{N}$):

(0) $\forall XY \, (X \in^0 Y)$

(I) $\forall XY \, (X \in^{i+1} Y \to X \in^i Y)$

(N') $\forall X \, (X \notin^1 \emptyset)$

(W') $\forall XYZ \left(\begin{array}{l} X \in^{i+1} \{\!|\, Y \,|\, Z \,|\!\} \leftrightarrow (X = Y \wedge X \in^i Z) \vee \\ \qquad\qquad\qquad\qquad (X \neq Y \wedge X \in^{i+1} Z) \end{array} \right)$

(E') $\forall XY \, (X = Y \leftrightarrow (\forall i > 0) \forall Z (Z \in^i X \leftrightarrow Z \in^i Y))$

(R') $\forall X \exists Z \forall Y \, (Y \in^1 X \to (Z \in^1 X \wedge Y \notin^1 Z))$

With these axioms, and making use of arithmetic, we can define usual predicates and operators, such as subset, union, difference. Before doing that, some choices must be made. In [14], for instance, the authors propose two kinds of union symbols: $\cup$ and $\uplus$.[4] Both of them are useful in different contexts. In particular, the $\uplus$ is the more natural for performing bag-union. Consider the bags of the example above: a bag that is union of the two bags contains 3 apples and 5 oranges. $\cup$ of [14], instead, takes the maximum number of occurrences of items of the two bags, and $\cap$ the minimum. The axiom for $\uplus$ is the following:

$$(\uplus) \; X = Y \uplus Z \leftrightarrow (\forall i > 0) \forall W (W \in^i X \leftrightarrow$$
$$\exists m_1 m_2 (W \in^{m_1} Y \wedge W \in^{m_2} Z \wedge i = m_1 + m_2))$$

In Section 5.2 we also discuss the relationships between bags and infiniteness. We define Φ_{with}^{mn} as the set of open first-order formulae of $\mathcal{L}^{mn}$ involving $\emptyset, \{\!|\, \cdot \,|\, \cdot \,|\!\}, \in^i$, $=$ and $\Phi_{\uplus}^{mn}$ as the set of open first-order formulae of $\mathcal{L}^{mn}$ involving $\emptyset, \uplus, \in^i, =$.

---

[4] In [14] a bag constructor of any finite arity is used to build bags. However, the binary functor symbol $\{\!|\, \cdot \,|\, \cdot \,|\!\}$ adopted in this paper is sufficient to perform that task.

5.1   Multiset *with* vs union

From [9] one can deduce a result for canonical formulae in a multiset context similar to that described in Lemma 2. The only difference is that $\notin^1$ must be used in place of $\notin$. That result (outside the scope of the paper) can be obtained using multiset unification [13,7] and rewriting rules for constraints [9].

Lemma 5 ([9]). *Let $\varphi(\boldsymbol{X})$ be a formula of Φ^{mn}_{with}, then there is a formula $\varphi'(\boldsymbol{X},\boldsymbol{Y}) = \bigvee_i \varphi_i(\boldsymbol{X}_i, \boldsymbol{Y}_i)$ such that $\{\boldsymbol{X}_i\} \subseteq \{\boldsymbol{X}\}$, $\{\boldsymbol{Y}\} = \bigcup_i \{\boldsymbol{Y}_i\}$, and each φ_i is a conjunction of literals in canonical form. Moreover, $0IN'W'E'R \models \forall \boldsymbol{X} \, (\varphi \leftrightarrow \exists \boldsymbol{Y} \bigvee_i \varphi_i)$.*

Proving two lemmata analogous to Lemma 3 and Lemma 4 (the valuations defined for these lemmata holds in the multiset case, as well), we can prove:

Theorem 4. *Let $\mathbb{T}$ be any theory implying $0IN'W'E'R$. There is no formula $\varphi(X, Y, \boldsymbol{Z})$ in Φ^{mn}_{with} such that: $\mathbb{T} \models \forall XY \, (X \sqsubseteq Y \leftrightarrow \exists \boldsymbol{Z} \varphi)$.*

Proof. Same proof as Theorem 2, with references to the multiset versions of Lemmata 3 and 4.   □

Theorem 5. *Let $\mathbb{T}$ be any theory implying $0IN'W'E'R$. Then the function symbol $\uplus$ can not be existentially expressed by Φ^{mn}_{with} in $\mathbb{T}$.*

Proof. From Theorem 4 and the property that $X = Y \uplus Z$ implies $Y \sqsubseteq X$.   □

5.2 $\uplus$ vs Multiset `with`

In the previous paragraph we have shown that the operator $\uplus$ is a natural *union* operator for multisets (bags). In this section we show that $\Phi^{mn}_{\uplus}$ can not existentially express the *with* operator $\{\!\!|\cdot|\cdot|\!\!\}$. The proof technique used for sets in Lemma 1 can not be applied to the case of multisets. There are two main problems. The first is that a constraint $X = Y \uplus Z$ is no longer satisfied when we add one element to any multiset occurring in it. The second is that it is too restrictive to analyze hereditarily finite solutions only, since simple constraints (e.g., $X = Y \uplus X \wedge Y \neq \emptyset$ or $X = \{\!\!|Y|X|\!\!\}$) have only infinite solutions. This is a very interesting feature of multisets: infiniteness can be expressed by a single equation, while nesting of quantifiers is needed in set theory (cf. [18]). Thus, we introduce axiom (S^ω) ensuring that, given an object c, there is the infinite multiset $\{\!\!|c, c, c, \ldots|\!\!\}$:

$$(S^\omega) \quad \forall Y \exists X (X = \{\!\!|Y|X|\!\!\} \wedge (\forall Z \in X)(Z = Y))$$

Observe that the universal quantification is not needed to ensure the existence of an infinite multiset (the first equation is sufficient). It is introduced to ensure the existence of the particular multiset containing only one element repeated an infinite number of times. However, $|\omega|$ is the only infinite cardinality we are interested in in this work. Thus, all models have domains in which bags contain at most $|\omega|$ elements. Since we are interested in infinite bags, in all axioms the numbers i must be intended to range over $0, 1, 2, \ldots, \omega$ and the symbol $\in^\omega$ to belong to the language. Let $\mathbb{T}_m = (0)(I)(N')(W')(E')(R)(\uplus)(S^\omega)$.

Lemma 6. *For each satisfiable formula φ in $\Phi_{\uplus}^{mn}$ it holds that $\mathbb{T}_m \not\models \forall(\varphi \to X = \{\!\{ Y \}\!\})$.*

Proof. As done in Lemma 1, without loss of generality we can consider a formula φ in flat form in DNF, namely a disjunction of conjunctions of literals of the form:

$$V = W \uplus Z, V = \emptyset, V = W, V \neq W, V \in^i W, V \notin^i W$$

Let $\mathcal{M} = \langle M, (\cdot)^{\mathcal{M}} \rangle$ be a model of T_m and assume there is a valuation σ of the variables of φ on M such that $\mathcal{M} \models \varphi\sigma$; we build a valuation σ' on a domain $\mathcal{M}'$ (possibly, on $\mathcal{M}$ itself) such that $\mathcal{M}' \models \varphi\sigma'$ and $\mathcal{M}' \not\models (X = \{\!\{ Y \}\!\})\sigma'$. Let ψ be a disjunct of φ that is satisfied by σ and involves variables X and Y (if there are no disjuncts of this form the result holds trivially).

If $\sigma(X)$ is $(\emptyset)^{\mathcal{M}}$ or a bag containing at least two elements, then choose $\sigma' = \sigma$.

Otherwise, consider a formula ψ' defined as follows: replace each variable V such that $\sigma(V) = (\emptyset)^{\mathcal{M}}$ with $\emptyset$ and retain only one representative for each set $V_1, \dots, V_n$ of variables such that $\sigma(V_i) = \sigma(V_j)$. Let X be the representative of its class. After performing simple rewrites on ψ modified as above it is easy to obtain a formula ψ' as conjunction of literals of the form:

$$V = W \uplus Z, V \neq W, V \neq \emptyset, V \in^i W, \emptyset \in^i W, V \in^i W, \emptyset \in^i W.$$

Moreover, for $V \in FV(\psi')$ it holds that $\sigma(V) \neq (\emptyset)^{\mathcal{M}}$ and if V, W are distinct variables, $\sigma(V) \neq \sigma(W)$. Observe that X belongs to $FV(\psi') = \{V_1, \dots, V_k\}$.

Let $c = \sigma(V_1) \uplus \cdots \uplus \sigma(V_k) \in M$ and c_ω be the solution on $\mathcal{M}$ of the formula $K = \{\!\{ c \,|\, K \}\!\} \wedge (\forall Z \in K)(Z = c)$. The existence of c_ω in M is ensured by axiom (S^ω) (informally, c_ω is the infinite bag $\{\!\{ c, c, c, \dots \}\!\}$). Observe that since $\mathcal{M}$ is a model of (R), it can not be the case that $c \in \sigma(V)$ for $V \in FV(\psi')$.

We define a Mostowski collapsing function [16] $\beta : M \longrightarrow M'$, as follows: $\beta(m) = m$ if $m \neq \sigma(V)$ for all $V \in FV(\psi')$; otherwise $\beta(m)$ is

$$\underset{m' \in^h m \wedge m' \notin^{h+1} m}{\biguplus} \underbrace{\{\!\{ \beta(m'), \dots, \beta(m') \}\!\}}_{h} \uplus \underset{m' \in^\omega m}{\biguplus} \underbrace{\{\!\{ \beta(m'), \beta(m'), \dots \}\!\}}_{\omega} \uplus c_\omega$$

The domain M' is such that $M' = M$ as long as there is in M $\beta(m)$ for all $m = \sigma(V)$. Otherwise, obtain $\mathcal{M}' = \langle M', (\cdot)^{\mathcal{M}'} \rangle$ expanding the model $\mathcal{M}$ so that to guarantee the presence of those elements. We prove that

$$m_1 \neq m_2 \to \beta(m_1) \neq \beta(m_2) \tag{1}$$

for all m_i such that $c \notin m_i$. Since M is a well-founded model of membership, we prove this fact by contradiction in this way. Let m_1 be a $\in$-minimal element such that there is m_2 such that $m_1 \neq m_2$ and $\beta(m_1) = \beta(m_2)$. The following cases must be analyzed:

$\beta(m_1) = m_1, \beta(m_2) = m_2$: then $\beta(m_1) \neq \beta(m_2)$ by hypothesis: absurdum.
$\beta(m_1) = m_1, \beta(m_2) \neq m_2$: This means that $c \in \beta(m_2)$. By hypothesis $c \notin m_1$: absurdum.

$\beta(m_1) \neq m_1, \beta(m_2) = m_2$: similar to the previous case.

$\beta(m_1) \neq m_1, \beta(m_2) \neq m_2$: by hypothesis $c \notin m_1$ and $c \notin m_2$; this means that $c \in^\omega \beta(m_1)$ and $c \in^\omega \beta(m_2)$ and c is not the element that made the two bags equal. Thus, for some elements $m_1' \in^1 m_1$ and $m_2' \in^1 m_2$ $m_1' \neq m_2'$ and $\beta(m_1') = \beta(m_2')$. This is absurdum since m_1 is a $\in$-minimal element with this property.

We are ready to define the valuation σ' on $\mathcal{M}'$: $\sigma'(V) = \beta(\sigma(V))$ for each $V \in FV(\psi')$. It is clear that for all $V \in FV(\psi')$ $\sigma'(V)$ is an infinite bag. Thus, clearly, $\mathcal{M} \not\models (X = \{\!\{ Y \}\!\})\sigma'$.

To complete the proof we must prove that $\mathcal{M} \models (s \ op \ t)\sigma'$ for each literal $s \ op \ t$ of ψ'. By case analysis:

$V \neq \emptyset, \emptyset \in^i W, \emptyset \notin^i W$: These literals remain true trivially.

$V = W \uplus Z$: We know that $\sigma(V) = \sigma(W) \uplus \sigma(Z)$. The function β extends the common elements in the same way. We only need to notice that c is inserted ω times on the l.h.s. and $\omega + \omega = \omega$ times on the r.h.s.

$V \in^i W$: Again, if $\sigma(V) \in^i \sigma(W)$ then $\beta(\sigma(V)) \in^i \beta(\sigma(W))$ by construction.

$V \neq W$: It derives from property (1) above.

$V \notin^i W$: If $\sigma(V) \notin^i \sigma(W)$ and $\beta(\sigma(V)) \in^i \beta(\sigma(W))$ this means that $\sigma(V)$ is a multiset without c among its elements that collapses using β with another multiset with the same property. This is absurdum by property (1) above.

Assign $(\emptyset)^{\mathcal{M}}$ to all variables of φ occurring only in the other disjuncts. $\qquad\square$

Theorem 6. $\{\!\{ \cdot \,|\, \cdot \}\!\}$ *can not be existentially expressed by $\Phi_\uplus^{mn}$ in $\mathbb{T}_m$.*

Proof. Immediate from Lemma 6 since $X = \{\!\{ Y \}\!\}$ is equivalent to $X = \{\!\{ X \,|\, \emptyset \}\!\}$. $\qquad\square$

Remark 1. The proof of Lemma 6 can be repeated on finite models. However, there is an interesting technical point. Suppose to have $X = Y \uplus Z$ and a multiset solution σ. If we add one element, say c, to all variables, this will be no longer a solution, since X should contain two occurrences of c instead of 1. Intuitively, we have to fulfill an integer linear system of equations, obtained from the formula ψ. But we can use the fact that the system is already fulfilled by σ. Thus, modifying the definition of β, second case, with: $\{\!\{ \underbrace{c, \ldots, c}_{|\sigma(V)|} \}\!\}$ we can prove the result for finite models of bags.

5.3 Independence Results for Bag Equational Theories

As for sets, the expressiveness results proved in the two previous subsections have a consequence on the class of formulae that can be expressed using multiset unification and constraints. The symbols $\emptyset, \{\!\{ \cdot \,|\, \cdot \}\!\}$, and $\uplus$ fulfill the equational properties $(C_\ell), (A), (C), (1)$ as in Fig. 1, while properties (A_b) and (I) are no longer true.

In [13,7] a general unification algorithm is presented for the theory (C_ℓ). In [9] constraint solvers for the theory are presented. For links to $AC1$ unification algorithms, see [4]. Similarly to what holds for sets (cf. Sect. 4), it is an immediate consequence of Theorems 5 and 6 that $AC1$ unification with constants can not express all general (C_ℓ) unification problems and general (C_ℓ) unification can not express all $AC1$ unification with constants problems. General $AC1$, instead, can deal with any general (C_ℓ) unification problem using the usual encoding of with (cf. Sect. 4).

6   Conclusions

We have analyzed the relationships between the expressive power of two very common set and multiset constructors. In particular, we have proved that *union*-like and *with*-like symbols are not mutually expressible without using universal quantification. This has many consequences, such as the fact that testing satisfiability of with-based formulae (constraints) is easier than for formulae of the other case. This is a criterion for choosing the admissible constraints ([15]) of a CLP language. The conjecture of this expressiveness result has been used in [11] to enlarge the class of admissible constraints of $CLP(\mathcal{SET})$ [8]. However, the main consequence is perhaps an independence result of two very common equational theories for handling sets $((A_b)(C_\ell)$ and $ACI1)$. The same results can be obtained for multisets theories. In particular, for proving the results we have pointed out that

- $\in$ is sufficient, for sets, for giving (clean) axioms for equality.
- For multisets, the symbols $\in^i$, $i \in \mathbb{N}$ with the meaning 'to belong at least i times' and integer arithmetic is needed to perform the same task.
- The result could be extended for lists as well, but a complex axiomatic based on the notion of 'to belong in the list at the position ith' is needed.

Observe that the canonical form of this paper is very "explicit" and it might require a lot of time to be computed. To find more implicit but efficient normal forms is crucial when developing a CLP language but not for the aim of this work in which we have used the *existence* of a solved form for a theoretical proof of expressiveness.

Acknowledgements. We thank Domenico Cantone, Enrico Pontelli, and Gianfranco Rossi for their active contribution during our discussions and the referees that greatly helped to clarify the presentation. This research is partially supported by MURST project: *Certificazione automatica di programmi mediante interpretazione astratta.*

References

1. D. Aliffi, A. Dovier, and G. Rossi. From set to hyperset unification. *Journal of Functional and Logic Programming*, 1999(10):1–48, September 1999.

2. P. Arenas-Sánchez and A. Dovier. A minimality study for set unification. *Journal of Functional and Logic Programming*, 1997(7):1–49, December 1997.
3. F. Baader and W. Büttner. Unification in commutative and idempotent monoids. *Theoretical Computer Science*, 56:345–352, 1988.
4. F. Baader and K. U. Schulz. Unification theory. In Wolfgang Bibel and Peter H. Schmidt, editors, *Automated Deduction: A Basis for Applications. Volume I, Foundations: Calculi and Methods*. Kluwer Academic Publishers, Dordrecht, 1998.
5. P. Bernays. A system of axiomatic set theory. Part I. *The Journal of symbolic logic*, 2:65–77, 1937.
6. D. Cantone, A. Ferro, and E. G. Omodeo. *Computable Set Theory, Vol. 1*. Int'l Series of Monographs on Computer Science. Clarendon Press, Oxford, 1989.
7. E. Dantsin and A. Voronkov. A nondeterministic polynomial-time unification algorithm for bags, sets and trees. In W. Thomas, editor, *FoSSaCS'99*, volume 1578 of *Lecture Notes in Computer Science*, pages 180–196. Springer-Verlag, Berlin, 1999.
8. A. Dovier. A Language with Finite Sets Embedded in the CLP Scheme. In R. Dyckhoff, editor, *Int'l Workshop on Extension of Logic Programming*, volume 798 of *Lecture Notes in Artificial Intelligence*, pages 77–93. Springer-Verlag, Berlin, 1994.
9. A. Dovier. *Computable Set Theory and Logic Programming*. PhD thesis, Università degli Studi di Pisa, March 1996. TD–1/96.
10. A. Dovier, E. G. Omodeo, E. Pontelli, and G. Rossi. {log}: A Language for Programming in Logic with Finite Sets. *J. of Logic Programming*, 28(1):1–44, 1996.
11. A. Dovier, C. Piazza, E. Pontelli, and G. Rossi. On the Representation and Management of Finite Sets in CLP-languages. In J. Jaffar, editor, *Proc. of 1998 Joint Int'l Conf. and Symp. on Logic Programming*, pages 40–54. The MIT Press, Cambridge, Mass., 1998.
12. A. Dovier, C. Piazza, E. Pontelli, and G. Rossi. ACI1 constraints. In D. De Schreye, editor, *ICLP'99, 16th International Conference on Logic Programming*, pages 573–587. The MIT Press, Cambridge, Mass., 1999.
13. A. Dovier, A. Policriti, and G. Rossi. A uniform axiomatic view of lists, multisets and the relevant unification algorithms. *Fundamenta Informaticae*, 36(2/3):201–234, 1998.
14. S. Grumbach and T. Milo. Towards tractable algebras for bags. *Journal of Computer and System Sciences*, 52(3):570–588, 1996.
15. J. Jaffar, M. Maher, K. Marriott, and P. Stuckey. The semantics of constraint logic programs. *Journal of Logic Programming*, 1–3(37):1–46, 1998.
16. K. Kunen. *Set Theory. An Introduction to Independence Proofs*. Studies in Logic. North Holland, Amsterdam, 1980.
17. E. G. Omodeo and A. Policriti. Solvable set/hyperset contexts: I. some decision procedures for the pure, finite case. *Communication on Pure and Applied Mathematics*, 9–10:1123–1155, 1995.
18. F. Parlamento and A. Policriti. The logically simplest form of the infinity axiom. *American Mathematical Society*, 103:274–276, 1988.
19. O. Shmueli, S. Tsur, and C. Zaniolo. Compilation of set terms in the logic data language (LDL). *Journal of Logic Programming*, 12(1-2):89–119, 1992.
20. J. R. Shoenfield. *Mathematical Logic*. Addison-Wesley, Reading, 1967.
21. F. Stolzenburg. An algorithm for general set unification and its complexity. *Journal of Automated Reasoning*, 22, 1999.
22. R. L. Vaught. On a Theorem of Cobham Concerning Undecidable Theories. In E. Nagel, P. Suppes, and A. Tarski, editors, *Proceedings of the 1960 International Congress*, pages 14–25. Stanford University Press, Stanford, 1962.

Author Index

Lecture Notes in Artificial Intelligence (LNAI)

Lecture Notes in Computer Science